THE FUNDAMENTALIST CITY?

Religiosity and the remaking of urban space

The relationship between urbanism and fundamentalism is a very complex one. This book explores how the dynamics of different forms of religious fundamentalisms are produced, represented, and practiced in the city. It attempts to establish a relationship between two important phenomena: the historic transition of the majority of the world's population from a rural to an urban existence; and the robust resurgence of religion as a major force in the shaping of contemporary life in many parts of the world.

Employing a transnational interrogation anchored in specific geographic regions, the contributors to this volume explore the intellectual and practical challenges posed by fundamentalist groups, movements, and organizations. They focus on how certain ultra religious practices of Christianity, Hinduism, Islam, and Judaism have contributed to the remaking of global urban space. Their work suggests that it is a grave oversimplification to view religious orthodoxies or doctrines as the main cause of urban terrorism or violence. Instead they argue that such phenomena should be understood as a particular manifestation of modernity's struggles.

AlSayyad and Massoumi's book provides fascinating reading for those interested in religion and the city, with thought provoking pieces from experts in anthropology, geography, sociology, religious studies, and urban studies.

Nezar AlSayyad is Professor of Architecture, Planning, and Urban History and Chair of the Center for Middle Eastern Studies at the University of California, Berkeley. He is also President of the International Association for the Study of Traditional Environments. He has published fourteen books including *Muslim Europe or Euro-Islam* (2002); *Urban Informality* (2004); *The End of Tradition?* (2005); and *Cinematic Urbanism* (2006).

Mejgan Massoumi is an urban planner and manager at the Center for Middle Eastern Studies at the University of California at Berkeley. She obtained her degrees in Architecture and City Planning from the University of California, Berkeley. She is co-editor of *Urban Diversity. Space, Culture and Inclusive Pluralism in Cities Worldwide* (forthcoming).

THE FUNDAMENTALIST CITY?

Religiosity and the remaking of urban space

Edited by Nezar AlSayyad and
Mejgan Massoumi

LONDON AND NEW YORK

First published 2011
by Routledge, 2 Park Square, Milton Park, Abingdon, Oxon, OX14 4RN

Simultaneously published in the USA and Canada
by Routledge, 270 Madison Avenue, New York, NY 10016

Routledge is an imprint of the Taylor & Francis Group, an informa business

Typeset in Helvetica and Aldine by Hope Services (Abingdon) Ltd
Printed and bound in Great Britain by CPI Antony Rowe, Chippenham, Wiltshire

British Library Cataloguing in Publication Data
A catalogue record for this book is available from the British Library

Library of Congress Cataloging-in-Publication Data
AlSayyad, Nezar.
The fundamentalist city? : religiosity and the remaking of urban space / Nezar AlSayyad and Mejgan Massoumi.
p. cm.
Includes bibliographical references and index.
1. Religion and geography. 2. Cities and towns--Religious aspects. 3. City planning--Religious aspects. 4. Religion and sociology. I. Massoumi, Mejgan. II. Title.
BL65.G4A47 2011
201'.630776091724--dc22
2010007903

ISBN13: 978-0-415-77935-7 (hbk)
ISBN13: 978-0-415-77936-4 (pbk)
ISBN13: 978-0-203-84459-5 (ebk)

Contents

Preface

The idea of a 'fundamentalist city' came to me some years ago while walking in the streets of Cairo. I grew up in what can be described as secular Cairo in the 1960s, and left Egypt by the end of the 1970s. However, between the 1970s and the end of the century, the number of mosques in the city almost quadrupled, while the population of the city only doubled. Of course, many of the new mosques were not really mosques, but garages and basements in apartment buildings which were converted into mosques. A series of tax exemptions passed by the government and a rapid increase in ritual practices may explain this change. As an urbanist who studies change in cities, I celebrated such transformations and saw them as a fertile arena for further interrogations of the nature and content of urbanism in the Middle East.

Two years ago, I was in Cairo on a Friday during one of the regular Friday prayers. I had some minor errands to run that morning in the neighbourhood that had been my childhood home. And when I went out to do them, I found that the city's streets were full of worshippers, as these small mosques expanded into the public space, literally taking over the city. Since, in Islam, any space, including a room, an open courtyard, or a street, can serve as a mosque, I realized that for an hour or so that morning the entire city had become one. Preachers with loudspeakers, often preaching from below ground level, were competing for attention with their sermons, while their flocks, composed of neighbourhood residents, spilled out into the adjacent streets, now covered with carpets and individual prayer rugs. Returning from my morning errands, I eventually discovered that the street which led to my family's house was blocked by carpets being laid in preparation for the prayer. The house could be accessed from another street, but when I circled around to approach that way, I was surprised to find that it, too, was blocked by people sitting on prayer rugs waiting for the sermon to end.

When a phenomenon is too powerful, one must surrender to it. When my second attempt failed to get me to where I wanted and despite being a non-religious and non-observant individual, I yielded. I took off my shoes and sat with the crowd until the sermon was over and I could pick my shoes up again and walk fast through the crowd before the actual prayer started.

This experience made me realize that many people of other religious orientations or beliefs, including Cairo's large Christian minority, had an increasingly diminishing right to the city. In recent decades subtle forms of spatial restriction and exclusion had emerged and become normalized in everyday life, so that they no longer warranted questioning by the very people who were being excluded. I was reminded of Henri Lefebvre's work on 'The Right to the City', and of Don Mitchell's application of it, arguing for the right of the homeless to public space in American cities. Was the analogy a useful one? It occurred to me then to raise the question of who had the right to public space in the city, and at whose expense. Clearly, the right to walk the city is a fundamental right of urban citizenship. To what extent should the surrender of this right, even for a short time, not only by minority believers but also by those who did not care to engage in the religious ritual, be justified or tolerated? This was the genesis of the question which led to this research project, and later to the book.

The idea behind this book goes back to a discussion I had with Mejgan Massoumi, who was at the Woodrow Wilson International Center for Scholars while the two of us were part of the 'Inclusive Cities' project. It became clear to us that patterns of segregation and exclusion which emerge in some cities are often based on specific practices of religiosity and religious activism. We also realized that these have not received their fair share of attention from scholars either of the city or of religion. Aware that the Wilson Center has a tradition of assembling working groups of scholars – think tanks of sorts – who work for a year on a given topic, and recognizing the gap in the literature on the connection between religion and urbanism, we decided to organize a working group and a conference entitled 'Cities and Fundamentalisms', stressing the plural of both words. Funded by the Wilson Center, the group was constituted of scholars who had done interdisciplinary work but who were also firmly anchored in disciplines such as anthropology, architectural history, cultural geography, planning, political science, sociology, and urban history.

The group met for the first time in Washington DC, in the summer of 2007. Our intent was to facilitate an exchange between these scholars about the challenges posed by fundamentalist movements and organizations (with a special, but not exclusive, focus on religious ones) with the aim of understanding how they are affected by the urban condition and how they affect the urban landscape. The relationship between Fundamentalisms and Cities is a very complex one, and our goal was to explore how the dynamics of fundamentalism – in its urban, social, religious, and political forms – are produced, represented, and practiced in the city. Working on the assumption that fundamentalism is not only an urban phenomenon, but that it also operates within institutional structures in cities across the globe, we decided to emphasize clearly and early on that we were not dealing with orthodox practices of religion at the end of the spectrum or extremist ideologies used to justify terrorism. In particular, our aim was to explore the following question: When do certain religious rituals/customs turn into exclusionary practices which ultimately lead to fundamentalist positions? We hope the different contributions to this book will offer some insights into this.

There are several institutions and individuals we need to acknowledge. First, we would like to thank the Woodrow Wilson International Center for Scholars (WWICS) in Washington DC, for programmatic and financial support, especially from its President, Lee Hamilton, and its Executive Vice President, Michael Van Dusen. The entire Comparative Urban Studies Project (CUSP) team provided major logistical support, specifically Blair Ruble, the project's Chair, whose early encouragement allowed the project to happen. The support of Allison Garland and Lauren Herzer was also invaluable. Co-sponsorship was also provided by Haleh Esfandiari and Azucena Rodriguez of the Middle East Program at the WWICS. At Berkeley, the International Association for the Study of Traditional Environments (IASTE) served as the intellectual origin for articulating the project, while the Center for Middle Eastern Studies (CMES) served as the base from which it was run and this book was assembled. We would like to acknowledge the support of Lily Cooc, Priscilla Minaise, Aashika Damodar, and Joe Gouig, who helped with the logistics of meetings held in Berkeley. Finally, a number of distinguished scholars have contributed papers or commentaries to some of the group's meetings. Among them have been Peter van der Veer, Salwa Ismail, Emily Gottreich, Minoo Moallem, Raka Ray, Arvind Rajagopal, Ananya Roy, John Voll, Haim Yacobi, Salwa Ismail, and Nora Colton.

Let me conclude by stating clearly that this book is not a treatise along the lines of the Cities of God paradigm; nor is it a detailed discussion of fundamentalist urbanism. Instead, it is an interrogation of the relationship between specific cities and certain practices of fundamentalisms. It simply raises the question: *Is a fundamentalist city possible*?

Nezar AlSayyad
Berkeley

The Contributors

Renu Desai is a Postdoctoral Research Associate at the Department of Geography, Durham University. Her research interests include urban planning and governance, urban informality, housing struggles, and questions around identity, violence and urban space.

John Eade is Professor of Sociology and Anthropology and Executive Director of the Centre for Research on Nationalism, Ethnicity and Multiculturalism at Roehampton University and the University of Surrey. His recent publications include two co-edited books: *Advancing Multiculturalism, Post 7/7* (2008); and *Transnational Ties: Cities, Migrations and Identities* (2008).

Omri Elisha is Assistant Professor of Cultural Anthropology at Queens College, City University of New York. He is currently finishing a book based on ethnographic fieldwork among megachurches and faith-based organizations in Tennessee.

Inger Furseth is Professor and Research Associate at KIFO Centre for Church Research, Norway and Research Fellow at University of Southern California. Among her books are *A Comparative Study of Social and Religious Movements in Norway 1780s–1905* (2002), and *From Quest for Truth to Being Oneself* (2006).

Francesca Giovannini is a Lecturer in International and Area Studies Program at the University of California, Berkeley. She has worked for the United Nations Development Program in Beirut, Lebanon and in Gaza.

Mona Harb is Associate Professor of Urban Planning and Policy at the American University of Beirut. She has contributed to many edited books and has published several articles in *Third World Quarterly, The Arab World Geographer, Annales de la Recherche Urbaine and Genèses*.

James Holston is Professor in the Department of Anthropology at the University of California at Berkeley. His publications include *Insurgent*

Citizenship: Disjunctions of Democracy and Modernity in Brazil (2008); an edited book entitled *Cities and Citizenship* (1999); and *The Modernist City: An Anthropological Critique of Brasilia* (1989).

Mrinalini Rajagopalan is Assistant Professor at New York University where she teaches in the John W. Draper Interdisciplinary Masters Program in Humanities and Social Thought. She has contributed to the *Encyclopedia of Urban Studies* (2009) and the *Handbook of Architectural Theory (forthcoming).*

Batya Roded is a Researcher in the Department of Geography at Ben Gurion University of the Negev in Beer-Sheva, Israel. Her fields of research deal mainly with the ethnic frontier and the processes and power relations which mould this territory.

Rhys H. Williams is Professor in the Department of Sociology at the University of Cincinnati. His recent publications include contributions to *Religion and Social Justice for Immigrants* (2007) and *Social Movement Studies* (2006).

Oren Yiftachel is Professor of Geography at Ben Gurion University of the Negev in Beer-Sheva, Israel. His recent publications include *Ethnocracy: The Politics of Judaizing Israel/Palestine* (2006) and *Planning a Mixed Region in Israel: The Political Geography of Arab-Jewish Relations in the Galilee* (1992).

Part I: Fundamentalisms: Between City and Nation

Chapter 1

The Fundamentalist City?

Nezar AlSayyad

Globalization and the compression of space and time have fundamentally changed the standard relationships between peoples and places. Across the world, despite these changes, national, communal, and religious allegiances have often only become stronger. Some scholars argue that these strengthened ties are an important means of resistance against the hegemonic forces of globalization. Others interpret the rise of fundamentalist practices as articulating alternative forms of non-Western modernity. Whatever the case, the unanticipated resurgence of religious and ethnic loyalties has given new meaning to religion in the public life of many communities. Christianity, Hinduism, Judaism, and Islam have all experienced such surges in commitment and practice. And in many instances radical groups espousing essentialist religious positions have spawned new political movements. The spread of global terrorism has often been explained in terms of the inability to understand these groups. But such a connection is often unjustifiable, and it would be a grave simplification to view all religious orthodoxy or fundamentalist doctrine as a basis for terrorist violence.

Many observers believe that the most powerful cause of this religious resurgence was the social, economic, and cultural modernization that swept across the world in the second half of the twentieth century, challenging deep-rooted understandings of modernity, tradition, and identity. In particular, some scholars argue that at the national level, people who migrate from the countryside to the city are often separated from older social networks. As they battle for economic survival, they may become exposed to new sets of relationships and experiences, and become caught up in the trauma of modernization and the search for identity. Religion, be it mainstream or not, may meet some of their needs. At the international level, the same argument has been made to explain the status of people who emigrated from former colonized regions (often including educated, middle-class individuals) to Europe and North America, and who achieved the status of nationals, forming substantial ethnic minority communities in these countries.

Religious groups, and more recently those affiliated with orthodox ideologies, have also been increasingly called upon to provide social services in these marginalized communities, as their needs have been left unattended by state bureaucracies. These services may include medical and hospital care, kindergartens and schools, programmes for the elderly, prompt relief after natural and other catastrophes, and welfare and social support during periods of economic deprivation (Appleby and Marty, 1993*a*). Indeed, the breakdown of order and state power under the neoliberal economic paradigm that swept the world in the late twentieth century has often created a vacuum that can only filled by religious groups. And when traditional religious ideas do not meet people's needs, they frequently turn to more radical and reactionary ones. Participants in new religious movements usually come from all walks of life, but they have overwhelmingly been residents of urban areas. The religious revival and the subsequent radicalism that has grown from it may thus be seen as an urban phenomenon. It appeals to people who are educated, oriented to the modern world, and holders of successful careers in all job markets. Doctors, lawyers, university professors, businessmen, and administrators are just a few constituents in certain regions.

The 'Cities and Fundamentalisms Project' from which this book grew had the aim of facilitating an interdisciplinary exchange between scholars and practitioners in the social sciences, humanities, and professions at the intersection of urban and identity studies. Each participant who contributed to this book has attempted to explore some aspect of the complex of issues discussed above. As a collaborative effort, the book thus aims to understand the urban processes by which religious movements transform into fundamentalist ones, possibly engaging in tactics of control that reshape the life and form of cities. In particular, we have attempted to understand how and where religious beliefs and the performance of religious rituals turn into exclusionary spatial practices that ultimately become fundamentalist. We never fully answered this question, but we hope we have identified and explained some of its manifestations. Of course, this task has been complicated by the current moment. Globalization, religious and ethnic racism, and the 'war on terror' have all intensified interest in radical religious groups – but frequently as a result of tainted assumptions, misconceptions, and one-dimensional views.

The relationship between fundamentalism as a concept and urbanism as lived reality is very complex. A proper study of it requires understanding each of these domains and their individual meanings. Embodying the form and culture of urban societies, urbanism is an important arena in which to observe how ethnic and religious subcultures mediate global forces and international currents at the local or national levels. By contrast, few would dispute the present negative connotation of fundamentalism, associated mainly with militancy and regressive politics. However, the *Oxford English Dictionary* simply defines fundamentalism as the 'strict maintenance of orthodox traditional beliefs or doctrines'; and the *American Heritage Dictionary* defines it as strict adherence to any set of basic ideas or principles.[1] A quick search on the Web, however, makes clear that in common usage fundamentalism is defined mainly in religious terms and is often understood as embodying a pathological condition of violence. In this regard, a Google-scholar search leads one to articles mainly from psychology and psychiatry which equate fundamentalism with terrorist acts committed by those who have been socialized into altered states of mind. Interestingly enough, some of these websites also contain listings for 'Capitalist or market fundamentalism', which they describe as a form of excess or deviation that results from political and economic hegemony. Meanwhile, the supposedly unscholarly Wikipedia stresses the subjectivity of the term. Such dictionary definitions never capture the complex nuances of a concept, but they can serve as a starting frame of reference. Thus, what we tend to classify today as fundamentalism is often construed as a form of religiosity in direct engagement or conflict with secularism.

In reality, the concept of fundamentalism has a Christian origin. The first use of the term in the manner we now use it came in relation to early-twentieth-century evangelical Protestants who stressed the inerrancy of the Bible and mandated its use as a binding historical document. The term appears to have been used first in connection with the American Northern Baptist Convention of 1920 to describe more conservative delegates, who desired 'to restate, reaffirm, and re-emphasize' the fundamentals of the New Testament.[2] A *Time* magazine article from 1923 explained 'Who is a fundamentalist?' and went on to detail their beliefs.

Whatever the origin of the term, however, the underlying phenomenon is not new. According to Emerson and Hartman (2006, p. 139), 'In the broad sweep of history fundamentalists are normal. There is nothing unusual in people taking religion very seriously. What we now regard as religious 'extremism' was commonplace 200 years ago in the Western world and is still commonplace in most parts of the globe'. Hence, fundamentalism has not always been viewed as an oppressive system of belief or a negative paradigm for action. Indeed, both fundamentalists and fundamentalisms have been with us since the beginnings of civilization. What has changed in modern life is not fundamentalism as a way of being but the introduction of a new value system offering a different vantage point from which to assess it.

This new viewpoint emerged as a result of the colonial project of the eighteenth and nineteenth centuries, which spread the ideas of the European Enlightenment, with its focus on the individual, to parts of the world that for centuries had abided by totally different ideological systems. These European ideas advocated a new, universal frame of reference which assumed the adherence of all human beings to a common system of values. And in the early twentieth century the peoples of many colonized regions invoked this same value system to justify their rebellion against their colonial masters, to engage in independence struggles, and to establish themselves as new nation-states. Their various forms of resistance ranged from appeals to the humanity of their oppressors, to invocations of religious principles, to sometimes desperate extremist violence. But during this early period no one described such movements as fundamentalist. And after the dust from independence struggles settled, many of these new nations adopted modernist governance structures. As such, however, they soon found themselves having to battle some of the same groups which had helped them achieve independence, but which were now rebelling against the European-inspired universal value system. These groups usually advocated a return to real or imagined former value systems and the revival of mechanisms of decision-making from the past.

The rise of information technologies and globalization, coupled with the re-emergence of religious activism, aggravated this situation. As Manuel Castells argues in *The Power of Identity* (1997), the space of flows, which replaced the space of places, facilitates the rise of such fundamentalist movements not only in newly independent states but also in the lands of the former colonizers. And, as Benjamin Barber points out in *Jihad vs. McWorld* (1996), many nations in this new global order want it all. Despite obvious contradictions, they want Coca Cola, McDonald's, and the Internet, while they insist on restricting freedom of speech, movement, and communication. Timothy Mitchell (2002), on the other hand, argues that *jihad* is not antithetical to the development of McWorld, and that McWorld is really 'McJihad', a necessary combination of a variety of social logics and forces. Indeed, he points out that *jihad* seems to have ridden on the back of the information revolution, and has done extremely well by it. Hence, developments in many cities and places that we refer to today as tolerating fundamentalism are in fact a product of a specific moment in time, facilitated by the space of flows. In one of the earliest works on fundamentalism, *Defenders of God: The Fundamentalist Revolt Against the Modern Age* (1989), Bruce Lawrence made some of these same connections. According to Emerson and Hartman (2006, pp. 132–33), Lawrence argued that 'fundamentalism is an ideology rather than a theology and is formed in conflict with modernism. His study set the groundwork for sociology because ... [it viewed] fundamentalism ... as a transcultural phenomenon located in a developmental historical framework'.

Martin Marty and Scott Appleby were among the first to recognize the upsurge of fundamentalism in the early 1990s. In an initiative supported by the American Academy of Arts and Sciences entitled 'the Fundamentalism Project', which lasted from 1988 to 1993, they convened ten conferences involving some one hundred experts, some of whom have also contributed to this volume. On the basis of the five volumes of case studies and essays that came out of these scholarly exchanges, Appleby and Marty (1993*b*, p. 2) argued that there is 'a family resemblance within Fundamentalism that, to a greater or lesser degree, unites movements within the religious traditions of Christianity, Islam, Judaism, Hinduism, Sikhism and Buddhism'. In their definition, present-day fundamentalists may be against modernity, but they are also modern, a condition supported by their ability to master technology in a drive to attract adherents and maintain their power base. Marty and Appleby also concluded that fundamentalist practices are not simply conservative or regressive; on the contrary, they are mainly 'reactive', and may be seen as the way some peoples and groups 'fight back' against modernity, relativism, and pluralism.

Thus, the question arises: What makes a movement or a practice 'fundamentalist'? To our way of thinking, fundamentalist movements may first be viewed as a direct and self-conscious response to a particular form of modernity. Thus, some scholars see the rise of fundamentalist movements as embodying a bold disenchantment with specific Enlightenment ideas, particularly those that dilute the role of religion as a public force. Others view it as a rejection of specific colonial practices that attempted to contain religious influences in the public sphere in much of the formerly colonized world – specifically areas of the Muslim world. In general, fundamentalism is often perceived in opposition to three basic tenets of the modern: the preference for secular rationality; the privileging of individualism; and the adoption of religious tolerance and relativism (Appleby and Marty, 1993*b*).

It would be naive to think that fundamentalism exists in complete opposition to the modern world or its various manifestations. Indeed, the category itself can be defined less by a rejection of modernity than by social, ethnic, and nationalistic grievances. For example, many of those who self-define or are viewed as fundamentalists actively employ such products and processes of the modern as industrial technology, new information and communication media, and scientific tools. One reading of this paradox might be that the relation between modernity and fundamentalism is less 'antagonistic' than 'transactional'. In this respect, fundamentalism may be considered 'thoroughly modern and change-oriented' (Appleby and Marty, 1992, p. 34).

According to Bruce (2000, p. 117), 'fundamentalism is the rational response of traditionally religious peoples to social, political and economic changes that downgrade and constrain the role of religion in the public world…'. Thus, although they conceive their activism in the public sphere or in civil society as part of a necessary mission, religiously orthodox individuals who challenge the state, particularly if they form a shadow state, are usually designated as fundamentalists. This case is best illustrated by the challenge to the Israeli state by the Israeli settler movement in the occupied West Bank – and, more recently, by the challenge to the Palestinian Authority with the takeover of Gaza by Hamas. In some of these cases, the conflict is between two, or more, usually antagonistic, understandings and regimes of knowledge and statehood. But here we must recognize that such conflicts are not just battles between fundamentalism and secularism; they are often between two irreconcilable forms of fundamentalism.

But should fundamentalisms really be understood as a new type of modern resistance? As has been discussed, while fundamentalists often reject many of the values brought about by modernization and modernity, they seldom reject modernity's means – particularly technological ones. It is true that most fundamentalist movements invoke an invented history to justify their claims, but almost all nationalist movements have done the same. The only difference may be that fundamentalists always invoke an essentialist history based on belief in the inerrancy of a text or texts. Hence, they justify a nation of God or a city of God by invoking scriptural truth, while the nation-state invokes only the apparent truth of its own modernity.

Identifying and Critiquing Conceptions of Fundamentalism

Before turning to a discussion of how religious activism affects urban form and contributes to the making of fundamentalist space, it is necessary to examine who the so-called 'fundamentalists' are and what common attributes, if any, they may share. Emerson and Hartman (2006, p. 134) have recently suggested that there are specific ideological and organizational characteristics that fundamentalists of different orientations share. Among the ideological characteristics are the following: reactivity to the marginalization of religion; selectivity in defending certain religious traditions in a manner that differs from mainstream religious practices; a dualistic view dividing the world into good and evil, righteousness and unrighteousness; absolutism based on the belief in an inerrant text; and millennialism or messianism in expectation of a holy end to history. Among the organizational characteristics of fundamentalist groups are the following: a view of themselves as called to a mission; a tendency to create sharp boundaries requiring other groups to be with them or against them; organization around charismatic authoritarian leaders; and the enforcement of behavioural rules regarding speech, dress, eating, entertainment, children, and family formation. It is interesting to note that some of these characteristics apply equally to many nation-states, political communities, and social groups.

Many are sceptical of this conception of fundamentalism, and, indeed, researchers must address critiques of the concept as articulated by Emerson and Hartman. Most importantly, those wishing to employ the concept must demonstrate why its use will help advance knowledge of religion and society. While the contributing authors to this volume have done so in different ways, I will examine some of these critiques below. I do so not to advocate the concept of fundamentalism but to set the stage for some of the other articulations in this book.

The first charge against the concept of fundamentalism is that work on it often conflates conservative religious movements with postcolonial national religious movements. Some argue, for example, that significant differences must be 'brushed aside to view, for example, U.S. Protestant fundamentalists and Hindu national fundamentalists as the same conceptually' (Emerson and Hartman, 2006, pp. 130–131). While this may be a valid criticism, the contributors to this volume argue that such differences are not irreconcilable, or of as great significance as they might initially appear.

A second critique (and one that has been levelled particularly at Marty and Appleby's approach) is that it is inappropriate to use the term fundamentalism (which as noted above was originally a reference to a movement in American Protestantism) to describe movements in other religions, particularly non-Western ones. This practice has been denounced as a kind of Eurocentric conceptual imperialism (an especially sensitive charge in the Islamic world, where those designated as fundamentalists are usually upset by Western political, economic, and cultural domination). Again, while it is important to be sensitive to the use of terminology, this critique is mainly based on linguistic literality, and not the importance of the content it describes. In the end, fundamentalism, as it is addressed in this book, is about the spatial practices declared and enforced by some, and applied and accepted by others, to manage urban life.

A third critique is that the term fundamentalism has acquired such narrow negative connotations – ones that have often aligned it unnecessarily with terrorism – that it can no longer be used to explain more broad-based religious activism. Indeed, significant negative connotations of the term – usually including bigotry, zealotry, militancy, extremism, and fanaticism – do raise questions about its suitability as a frame for scholarly analysis (Emerson and Hartman, 2006). Furthermore, this type of labelling runs the risk of singling out and negatively tainting certain racial, ethnic, or religious groups.

Again, there is considerable truth in this view, particularly in the Islamic context because of its implied discriminatory stance. However, it is also important to offer a more nuanced understanding of the communicability of the concept. For example, no exact correlate for the term 'fundamentalist' existed in Arabic before the present discussions. When the concept had to be translated from English by the Arabic press, a term had to be invented. One of the earliest was *mutatarefeen*, which meant 'marginal extremists'. But non-militaristic religious activists who rejected this name came up with their own, *ussulleen*, which literally meant 'those who are returning to the origins'. What is interesting is that both terms were a reaction to the English term, and not terms that in and of themselves described an existing conscious understanding of the transformation of religious activism into religious extremism. Of course, this is a form of self-Orientalization that cannot be seen independently of the larger legacy of colonialism.

With the passage of time, however, the name distinctions between types of religious activists started to disappear, lumping the orthodox with the extremist, but also absolving the former of the responsibility to stand up to the hegemony and the violence often advocated by the latter. Hence, scholars of Islam and the Middle East have backed away from using the term 'fundamentalism' and have instead introduced terms like 'political Islam', and more recently, 'Islamist politics'. But we have to recognize here that any classification is based on the politics of the person making it; that no term is ever innocent of implication; and that all terms at some level are laden with unintended consequences. In one sense, then, these valiant attempts to be more accurate in naming the phenomenon have avoided a serious confrontation with it. Similar problems existed when terms like 'racism' were first introduced to scholarship to describe an exclusivist phenomenon in the nineteenth century, before the term settled into a comfortable meaning. Thus, in the end, this critique also remains of limited utility to those interested in studying how space is shaped by extremist ideology.

Some scholars, like Iannaccone, for example, argue that we should study sectarianism instead of fundamentalism. He writes that 'Sectarian religion is high-powered religion rooted in separation from and tension with the broader society'. And he goes on to suggest that 'studying sectarianism avoids problems of trying to explain why fundamentalism seems only to have appeared on a global scale since the 1970s, long after modernity had appeared on the scene in some places, and only shortly after it had appeared on the scene elsewhere' (Iannaccone, 1997, p. 114). This may, indeed, be true, but it may also be an escapist position, as there are few situations around the globe where sectarian conflicts do not involve fundamental positions and non-negotiable stances. And from a sociological perspective, Riesebrodt (2000) has suggested that many of the resurgent conservative religious movements share common features, and that despite the often significant differences between them, knowledge about them can be advanced by considering fundamentalism as a sociological category in need of theoretical development and empirical study.

In the end, the scholars who participated in our project and the contributing authors to this volume decided that having one single, generic definition of fundamentalism may not be very useful, as the urban context of each setting studied offers different, and possibly new, attributes to the definition of fundamentalism. Hence, in this book, we speak of fundamentalisms in the plural, with an 's'. Our use of the term is not an endorsement of the concept, but a pragmatic acceptance of its currency. It also embodies a recognition that our inability to agree on a single definition of a specific phenomenon, or to invent a new term to describe it, should not hinder our ability to discuss its implications for cities. Theoretically, a fundamentalist, according to one group, may be hero or a leader, while to another he or she may be a criminal. Indeed, some may consider the motivation of fundamentalists to be moral activism and its manifestation in the city to be religious urbanism.

The Territorial Character of Fundamentalism

Here one has to ask: If fundamentalism is a categorization based on religion, then what does it have to do with cities? Riesebrodt (1993, p. 9) defines fundamentalism as 'an urban movement directed primarily against dissolution of personalistic, patriarchal notions of order and social relations and their replacement by depersonalized principles'. The city may be important to the formation of fundamentalism as an ideological framework, and urban space needs to be considered in our understanding of fundamentalism. However, the purpose of this book is not to probe whether fundamentalism is essentially an urban phenomenon, because, as the contributors demonstrate, its practices change from place to place. Instead, the book is an attempt to spatialize and territorialize the logic of various types of fundamentalism and to cast the discourse in terms of cities and citizenships.

In a sense, our project is also about the intersection of the territorial manifestation of various fundamentalisms with the exclusionary and often hegemonic nature of the city. One might hypothesize that there are two versions of the fundamentalist city: the parochial and the global. The parochial fundamentalist city is one where mainly local forces have shaped the fundamentalist, exclusionary nature of space (for example, the cases of Peshawar and Hebron covered in this book). The global fundamentalist city is one where networks of global interaction become the reason for exclusionary practices (examples here include ethnic neighbourhoods both in the cities of the Islamic global South as well as the cities of the global North that have sizable minority populations).

Since the start of our project, our object of analysis has shifted from fundamentalisms to the city. Because of that shift, we have become much more interested in the territorial character of fundamentalisms and less in whether they are urban or not. Scholars have debated this issue and have come up with various viewpoints. Hence, as much as it is important to examine the territorial character of various fundamentalisms, it is equally important to interrogate the fundamentalist potential latent in the contemporary city. In the case of Islam, Asaf Bayat has argued that, although Islamist politics may be considered a social movement, it is a movement that does not always express itself in the urban realm. 'The identity of Islamism does not derive from its particular concern for the urban disenfranchised. It has never articulated a vision of an alternative urban order around which to mobilize the community members, whom the Islamists see as deserving welfare recipients, to be guided by leaders. These members are rarely expected to participate actively in making their communities' (Bayat, 2002, p. 13). On the other hand, invoking her work on Cairo, Salwa Ismail has suggested that the urban may indeed play a bigger role, and that militant Islamism has developed as an appropriation of spaces that emerged through the processes of urban growth and expansion. 'There are indications that the new landscape does not represent rupture with social traditions and popular modes of life, but is, rather, a variation on the historically evolving urban tapestry…' (Ismail, 2000, p. 264).

The relationship between the city and its hinterland in the fundamentalist equation is an important one, and one that is relatively unexplored by scholars of either religious or urban studies. This core-periphery relationship, which in certain cases describes the connections between city and desert, and in others between town and village, may explain much about the roots of various fundamentalisms. Two examples of the first type are eighteenth-century Wahhabism in central Arabia – which was reincarnated in the twentieth century during the events that accompanied the siege of Mecca in 1979 (and which emerged again with 9/11), and the cultish Jamaat al-Takfir wal-Hijra in Egypt – whose activities culminated in the assassination of the Egyptian President Anwar Sadat in 1981. Both of these movements played on a special and mutually reinforcing relationship between the city and its desert hinterland, in which plans concocted in the former, assumed to be a space of virtue and purity, were executed in the latter, condemned as a place of vice and infidelity.

Wahhabism, as a sect, is attributed to Muhammad Abd-al-Wahhab, an eighteenth-century figure from what is today Saudi Arabia. Wahhabi theology treats the Quran and the Hadith as the only fundamental and authoritative texts. Some thirty years ago, a group of Wahhabi militants laid siege to Islam's holiest mosque in Mecca. Led by a young man who believed himself to be the promised Mehdi, they came to cleanse the earth of sin and bring the Kingdom of God to man. In his book *The Siege of Mecca*, Yaroslav Trofimov has described how this event, which occurred on 30 November 1979 – the first morning of a new Muslim century – launched the first global *jihad* and later contributed to the founding of al-Qaeda. It also shows the interesting connection between the city, in which the action occurred, and the desert, which served as a training ground. The practices of Wahhabism may explain why Saudi Arabia as a country is as orthodox as it is today. They also may explain much of the discourse identified as fundamentalist in the Muslim world today.

Jamaat Takfir wal-Hijra was originally founded in Egypt, emerging in the 1960s as an offshoot of the Muslim Brotherhood. Members of the group were Islamists who claimed not to be bound by the usual religious constraints. Although they were originally ambivalent about the city, and migrated to the desert to perform Hijra in the tradition of Prophet Mohammad, they were always interested in returning to the city. The opportunity came with the plot to assassinate Anwar Sadat. They believed that ends could justify means, and that killing other Muslims was acceptable to further their cause. They were also willing to adopt non-Islamic appearances to blend into crowds and make themselves hard to detect once they returned to the city.

Another example that demonstrates this relationship between the city and the countryside may be the role of the village and the small town in the events in India around the Godhra train burning which led to the Gujarat riots of 2002. The Gujarat riots and their repercussions, such as the communal riots in Bombay, have been represented worldwide in media accounts as a clash between Hindu and Muslim fundamentalists – a predictable outcome of fundamentalist ethos, motivated by religious zeal. However, the question remains: Why did events in the small and unimportant towns of Godhra and Ayodhya find their most violent repercussions in some of the larger cities of India? Ordinarily, the question of space (particularly urban space) has been peripheral to the understanding of fundamentalism as a phenomenon. In fact, when urban space is referred to, it is usually only as a tableau against which fundamentalism is played out. Yet, in an essay reflecting on the rioting in Bombay following the destruction of the Babri Masjid in 1992, Arjun Appadurai (2002) claimed that the violence perpetrated against Muslims was in fact motivated by the desire of Hindus to claim valuable urban resources such as property and water, and that this could only be achieved through the elimination of Muslim bodies. In other words, fundamentalism served as the pretext for a larger campaign of land-grabbing, in which Muslims were seen as undeserving of urban space or resources.

Much of the work on fundamentalism and space to date either abstracts the urban as a micro-site of the nation or reduces it to a set of resources that provides a subsidiary battleground for ideological conflict. Another way to understand the fundamentalist city beyond the lens of religious activism, however, is to interrogate it through practices of exclusion. In this regard, one can use the frameworks of the apartheid city, the ghettoized city, or the ethnically, racially, or religiously divided city to understand fundamentalist urban practices. Oren Yiftachel (2006) articulates this point using the concept of ethnocracy in his most recent book. There, he suggests that the urbanization of Israel can be seen as a social, cultural, and political engine through which the nation has been able to maintain its hegemonic control over the Palestinian territories, and thus fulfil its mandate as a Jewish state. In an earlier work (2003), he used the case of the city of Beer Sheva to argue that Israel has employed planning as an instrument of control to establish an ethno-nation. Urban informality was deliberately created in Beer Sheva to exclude and impede the development of an entire ethnic population: the city's Arabs. Despite the fact that these Arabs hold Israeli citizenship, most of them have lost their right to the land through a process of legalized expropriation by the state.

Mike Davis (2007) has approached these issues from another direction in his writings about the rise of Pentecostalism in squatter settlements in Latin America and elsewhere. He has argued that this form of religious affiliation, which many would actually consider fundamentalism, has been a powerful negotiating tool for the marginal urban poor all over the world. Finally, Loic Wacquant (2008) has argued that generic macroeconomic forces have interacted with the particular race/class structures of cities and national-state strategies to produce various modalities of advanced marginality in post-industrial societies. These studies provide valuable insight into the discussion of space and fundamentalism and articulate how the urban often becomes a micro-site of the nation and a site of religious contestation.

Defining the Fundamentalist City

The main question this book attempts to answer is this: What makes a city a fundamentalist, and in whose eyes? From this, of course, follow a number of ancillary questions, including: Is a city which deprives a particular gender of access to its streets a fundamentalist city? Is a city which requires religious minorities to observe and obey the practices and dress codes of its majority a fundamentalist city? Is a city which bans smoking and renders smokers a banished group a fundamentalist city? When is this line crossed?

These questions are not simply rhetorical, as these conditions exist in many cities around the world – from Peshawar, Pakistan; to Cairo, Egypt; to Berkeley, California. For example, the current Saudi government restricts access to Mecca to Muslims. Some interpret the Quran as having specified this prohibition, but does that make Mecca a fundamentalist city? What if this same quality were applied to entire regions or nations? Would we then consider them fundamentalist? Following the events of 9/11, the United States adopted entry requirements of a similar nature. Does that make the U.S. during that period, a fundamentalist nation? Who has the right to make this designation, and to what end?

While cataloguing so-called fundamentalist cities may be a useful exercise for agencies concerned with national security, who often confuse fundamentalism with terrorism, this is not the project of the contributors to this book. Therefore, as a working hypothesis to flesh out the discourse, let me suggest that the fundamentalist city has a few hypothetical attributes.

1. It is a city which categorically excludes by law, tradition, declared policy, or latent practice individuals who are adherents of another religion or who belong to a different ethnicity than those of the ruling power or majority population. It is a city whose minority residents, belonging to different ethnic or religious groups, are denied access to basic urban services or specific public spaces.

2. It is a city whose religious majority mandates, or practically expects, that its minority groups will conform to all the rituals of public behaviour prescribed in the religious code of the majority.
3. It is a city that is so gender segregated, as mandated by a male-dominated society, that women have little or severely restricted access to public space.
4. It is a city which normalizes most of the above-mentioned forms of control or oppression in everyday life to the extent that the minority ceases to question them.

Of course, all these are attributes of a theoretical, generic fundamentalist city. And some of them may, in fact, describe one actual city or another. However, it would be a grave error to use them as test criteria. Indeed, any discussion of fundamentalism in a given place needs to take into account the specificity of local culture and its location within a web of global interconnectedness. Hence, a number of serious qualifications need to be made.

The first is that we should not lose sight of the main goal, which is exploring the specific urban conditions that contribute to the rise of religious extremism, and, hence, fundamentalism. Many of the contributors to this book have attempted to understand how cities might provide a particular utility to the larger mission of religious fundamentalisms or political extremisms and the various ways these have been manifested around the world. This issue is also historical: it requires close examination of the particular nature of the urban at this moment in time, as well as the specificities of contemporary fundamentalisms and how they might be forged as the result of a mutually constitutive dialogue.

The second qualification is that we believe that cities, as well as fundamentalisms, are not monolithic, but range across a wide spectrum. Quite simply, there are as many different types of fundamentalism as there are categories of cities. As we accept this qualification, we will begin to identify urban trends around fundamentalisms, and vice-versa. For example, almost every type of religious fundamentalism has been associated with an iconic city, based on historical, cultural, or religious importance. A closer examination of these cities might help in identifying particular types of urbanity – or at least the particular urban imaginaries that underpin each of their respective fundamentalisms.

A third issue is that fundamentalism has always been linked to authority, whether manifested in a single person (ayatollah, pope, etc.) or in a literary manuscript (Quran, Veda, etc.). Often fundamentalist principles stress the importance of such a text as an inerrant historical record. This is a crucial qualification to the idea of space and the city. Often the imposition of fundamentalism onto a given landscape requires a great deal of creativity, because the literal texts remain ambiguous about cities as they exist in the modern world.

A fourth qualification relates to the connections between modernity and fundamentalism in cities where there is strict public adherence to religion (and there are many such cities in the world today). For example, when the practices of modernity are not the central concern or target of a fundamentalist movement, other fundamentalisms may be invoked to play that role. Indeed, one may argue that almost all fundamentalisms are defined against one another, and often in response to one another. For example, Hindu fundamentalists in India view their enemy as Muslim fundamentalists, whether the threat of Muslim fundamentalism is serious or not. Such fundamentalisms are conceived in response to a threat – real, exaggerated, perceived, or imagined. A similar situation exists in terms of the Muslim world and its postcolonial relationship with Europe; in this case, a Crusader Christianity is perceived as being the prime motivator of Western interventions in the Middle East.

A final issue relates to the supposed difference between ultra-traditionalists and fundamentalists. Is there really a difference, as some have claimed, or is the difference one of space and location? It has been suggested, for example, that the Amish of Pennsylvania are ultra-traditionalists, not fundamentalists like the Muslim Brotherhood of Egypt. This is supposedly so because the former mind their own business, hope to be left alone in their insular rural setting, do not proselytize, are not antagonistic to the state, and do not seek to change the nation or enforce their views on it. But some may argue that the real difference is that the former is principally rural, while the latter is mainly urban.

Fundamentalisms and the Making of Urban Space

As I indicated earlier, the first section of this book deals with the larger discourse on the city, citizenship, and fundamentalisms. In the next chapter, Inger Furseth picks up this theme by reminding us that until the late 1970s and 1980s, the operative assumption of many social scientists was that the growing secularization of Europe during the twentieth century would spread to the rest of the world and become the rule for public life, particularly in urban areas. Since then, however, a great amount of research has shown this has not been the case.

Furseth distinguishes two major approaches to the concept of fundamentalism: demand-side approaches, such as crisis theory, examine fundamentalism as a response to external cultural and socioeconomic forces; supply-side approaches, such as resource-mobilization theory and political-process theory, attempt to understand internal motivating factors such as ideology, organization, pre-existing social networks, and gender. Research also indicates how cities provide a wealth of resources that fundamentalist groups may use to advance their purposes. The compact quality of urban space is conducive to recruitment; there are advantages gained from working in close proximity with other social groups; broad ideological messages can be framed within the context of local grievances; pre-existing social networks can be leveraged for new purposes; and gender distinctions can be exploited. Furseth also illustrates how the modern city has proved to be fertile ground for the spread of new religious movements, the growth of Protestant evangelicalism, and the rise of political Islam.

Coming at the subject mainly from the point of view of fundamentalism and citizenship, James Holston then examines the internal moral basis for fundamentalist regimes of power. Civility is the ideological consensus that underlies all regimes of citizenship, he argues; and since the Enlightenment, Western social theories have tended to equate it with values such as 'civilization, liberty, trust, generosity, inclusion, secularism, and democracy'. But all citizenships are based on modes of exclusion. And Holston points out that fundamentalist religious regimes are only an extreme example of the ability to generate consensus on the basis of a denial of rights.

The chapter then examines the conditions that sustain regimes of national citizenship whose very basis is the differential apportionment of rights and privileges. Using the example of the genocide practiced against Native Americans by fundamentalist Spanish colonial authorities, Holston suggests that 'citizenship remains overwhelmingly a means for distributing and legitimating inequality, and therefore for creating both marginal (national) citizens and stateless non-citizens'. However, what most distinguishes fundamentalist regimes is their use of transcendental truth as a point of reference to determine standards of conformity. They admit no rational debate, assigning both the right and the duty of realization to the believer, and render the views and interests of the non-believer irrelevant.

The second section of the book turns to the relation between forms of fundamentalism and urbanism, using case studies that cover Christian, Muslim, and Hindu practices. Rhys Williams's chapter focuses on the birth of Protestant fundamentalism in the United States – 'a forceful reassertion of theological orthodoxy that grew into a significant religious, and eventually social, movement'. In particular, the chapter distinguishes fundamentalism as it emerged in the early twentieth century from older American religious practices of traditionalism and evangelicalism. In the process, Williams shows how American fundamentalism shares traits and attitudes with fundamentalist movements elsewhere.

This is a story, Williams writes, 'with obvious implications for the contemporary United States, particularly as American cities are again being transformed by immigration and incorporation into global markets'. In the United States, Protestant fundamentalism rose in response to modernity as a doctrinal reaction both to secularism and to the attitudes and practices of established churches that sought to accommodate themselves to it. As Williams points out, it is sometimes thought of as antithetical to cities; however, it is better thought of as a response to the changed urban context in the U.S. at the beginning of the twentieth century, which was characterized by an uncomfortable mixing of populations and religious ideas. The association of fundamentalism with cities in America is therefore 'both foundational and deeply engrained', Williams suggests.

Renu Desai then provides a view into the divisive and exclusionary Hindutva movement, which has attempted to promote a view of India as an exclusively Hindu nation. Her chapter uses the case of Ahmedabad, one of India's largest cities, to reveal how such fundamentalist movements rely on the structuring capacity of urban areas to reinforce their ideological hold over target populations. The chapter examines the practices through which Ahmedabad has been reconfigured by Hindutva politics and the ways in which the ideology has been embedded in the life of the city and its residents. Desai interrogates what she refers to as the 'communalized city' as an instrument of this ideological campaign. Ahmedabad has long been characterized by patterns of residential segregation by caste and socioeconomic status. But in recent decades the patterns of exclusion have been deliberately altered through organized campaigns to create suspicion, fear, and exclusion of an imagined Muslim Other among Hindus. One of the purposes of this violence has been to fix separate and antagonistic notions of identity in urban space, and create separate Muslim and Hindu neighbourhoods. These new enclaves in turn emphasize a series of internal urban 'borders', where 'communal hostility is displayed to reinforce separateness, antagonism and irreconcilability'.

As Desai writes, in the secular nation-state of India such a campaign also 'embodies a militant rejection of secular modernity'. And these practices have been both reinforced and undercut in recent years by the forces of globalization. To participate in the global economy, the city has attempted to shed its image as a site of destructive communal violence. But both government and business have reacted to recent outbreaks of rioting not by addressing its dehumanizing aspects, but by seeking to normalize the new patterns of urban exclusion, thus silently reproducing the 'communalized city'.

Mona Harb's chapter then investigates the impact of the Iranian-backed Hezbollah, or Party of God, on the structure of metropolitan Beirut. Contemporary characterizations to the contrary, Hezbollah may best be understood as a broad social movement representing the social, political, and cultural aspirations of Lebanon's Shi'a population in the face of historically stronger powers in the region. Like many other organizations that emerged from years of civil war in Lebanon, Hezbollah is fundamentalist both in terms of ideology and organization. But Harb describes how Hezbollah has benefited greatly from its ability to control and structure a physical territory to reflect its values and socio-political positions. In particular, it has embedded and inscribed itself into the social and spatial setting of al-Dahiye, the southern suburb of Beirut which now symbolizes its posture of defiance. Harb describes this suburb as a space of religious ideology, an 'Islamic milieu', monitored and administered by a series of Hezbollah-related NGOs that deliver social and economic services to loyalists. In doing so, Hezbollah has also adopted a distinctly modern, rational approach that highlights technical competence and accountability based on modern systems of organization, monitoring, and control as well as a centralized administrative structure.

Another component of Hezbollah's spatial strategy has been the inscription of religious symbols and the imposition of its own system of territorial demarcations. Harb writes that 'Al Dahiye has thus been (re)produced by the combined actions of Hezbollah's networks, private and associative stakeholders, and ordinary residents. These actions have produced a sense of territoriality which allows the Islamic milieu to be embedded in the daily lives of al-Dahiye's residents while simultaneously providing them with a "stage" through which their socio-cultural identities can be displayed, gazed at, expressed, and renegotiated'.

In her chapter, Francesca Giovannini turns to the triumph of Hamas among Palestinians in the Gaza Strip. The success of Hamas in the 2006 elections marked a transition in the struggle for Palestinian statehood, as a predominantly secular, political effort was seemingly transformed overnight into a struggle over religious identity. Giovannini writes, however, that this transition was a long time coming, and owed much to the tireless work of Hamas among the residents of Gaza's eight refugee camps. These camps have now become 'the bedrock of Hamas's political and social support and a privileged base for recruitment into its *jihadist* operations'. Giovannini explains how the spatial isolation of the camps has been crucial to this success. Separated from the rest of the urban environment, the camps have provided an ideal place to consolidate and actualize Hamas's message and system of governance.

Hamas's charter is fundamentalist in its core, as it claims the land of Palestine as a religious endowment, obviating any possibility of negotiation. Ironically, this position may be what Hamas shares most literally with its arch enemy, the state of Israel. But, according to Giovannini, Hamas's strategy inside the camps has involved 'the creation of Islamic spaces and symbols, a skilful and methodical use of these religious spaces for humanitarian purposes, and the manipulation of vulnerable groups such as women and children'. Their effect has been to create 'trapped spaces' that express the very ideological foundation of the organization. These operate both at the physical and the symbolic levels, allowing Hamas to connect the personal struggle of the refugees for daily sustenance and meaning to its own Manichean efforts to assert an ultimate, divine truth.

The third part of the book contains five chapters that address the intersection of issues of identity, tradition, and heritage as essential attributes of the fundamentalist equation, particularly in relation to space. It begins with Oren Yiftachel and Batya Roded's chapter on the political geography of 'religious radicalism' in Israel. They prefer to use that term in the Israeli context as an equivalent to fundamentalism, arguing that such radicalism is closely related to ethno-national conflicts in general, and to the recent development of urban colonialism in particular.

The chapter highlights two related phenomena: the typical transformation of relations between ethnocratic and theocratic mobilizations from mutual reinforcement to intensifying conflict; and the propensity of state-supported urban colonialism to generate counter-religious mobilizations. It further reminds us of the importance of bringing conditions of material life into abstract discussions of religious fundamentalism, particularly in relation to issues of access to and control of space. To illustrate the argument, Yiftachel and Roded analyze the dynamics of ethno-religious relations in three sacred cities that are sites of contestation between Jews and Muslims – Hebron, Jerusalem, and Beer Sheva. They argue that the state's ethnocratic geopolitical policies remain the main (although not the sole) cause of religious radicalism. They claim that religious radicalism often derives from the very identity projects instigated by modern nation-states, and by the social and economic conditions they create.

Mejgan Massoumi next examines the resurgence of the Taliban movement in Pakistan, where cities near the Afghan border have been transformed into sites of strategic importance both locally and globally. Here, Islamic fundamentalism has been reinvented in its most insidious form, and in ways that reveal its conflicted relation with modernity. But Massoumi suggests that the rise of the Taliban can be viewed both as part of a broader resurgence of religious fundamentalism since the 1970s and as the expression of an 'imagined' form of Islam grounded in the culture of a single ethno-linguistic group: the Pashtuns.

Over the course of three decades, as the Taliban have used Peshawar as a base of operations, they have gradually enforced strict new codes of religious expression in public life. The chapter identifies its specific forms of urban exclusion and surveillance as 'strategies of performance and practices of oppression'. These include the deprivatization of religion in public space, street spectacles, a campaign of urban violence, and the spatialization of fundamentalist knowledge through various local religious institutions. Together, these activities implant and enforce the Taliban's values, norms, belief system, and order on the city and its people. Massoumi also suggests that Peshawar can be seen as a fragmented and unregulated informal city – a city in which a *de facto* sovereign power has established both a distinct form of governance based on an 'imagined' religious tradition and particular forms of negotiated citizenship based on its own regime of rule.

In his chapter, Omri Elisha discusses the growing participation of suburban evangelical mega-churches in philanthropic and civic enterprises seeking to promote social and physical revitalization in inner-city communities. He suggests that these new forms of evangelical engagement represent a departure from Christian fundamentalism's principled withdrawal from matters of social and public policy. But such activities are also consistent with longstanding missionary and revivalist sensibilities.

In Knoxville, a predominantly conservative Protestant community where 'white flight' has been especially detrimental to inner-city communities, Elisha explains how such white evangelical practices are also filled with cultural antagonisms, racial politics, and conceptual ambiguities. These include a tendency by suburban churchgoers to imagine themselves, according to biblical passages, as 'exiles' in their own world. Elisha analyzes the conceptual dimensions of this 'exilic frame' in the context of evangelical outreach and discusses the ways in which it reflects ideals rooted in the millennialist paradigm of the 'Kingdom of God on earth'. Thus, the broader mission seems less concerned with repairing social injustices than Christianizing regional cultures and establishing white evangelical institutions as the dominant engines of public morality and civic life. As socially engaged evangelicals become increasingly engaged with issues of urban revitalization, gentrification, and neoliberalization, Elisha concludes they may continue to be hampered by this older, antagonistic view of cities.

In her chapter, Mrinalini Rajagopalan questions the secular foundations of the urban in the contemporary world and conventional readings of religious fundamentalisms that are 'aspatialized', or vaguely defined, in terms of urban geography. Her specific focus is India's capital city, Delhi, and its contemporary position in the rhetoric and ideology of Hindutva, the resurgent religious and political movement discussed in Desai's earlier chapter. As Rajagopalan points out, the position of Delhi within the master narrative of Hindutva is precarious since it has a long history as a centre of Islamic power. The rich visual culture left behind by this history and its subsequent absorption into the framing of India as a 'secular' nation has proved a stumbling block to Hindutva's chauvinistic aspirations. Indeed, the city's surplus of Islamic architecture and heritage stands in sharp contrast to the conspicuous absence of any real 'Hindu' heritage.

The chapter further argues that attempts by agents of Hindutva to reclaim Delhi are not simply aimed at a renegotiation and re-presentation of the decades of Islamic culture. The project also attempts to recalibrate the secular foundations of modern India. Rajagopalan suggests that while the rubric of fundamentalism has rarely been used to describe the ideological underpinnings of Hindutva, its use puts into context the movement's appropriation and recalibration of secularism. She ends by examining various Hindutva interventions aimed literally at rewriting India's past by attempting to rewrite the history of its built environment.

In the final chapter of the book, John Eade examines the cross-cultural conflicts generated by this sense of mission in two urban contexts: the effort by the Muslim Tabligh Jamaat to build a congregational mosque at Abbey Mills in London's East End, and the effort by Polish Catholic leaders to create their own churches apart from the acculturating influences of British Catholicism. Eade describes the ways the Tabligh Jamaat and the Polish Catholic Mission have attempted to use migrant communities to promote their visions of resistance to Western secularism. But this effort has required a physical presence, which has galvanized opposition from neighbourhood activists, who have used the trope of fundamentalism and raised the spectre of a global campaign against Western values.

The cases are clearly different, but Eade points out that in each instance the 'need for a territorial space' has required negotiations with existing religious and secular authorities and in the process opened both missions to inclusivist forces. 'The impulse towards a purificatory exclusivism was countered, therefore, by the everyday force of cultural diversity and engagement with mainstream society', he writes. Thus, contrary to the common view that physical structures serve as sites of fundamentalist preaching and the spread of extremist ideas, he concludes they are also sites of struggle between inclusivist and exclusivist forces, where 'the pressure toward inclusion ... promises to overcome attempts to keep migrants within the walls (physical and mental) of embattled religious communities'.

The New Urban Frontier

All these examples lead to a basic position that the fundamentalist city is where certain categories of people or the religious other are rendered 'bare life' (Agamben, 2005). Is the fundamentalist city an approximation of Agamben's camp? Or is it a post-city, offering neither citizens nor citizenship? The fundamentalist city is a place where religious fundamentalisms do not see the urban as evil, but actually claim it as a new domain beyond the idea of the nation (in the case of Islam, the idea of the *Umma*). But we should also not forget that fundamentalisms rely on an ideology of exception and a culture of constant surveillance. Cities in the global era are turning into fragmented landscapes made up of spaces of exception. Hence, the relation of both fundamentalism and cities to issues of citizenship becomes very important.

Are we then returning to an era of medieval modernity? As I have argued elsewhere, 'If the "feudal" is a system of political, economic and social relationships and if the "medieval" is a system of ordering space, then the seemingly oxymoronic phrasing of "medieval modernity" indicates how the medieval lurks at the heart of the modern, how the feudal exists within capitalism' (AlSayyad and Roy, 2006, p. 16). In the fundamentalist city, the examples of citizenship indicate how the modern city functions through a medieval ordering of space. The oxymoronic phrasing of 'medieval modernity', used as an analytical concept not a historical period, thus reveals the inherent paradoxes of the modern: fiefdoms of democracy, the materialist immediacy of religious fundamentalism, and the simultaneity of war and humanitarianism. Here, modernity must be recognized as 'an inevitably fractured, divided and contradicted project' (*Ibid*., p. 17). It is clear that modern forms of national citizenship might be giving way to 'fractal and splintered territorialisation of citizenship in medieval enclaves' (*Ibid*.).

When the values of urban citizenship become irreconcilable with the intent of various fundamentalisms, the city itself becomes a meaningless fiction. The idea of the denial of existence altogether to groups of people, which often leads to ethnic cleansing at the level of nations, is the ultimate form of fundamentalism. The Nazi city was a product of the Nazi nation. Today we live in an increasingly urban world, where more than half of all human beings live in cities. It remains to be seen whether the so-called fundamentalist cities of today will make the fundamentalist nations of tomorrow.

Notes

1 American Heritage Dictionary online at http://dictionary.reference.com/browse/fundamentalism.
2 *The Fundamentals: A Testimony to the Truth* (1910–1915) was a series of twelve pamphlets that attacked modernist theories of biblical criticism. Their central theme was that the Bible is the inerrant word of God and should be taken literally. But, as a writer in the 19 May 1923 issue of *Time* magazine discovered, there are earlier sources for such viewpoints. For example, the Westminster Confession of 1643, which provides the constitution of Presbyterianism, substituted the authority of the Bible for the authority of the Roman Pope, and held the Bible to be 'the only infallible rule of faith and practice'.

References

Agamben, G. (2005) *Homo Sacer: Sovereign Power and Bare Life* (translated by D. Heller-Roazen). Palo Alto, CA: Stanford University Press.
AlSayyad, N. and Roy, A. (2006) Medieval modernity: on citizenship and urbanism in a global era. *Space and Polity*, **10**(1), pp. 1–20.
Appadurai, A. (2002) Spectral housing and urban cleansing: notes on millennial Mumbai, in Breckenridge, C.A. (ed.) *Cosmopolitanism*. Durham, NC: Duke University Press.
Appleby, S. and Marty, M. (1992) *The Glory and the Power: The Fundamentalist Challenge to the Modern World*. Boston, MA: Beacon Press.
Appleby, S. and Marty, M. (eds.) (1993*a*) *Fundamentalisms and Society*. Chicago, IL: University of Chicago Press.
Appleby, S. and Marty, M. (eds.) (1993*b*) *Fundamentalisms and the State: Remaking Polities, Economies, and Militance*. Chicago, IL: University of Chicago Press.
Barber, B. (1996) *Jihad vs. McWorld*. New York: Ballantine Books.
Bayat, A. (2002) Activism and social development in the Middle East. *International Journal of Middle East Studies*, **34**(1), pp. 1–28.
Bruce, S. (2000) *Fundamentalism*. Malden, MA: Blackwell.
Castells, M. (1997) *The Power of Identity*. Malden, MA: Blackwell.
Davis, M. (2007) *Planet of Slums*. London: Verso.
Emerson, M.O. and Hartman, D. (2006) The rise of religious fundamentalism. *Annual Review of Sociology*, **32**, pp. 127–144.
Iannaccone, L.R. (1997) Toward an economic theory of 'fundamentalism'. *Journal of Institutional and Theoretical Economics*, **153**, pp. 100–116.
Ismail, S. (2000) The popular movement dimensions of contemporary militant Islamism: socio-spatial determinants in the Cairo urban setting. *Comparative Studies in Society and History*, **42**(2), pp. 263–293.

Lawrence, B. (1989) *Defenders of God: The Fundamentalist Revolt Against the Modern Age*. San Francisco, CA: Harper & Row.

Mitchell, T. (2002) McJihad: Islam in the U.S. global order. *Social Text 73*, **20**(4), pp. 1–18.

Riesebrodt, M. (1993) *Pious Passion: The Emergence of Modern Fundamentalism in the United States and Iran* (translated by D. Reneau). Berkeley, CA: University of California Press.

Riesebrodt, M. (2000) Fundamentalism and the resurgence of religion. *Numen*, **47**, pp. 266–287.

Trofimov, Y. (2008) *The Siege of Mecca: The Forgotten Uprising in Islam's Holiest Shrine and the Birth of al-Qaeda*. New York: Knopf.

Wacquant, L. (2008) *Urban Outcasts: A Comparative Sociology of Advanced Marginality*. Cambridge: Polity Press.

Yiftachel, O. (2003) Control, resistance and informality: Jews and Arabs in the Beer-Sheva region, Israel, in AlSayyad, N. and Roy, A. (eds.) *Urban Informality in the Era of Liberalization: A Transnational Perspective*. Boulder, CO: Lexington Books, pp. 111–133.

Yiftachel, O. (2006) *Ethnocracy: Land, and the Politics of Identity in Israel/Palestine*. Philadelphia, PA: Penn Press.

Chapter 2

Why in the City? Explaining Urban Fundamentalism

Inger Furseth

The religious resurgence that began to appear in the late 1970s and the early 1980s in several parts of the world took most social scientists by surprise. They had anticipated that the secularization which had been growing in Europe for decades would spread to the rest of the world. Even more surprising was the fact that these 'fundamentalist movements', as they were soon called, seemed to appear and grow in urban areas. Sociologists of religion had long assumed that modernization with its institutional differentiation and urbanization would lead to the decline of religion and its disappearance into the private sphere. Instead, a growing number of fundamentalist movements within all world religions made claims on the public sphere. Some movements aimed at just being visible in public, whereas others demanded to have a voice in political matters. Again others attempted to overthrow the state and impose new regimes based on religious doctrine. These movements varied from smaller groups to large-scale global movements, and some were peaceful, whereas others engaged in violence.

Scholars from various fields, the social sciences as well as the humanities, have produced a great deal of research on fundamentalism. This chapter explores and discusses some theories of social and religious movements that may be useful in studies of urban fundamentalism. It discusses the term 'fundamentalism' and distinguishes between so-called demand-side and supply-side movement theories. The main argument offered is that in order to understand fully and to explain the growth of urban fundamentalist movements, attention must be given to both external and internal factors. The discussion centres first on two external factors: the urban condition, and the context of other movements. Four internal factors are then analyzed: movement ideology, organization, pre-existing social networks, and gender. In the discussion, examples will be taken from fundamentalist movements within various world religions.

Fundamentalism

The term 'fundamentalism' originated in the United States, where it appeared in the 1920s as a reference to Protestants who published a series on 'The Fundamentals' of the Christian faith. In the 1970s and 1980s, however, it became a broad concept used to describe most groups that took religion seriously, and especially those that made claims on public policy (Bruce, 2007, pp. 10–12). The term also often had negative connotations, implying that fundamentalists were irrational and emotional extremists, and that their religious interpretations and practices were backward.

Scholars have disagreed whether the term is useful: some have rejected it, while others have found it useful. Martin E. Marty and R. Scott Appleby, who led the Fundamentalism Project during the 1990s, argued that there were enough common features among different fundamentalisms to continue using the concept. Likewise, Steve Bruce (2007) outlined at least five features that many scholars think various fundamentalisms share. One is the claim that one or more sources of ideas, such as religious texts, are complete and inerrant. Second, fundamentalism presents an ideal past that demonstrates original religious conditions. Third, although fundamentalism is based on the past, it is reworked for present purposes. The fourth feature is that fundamentalism tends to find support in marginalized groups. Finally, many fundamentalisms adopt and use modern technology. Other scholars have stressed different aspects of fundamentalism – as, for example, psychological traits of the participants such as seeking authority, causing scandals, opposing ambiguity, creating divisions between 'us' and 'them', and being potentially or actually aggressive (Marty, 1992). Again others have stressed the mythical aspects of fundamentalist ideologies, such as the expectation of a utopian golden age (Nielsen, 1993) or opposition to the global institutional differentiation process (Shupe and Hadden, 1989).

A problem in using the strategy of outlining features in movements and searching for empirical cases that fit these features is the tendency to explain by 'definition', which means that concepts of fundamentalism contain assumptions that determine that any hypothesis in which the concept is used can only be confirmed. For example, fundamentalist movements can be defined in such a way that they only include people who seek authority and oppose ambiguity. These movements are often defined by labels, which include implicit sets of assumptions about people's motives for joining, their normality, and why these movements develop. Rather than studying empirically what types of motivations people have for joining, their reflections about participation, and the ways they were recruited, these issues are often left unexamined. For example, Bruce's idea that fundamentalisms mostly appeal to marginalized groups is not necessarily supported by empirical research. Studies of political Islam in the Middle East, for example, show that it has a relatively wide base. Hamas is supported both by the 'disinherited youth as well as among the traders and middle class' (Kepel, 2002, p. 324). The support for Islamist groups in this region has also changed over time. In the 1970s, they mostly found support among the urban poor and the devout bourgeoisie, whereas in the 1980s, the Islamist cause became significantly more middle class (*Ibid*., pp. 67, 80).

Yet another shortcoming is that concepts of fundamentalisms fail to explain why some movements are successful in mobilizing large groups of participants. Therefore, they cannot be used in explanations of fundamentalist growth. Nevertheless, concepts of fundamentalism, like the one outlined by Bruce, can still be fruitful as a sensitizing tool in distinguishing between different types of religious movements. This is the way the concept is used here.

Explaining Fundamentalist Movement Growth

If our aim is to explain fundamentalist movement growth, different sociological theories are available. The main distinction lies between the so-called demand-side and supply-side approaches. The demand-side approach emphasizes conditions external to the movements in question. It focuses on how the urban condition creates a demand for, or produces, fundamentalist movements. The supply-side approach stresses internal variables, analyzing the dilemmas facing movements attempting to mobilize. This, in turn, shifts attention to the allocation of resources and the different ways in which movements organize in order to reach their goals.

One example of the demand-side approach is crisis theory, which is frequently used implicitly in studies of fundamentalist growth (Furseth and Repstad, 2006, pp. 159–160; Sahliyeh, 1990, pp. 3–16). The argument here is that religious resurgence is a result of the modernization crisis. Different aspects of modernization are stressed – for example, a decline in traditional moral values, the loss of cultural identity, political oppression, and socio-economic inequality. One example is a study of India, where economic deprivation, social exclusion, and the political under-representation of Sikhs were used to explain the formation of militant religious movements (Singh,

1990). Studies of Islamic fundamentalism also often stress leadership and economic crises as explanatory factors. Thus, many Arab leaders during the 1960s and 1970s used socialist ideology to suppress opposition and maintain control of the state. But their efforts at governing were confounded by corruption, an inability to resolve issues of poverty, and the loss of the 1967 war against Israel (An-Na'im, 1999; Esposito, 2002; Kepel, 2002). Some scholars also emphasize negative reactions to modernization. The growth of the new Christian Right in the United States has often been interpreted from this perspective. This movement is seen as a conservative response by traditional Protestants, who oppose the decline in traditional values, especially sexual permissiveness (Hertzke, 1990). According to crisis theories, religion functions as a type of coping mechanism, helping people deal with difficulties in life by offering them comfort, support, discipline, and advice.

Nevertheless, crisis theories face several shortcomings in explaining the emergence and growth of fundamentalist movements. These theories assume uniformity among participants – for example, that they all have the same reasons, such as protest or discontent, for joining. Instead, it is reasonable to assume that participants in movements have different motives for joining, and that not all participants agree with everything for which a movement stands. Furthermore, crisis theories only invoke the conditions that may or may not result in a person's motivation to join. The emphasis is on identifying conditions for the existence of the movement, with the assumption that people act as they do because of external pressure. Yet, these external factors cannot in and of themselves explain the recent growth of religious movements.

During the 1960s resource-mobilization theory appeared, focused on changes in political opportunities or in the resources available to dissatisfied groups (Zald and Ash, 1966). Rather than asking 'why' people participate in religious movements, this approach asks 'how' religious movements are able to mobilize people into collective action. It also argues that participants in religious movements are just as rational or irrational as other social actors. Resource-mobilization theory emphasizes the supply side, i.e., the social networks and resources available for movements to mobilize. By using resource-mobilization theory, the analysis of fundamentalist movements benefits in several ways. It allows discontent to become a variable of investigation, and it enables analysis of how a movement organizes internally and externally to aggregate resources to pursue its goals. It also incorporates an interaction between the social movement and society at large.

Some resource-mobilization theorists focus on the impact of religious organizations upon the political system, as in the role of the African-American churches in the Civil Rights Movement (Morris, 1984). Nevertheless, the theory has been criticized for mainly addressing the internal dynamics of social organizations. Political-process theorists argue that social movements do not act as isolated units. Rather, they emerge in an environment which either facilitates or hinders their growth, and they exist within the context of other movements (Kitschelt 1986, p. 59; Tarrow 1988, pp. 422, 431). Political process derives from and is linked to resource-mobilization theory, but it tends to focus on environmental constraints to collective action, such as political opportunity structures, competing social and religious movements, as well as cycles of protest and reform.

Resource-mobilization theory and political-process theory have been used in several studies of social and religious movements (see Furseth, 1996, 2002). The theories argue that movements arise at a particular point in time as a result of the changing availability of resources, organization, and opportunities for collective action. Changing availability might include political and institutional changes that make mobilization easier – for example, legal changes that permit collective action. It could also mean that leaders have the ability to redefine old grievances in new terms – for example, that old claims of injustice might be redefined in new terms drawing upon the religious tradition. Thus, these two forms of theory combine both external and internal factors in order to explain movement formation, growth, and outcome.

It is perspectives taken from resource-mobilization theory and political-process theory that will be used in a discussion of urban fundamentalist movement growth here. Every fundamentalist movement appears in an environmental context that facilitates or hinders its growth. Which aspects of the city are conducive to fundamentalist movements? How does the existence of other movements affect the ability of any particular movement to be heard? Movements also face several dilemmas when trying to mobilize. What are their ideology and goals? How does the ideology, on the one hand, mobilize participants, and on the other, create conflict with the larger society? How do fundamentalist movements organize in order to mobilize resources such as people, money, and communication networks? What role do pre-existing social networks play in recruitment and building commitment? Are fundamentalist movements gendered in the sense that they primarily target a male audience, or do they include women? What roles do women have? These are some of the questions discussed here.

The Relationship between the Urban Condition and Religion: Different Trends and Perspectives

Among sociologists and historians, there are three primary interpretations of the relationship between the urban condition and religion: the orthodox, the revisionist, and the conflictual (McLeod, 1995, p. 9). The orthodox line of interpretation claims that cities are strongholds of irreligion. Inspired by Max Weber's view of modernization as a secularizing process and Karl Marx's emphasis on working-class consciousness and organization, many scholars in the 1960s and 1970s assumed a strong link between urbanization and secularization. Historians found evidence for this view in many European and American cities in the late eighteenth century, where industrialization, demographic upheaval, masses of mobile migrants, and a new pluralism led to a 'death of churches' (Callahan, 1995; Singleton, 1979; Wickham, 1957). Likewise, sociologists witnessed a decline in religious faith and practice in urban areas in the 1950s and 1960s, and assumed that there was an inevitable link between urbanization and secularization (Berger, 1967; Bruce, 1992; Wilson, 1966). The city as a whole was seen as a religious desert, a view challenged only within theology (Cox, 1990).

During the 1970s, the secularization thesis came under attack from sociologists who claimed there was a lack of empirical evidence to support it (see Hammond 1985). The revisionist sociologists questioned the assumption that urbanization automatically leads to the decline of religion. Some historians also critiqued the simple dichotomy between the rural and the urban which provided the basis for the orthodox view on religion in the city (Brown, 1992, 1995). The orthodox view assumed that religion flourished in the countryside, whereas the city was a hostile environment for it. However, historical studies from Europe suggested that nineteenth-century church leaders were quick to adapt and respond to social change. The results of these efforts varied: some cities saw religious growth, whereas others experienced decline in church attendance but stability with regard to rites of passage (Strikwerda, 1995; Otte, 1995). There was also more continuity between countryside and town than the orthodox view admitted, as migrants brought their traditional religious culture, strong sense of neighbourhood and community with them to the city. The social significance of religion was thus sustained or even increased in many European cities – for example, in Great Britain, during the transition from pre-industrial to industrial society (Gill, 2003; McLeod, 1995; Williams, 1995).

The change in perspective came with the transformation of American and European cities as a result of the development of post-industrial societies.

The movement of industrial production out of the city, the growth of the service sector, and the spread of the knowledge class have all had a profound effect on cities (Bell, 1976). In many cities, the emptied industrial space has been used for cultural goods and services. Since the 1980s and 1990s, culture has also played an important part in urban economic growth, a trend that is fuelled by tourism (DiMaria and Vergani, 2005). Indeed, the urban planning technique of 'cultural regeneration' has been widely used in Europe and the United States to promote urban development. Part of the new emphasis on culture includes religion, as religious sites and symbols are linked to local identity and included in the images of themselves that cities present to the world. Yet, a lot has happened in cities, including European and American ones, which is not good. Many have become sprawling concentrations filled with pollution, poverty, crime, and violence, as well as racial, class, and religious hostility. Millions of people in cities such as Manila, Mexico City, Jakarta, and Rio de Janeiro have attempted to escape misery in the countryside only to be locked into ghettos of poverty in the city. Modern cities are complex environments with contradicting religious trends. The city is a place for secularity and decline in traditional religion, but studies also show that modern cities remain spaces for religious growth (Hammond, 1985). Indeed, several examples of the emerging 'sacred in a secular age' (to use Hammond's title), such as new religious movements, fundamentalist Islam, and evangelism in America, grew out of urban areas.

A common feature among the orthodox and revisionists is the focus on secularization (see Bruce, 1992). Some historians and sociologists stress that the most important feature of religious development in cities is perhaps neither religious decline nor growth, but conflict. Historians claim that nineteenth- and early-twentieth-century European cities were characterized by polarization and antagonism between rival religious organizations, and between believers and unbelievers. Nineteenth-century cities were religiously heterogeneous, and urban religion had become a source of division rather than a communal bond. Urban religion was still a collective phenomenon, but it was tied to the identity of social classes and ethnic groups rather than the community as such (McLeod, 1995, pp. 23–24).

The issue of religious conflict is also apparent in sociological analyses of contemporary urban religion. Whereas sociologists previously tended to view religion as a primary source of social cohesion, religion's contribution to social conflict has become a more common theme, especially in analyses of religious fundamentalism (see Bruce, 2007; Marty and Appleby, 1991). The conflictual aspect is evident in analyses of movements across the world that use religion to promote political protest and challenge the secular state, especially those that use violence (Juergensmeyer, 2001, 2008). The new emphasis on conflict can also be related to the development of modern cities. This is especially evident in the West where there have been large increases in immigrant populations in recent decades. The religious heterogeneity in urban space is demonstrated in the building of new sites of worship such as mosques and Hindu and Buddhist temples. Religious conflict is now a common theme in studies of twentieth-century cities in the West, in particular conflicts between Muslim minorities and the Christian and secular majority populations (see Haddad and Smith, 2002; Rath *et al.*, 2001).

The orthodox interpretation has for long been out of date, and today the relationship between religion and urban society is evaluated in completely different terms. The view of a continuous decline of religion in modern cities can hardly be sustained in view of empirical evidence showing complex, and seemingly contradicting patterns of religious decline, growth, and conflict. The relationship between the urban condition and religion will vary according to the historical, social, political, and economic situation. Rather than defining this relationship in a specific metahistorical way, it must be evaluated within a specific context. With this in mind, it is important to look more specifically at some urban conditions that may help explain the rise of religious fundamentalism.

Why in the City? The Conduciveness of Urban Contexts to Fundamentalist Movement Growth

The city is a space and a social context that seems to facilitate fundamentalist movement growth. In fact, early North American Protestant fundamentalism emerged in the urban centres of industrialized Canada and the United States such as Toronto, Boston, New York, and Chicago (Ammerman, 1991, p. 17). The resurgence of religious fundamentalist movements since the 1970s has also tended to take place in urban areas. Likewise, the drastic growth in immigrant religions in the West, of which not all are fundamentalist, has primarily taken place in large cities, such as Los Angeles, Washington DC, New York, Amsterdam, London, Paris, and Berlin. What is it about the city that seems to provide opportunity structures to fundamentalist movement formation and growth?

Several studies from different parts of the world stress that urbanization, industrialization, and the migration of people into urban centres create a fertile ground for fundamentalist movements (see Marty and Appleby, 1991). Fast-growing cities tend to act as magnets for migrants from rural areas as well as immigrants from other parts of the world, who come from very different religious and ethnic backgrounds. The result is a heterogeneous population. The idea that the mixing of those from different religious and ethnic backgrounds would create a melting pot, which was prevalent in the United States during the first half of the twentieth century, was far too optimistic (Glazer and Moynihan, 1963). Instead, there has been an increasing religious heterogeneity in many cities, not only in the West, but also in other parts of the world.

Processes of industrialization and migration to cities seem to challenge traditional religion. In nineteenth-century European cities, these processes resulted in movements out of the majority churches and a growing religious diversity (McLeod 1995, pp. 25–27). In many instances, those who defected joined various forms of religious dissent, such as Methodists and Baptists. In other instances, people refrained from joining any religious association, creating a growing group of professed unbelievers, especially among middle-class men. At the same time, working-class and lower-middle-class secularism experienced growth.

Urbanization and migration seem to continue to present challenges to traditional religion. From 1950 to 1980, millions of peasants all over Latin America moved from the rural areas to escape poverty, as well as guerrilla and military violence, and seek a better future in the cities. The migration changed the population balance from rural to urban, so that by 1980 about 60 per cent of the population in Latin America lived in urban areas. As the migrants moved into the cities, which were characterized by poverty and crime, an exchange of old religious values for new ones often occurred. In particular, the urban poor became a target for Protestant proselytism. Traditional Roman Catholicism, which was increasingly perceived as a religion for the elite, failed to deal with the situation. The Roman Catholic clergy had been oriented towards a rural ministry, and there were few clergy trained to deal with the pastoral challenges in the city. These processes created a favourable situation for Protestant fundamentalism (Deiros, 1991, pp. 155–156).

Likewise, economic growth, industrialization, and urbanization in Thailand during the 1980s displaced traditional village-based values and presented a challenge to Thai culture. Among the disruptive new trends were an increase in landless, wage-dependent peasants, an expansion of higher education, and the development of a nouveau-riche commercial class in the cities. Increasing urbanization and the push towards national integration also undermined regional and local identities which were often based on religious traditions. Tourism and prostitution presented yet other profound challenges to Thai culture. These changes combined to erode the traditional symbol of Thai identity, loyalty to Buddhism and the king, and fundamentalist Buddhist movements, Dhammakaya and Santi Asoka, grew to address the situation of religious decline (Swearer, 1991, pp. 655–656).

Urban religious diversity is almost a recipe for conflict. Whereas relatively open and tolerant religious groups can live together in peace, more exclusivist and fundamentalist forms of faith may find this difficult, especially if they are seeking to achieve a dominant position. In addition, new sources of religious division relating to social class and ethnic differences may add to the intensity of a conflict. In nineteenth-century Europe and America, religious denominations were often identified with the values and interests of specific social classes (Niebuhr, 1975). In contemporary cities in the West, many new immigrant religions are profoundly tied to ethnic identity (Warner and Wittner, 1998). In some instances, religious and ethnic identity may be so intertwined in these faith communities that it can be difficult to separate them (Ebaugh and Chafetz, 2000, p. 94). When religious communities are closely identified with the values and interests of specific ethnic groups, members of other ethnic groups will often feel forced out. For example, in many European cities, mosques are established along ethnic lines, so that there are different mosques for Arabs, Pakistanis, Moroccans, and Turks. Such divisions may lead to a sharpening of differences, conflict, and a struggle for power. In contrast, many American cities, such as Los Angeles, are characterized by the formation of multiethnic mosques, which more likely will lead to cooperation among Muslims and less interethnic conflict.

Contemporary American and European cities contain a wide variety of religions. In the eyes of the majority populations this religious diversity may be a source of regret. Some look at the disintegration of old religious cleavages which for the most part distinguished between believers and non-believers, as well as active church goers and passive church members. Others long for a more homogeneous religious and secular culture that refers to Christianity and secular humanism, not all the world religions in one city. This may explain the conflict over issues relating to the role of religion in the public sphere, as demonstrated in issues such as the Muslim headscarf and the role of religion in public schools, which has caused debate in many European cities. Even if the legal toleration of religious minorities is protected, mental toleration is more limited. By not welcoming people of immigrant descent into the work sphere, the political sphere, middle-class neighbourhoods, and religious communities, many Western societies isolate a growing part of their populations. At the same time, different urban religions deal with diversity by building ghettos where their people are protected from close contact with people of alien faiths. Modern cities may combine a high degree of religious diversity with low levels of mutual acceptance.

So far, the focus here has been on several aspects of the demand-side of the urban condition. These conditions provide grounds for fundamentalist movement growth by creating a pool of potential recruits. As people try to adjust to the new urban conditions, fundamentalist movements offer 'alternative philosophies, structures, and institutions that would retain certain traditional values' (Marty and Appleby, 1991, p. 823). Nevertheless, these aspects of the city cannot in and of themselves explain fundamentalist movement growth. Examination of the supply side is also needed, particularly how movements capitalize on resources found in urban areas. Which aspects of the urban condition are particularly useful for fundamentalist movements as they attempt to mobilize a large following?

When a movement tries to recruit new people to join its ranks, it is dependent upon accessing a pool of potential members. In condensed urban space, fundamentalist movements have a larger chance of accessing such a pool than in more scarcely populated space. In fact, rural areas pose a particular set of challenges for movements to mobilize. For example, a movement must come in contact with potential joiners, who also need to receive information about it, in order for the movement's mobilization potential to be realized (Klandermans and Oegema, 1987, p. 520). This poses a challenge in rural areas, just because people tend to live far away from each other. In addition, some people are more available for movement participation than others. In particular, people who are young, single, and in a transitional role, such as students or the unemployed, are considered more open to collective action, because they possess unscheduled time and face minimal risks of sanctions (Snow *et al.*, 1980).

Urban space itself is also conducive to the recruitment process. Recruitment takes place in spatial settings of social life, in public and private places. In general, movements use four outreach channels for recruitment (*Ibid.*, pp. 789–790). First, many movements recruit among strangers in public places by face-to-face contact. This strategy is often used in larger public places, such as crowded streets, squares, train stations, and airports. In these social settings recruiters may contact strangers by giving them flyers, inviting them to meetings, or initiating conversations. Second, movements promote via the institutionalized mass media. Many movements attempt to gain worldwide attention by making the news – some using spectacular forms of violence. Such acts will normally be staged in a condensed urban space, because it is here that a violent act will affect the largest group of people (Juergensmeyer, 2001). Third, movements rely on recruitment among strangers in private, such as through door-to-door visits. This strategy can be efficient in urban settings, where people tend to live relatively close together in condensed neighbourhoods. Finally, movements recruit through the extra-movement networks of those who already participate. These pre-existing networks often prove to be the richest source of recruits (Klandermans and Oegema, 1987).

Cities provide a base of large numbers of organizations, local as well as national, and the more a movement's networks are woven into other organizations, the more people are reached. Indeed, to organize unattached individuals demands more resources and time than network recruitment. Urban space may facilitate the emergence of fundamentalist movements.

The Context of Other Movements

The opportunities for a group to engage in successful collective action vary over time. One important contextual factor is the structure of political opportunities (McAdam, 1982, pp. 40–58). Is it legal for the movement in question to mobilize, or do the participants face threats of persecution and imprisonment? Does the movement receive support or opposition from the political elites and other established organizations and movements? This support or opposition is often related to the political situation (Jenkins, 1983, p. 543). In his analysis of political Islam, Kepel (2002, p. 62) stresses the political opportunities to act in different countries as one of several important factors to explain its diffusion. The success or failure of a fundamentalist movement may be connected to shifts in political opportunity structures, which either facilitate the possibility for groups to access the polity or to exclude them.

In fact fundamentalist movements usually function in a context of other movements and, in some instances, draw on resources from them. Participants in social and religious movements commonly learn civic skills, such as giving speeches, raising money, disseminating the ideology, organizing groups of people, and developing successful strategies. Many movements draw lessons from other movements in their attempts at mobilization and organization (Furseth, 2002). This is also the case for some urban fundamentalist movements. The account of a former Islamist in Great Britain reveals how the participants in Hisb ut-Tahrir in London adopted successful strategies used by British radical socialists when it came to disseminating their ideology and co-opting other Muslim organizations into their fold (Husain, 2007). The context of other movements is a resource for new movements.

In other instances, movements compete with each other. They appeal to the same constituencies and thereby recruit from the same pool. They may also compete for other resources, such as money, influence in extra-movement organizations, or political representation. Competition may push moderate groups to stretch their rhetoric to be heard, or it may push already radical groups further to the edge to prove how committed they are to the cause. It is reasonable to believe that the 'theatre of horror' some violent fundamentalist movements perform (Juergensmeyer, 2001) has to do with competition for media coverage and worldwide attention. Competition with other movements may lead some fundamentalist movements to go to increasing extremes in order to be heard.

Movement Ideology

It has often been assumed that the goals of a particular movement more or less automatically reflect the interests of its participants. If this were true, adherents of political Islam would propose this solution because it would benefit their religious, political, and economic interests. However, this is a somewhat simplified view of movement ideology. Participants in social and religious movements do not only act upon the world or segments of it; they also frame or interpret the world in which they are acting (Snow *et al.*, 1986, p. 466). For example, movement leaders may interpret social and economic grievances that people have within a religious framework, and many of their speeches, public performances, symbols, and efforts to train participants are aimed at framing individual interpretive orientations. Religious movement leaders do not just express values, but form them. The process of 'frame alignment' (Snow *et al.*, 1986) can be seen in fundamentalist movements, such as Hamas. By using the delayed results of the peace process and the corruption of the leaders of the Palestinian Authority, they were able to introduce the idea of *jihad* as an alternative interpretive orientation that seemed to present a solution to the disempowerment of Palestinians in their own territories (Kepel, 2002, pp. 324–325). Many of Hamas's activities, such as ideological training and use of symbolism, were aimed at framing the thinking of Palestinians according to this idea.

Why, then, is movement ideology important in mobilization processes? Potential joiners need to target the source of what they perceive to be their grievances, interpret them, and find appropriate responses to them. Ideology makes grievances meaningful. To be able to act collectively, participants need to agree upon certain values, goals, and strategies (Heberle, 1951, p. 21). Movement ideology can be defined as 'a set of beliefs about the social world and how it operates, containing statements about the rightness of certain arrangements and what action would be undertaken in light of those statements' (Wilson, 1973, pp. 91–93). An ideology often consists of three elements: a definition of the source of the current problem; a goal or solution; and a motivation for joining and staying active in the movement.

Religion can provide these elements for a movement ideology. For example, all world religions offer an analysis of the moral situation of the world in terms of a divine rather than a secular standard. Religions tend to offer total 'packages' of interpretations, where the social and moral ills of the world are defined, and where the goals are clearly outlined. Furthermore, religion can provide several rationales for movement participation (Snow, 1987). In particular, it will define the mission as 'divinely' ordained and claim that it is the only 'proven' religion. Religion is also able to present itself as a continuation or fulfilment of older traditions. Furthermore, religion can offer greater rewards (for example, salvation) for those who participate actively than for those who remain passive, and it can give the active participant a special status, a sense of mission, and the sense of being part of an elect group. Religion offers powerful inducements for joining a movement and staying committed.

In fact, sociologist Mark Juergensmeyer (2001) shows the role that ideology plays in fundamentalist, violent movements. Different violent movements share some common, ideological features. One is the idea that their particular movement or religious community is under attack and that their actions are only responses to the violence that is inflicted upon them, especially by the state. Another common feature is that they use images of cosmic confrontation and warfare to interpret worldly conflicts and battles. Movements also tend to engage in processes of satanization of the 'others'. They believe that their group must engage in symbolic empowerment, which they do through violent acts. Juergensmeyer (2001, p. 10) claims that religion is not always innocent when it comes to violence. The violence associated with religion is not an aberration, but derives from the belief system of all major religions. There are aspects of all world religions that can be used to justify violence.

Nevertheless, a significant feature of movement ideology is that those aspects of the ideology that are effective in mobilizing people are often the same aspects that create conflict with the larger society. If the goal of a movement is to change individuals by changing their hearts and minds – which is often defined as the goals of religious movements (Zald and Ash,

1966, p. 329) – the reaction from society may be indifference. Yet, if the goal is to change social institutions or structures, which is commonly the goal of social movements, the opposition from the surrounding society, and especially the state, may be more severe. Indeed, many fundamentalist movements direct their efforts towards altering social structures, the legal system, and in some instances, overthrowing the state. This is especially true for political Islam, which is one reason why this movement has been met by such severe opposition and oppression in most countries (Kepel, 2002). The goals and tactics of a movement are important in understanding the mobilizing potential a movement has, the degree of threat it poses, and thereby its acceptance or opposition.

Movement Organization

Earlier studies of social and religious movements tended to downplay the importance of movement organizations, and assumed that organization plays a role later in a movement's history (Jenkins, 1983, p. 543). Resource-mobilization theory argues that organization is a fundamental factor in movement development. In order to aggregate resources, such as money and people, some form of organization is needed, and these resources must be controlled or mobilized before protests, rallies, or other forms of collective action can take place (McCarthy and Zald, 1977, pp. 1216, 1221). For example, Islamic groups in the 1970s needed to organize in order to access the wealth available to them as a windfall from the oil industry. Indeed, it was a complex Islamic economic system that helped build the so-called 'petro-Islam' (Kepel, 2002, pp. 61–80). Emerging movements require quite a bit of organization of practical matters before they can function.

Juergensmeyer's (2001) analysis of fundamentalist movements that use urban space to perform acts of terror shows the amount of organization required to carry out these acts. These movements select safe, public places, such as shopping malls and public transportation, and time their attacks so that as many people as possible are present – for example, during the rush hour. The explosives are also selected with a particular aim in mind, hurting as many people as possible. Juergensmeyer interprets these acts as a 'theatre of horror', where the purpose of violence is to make a symbolic statement. To promote the image of the group and demonstrate its power, they select symbolic days and choose acts that will draw the greatest possible media attention.

Another important aspect of movement organization is the role that pre-existing networks and organizations have in recruitment. Thus the violent anti-abortion movement in the United States has drawn upon existing institutions and organizational forms, namely Christian churches (Ginsberg, 1993). Indeed, some of its leaders have been ordained ministers. For example, Reverend Michael Bray and two other defendants were found guilty in 1985 of bombing seven abortion facilities. In 1994 Reverend Paul Hill, Bray's friend, killed Dr John Britton and his escort James Barrett as they arrived at an abortion clinic (Juergensmeyer, 2001, pp. 20–30). The Christian Right also used congregations as its focal point in its early phase before it developed more autonomous organizations (Williams, 2003, p. 319). By drawing on these pre-existing organizations, these movements were able to mobilize and grow.

The Role of Pre-existing Social Networks in Recruitment and Building Commitment

The above discussion illustrates how urban space is conducive to fundamentalist movement growth, because it facilitates the recruitment of individuals as well as social networks and organizations. The role of pre-existing formal and informal social networks for movement growth has been emphasized in several studies (see Furseth, 2002). These networks can range from structured organizations to informal associations or networks of family, friends, and neighbours. Pre-existing networks are extremely important in recruitment to all sorts of social and religious movements. In fact, most studies show that people do not join because they are convinced of an ideology, but because they have social networks, such as friends and family, within the movements (McCarthy and Zald, 1977). Through existing networks, potential joiners get to know about the movement and receive motivations to join.

Formal organizations often have their headquarters and sub-organizations located in cities, and it is within urban space that contact between rising movements and these types of organizations take place. If a movement has ties with other formal organizations, 'en bloc recruitment' may occur (Oberschall, 1973). This means that the entire pre-existing organization and its membership will join. In order to explain the growth of Islamist groups in the 1970s, Kepel (2002) analyzed several factors, including the numerous organizations that participated and the alliances they built. For example, one factor that accounted for Khomeini's success in Iran was his ability to mobilize various pre-existing networks. During the 1960s, Islamism in Iran found support in two groups in particular: young militants who reinterpreted Shiite doctrine within a Marxist and Third World interpretive scheme, and a section of the clergy who adopted an anti-modern ideology. Khomeini managed to combine these two groups, and thereby gain support from the educated urban middle class (Kepel, 2002, p. 37). Other types of networks, such as Quranic schools and mosques, have been important in the rise of Islamism in other parts of the world, as in Indonesia (*Ibid.*, p. 57).

Movements must not only gain new participants, but must maintain their commitment. Another important aspect of pre-existing networks is that they tend to produce loyalty. Recruitment among other formal organizations may have this effect, as these members are already committed to the cause. Recruitment among informal networks in urban neighbourhoods and among family and friends also creates a strong sense of solidarity among the participants, because affective ties create commitment. This may be one reason why many religious movements attempt to recruit entire families into their fold. The familial focus has, for example, been a strong feature in American Protestant fundamentalism, where the evangelical family has had an elevated status (Bartkowski, 2001, pp. 17–38; Gallagher, 2003, pp. 39–62). Likewise, the family is a central focus among many Arab Islamists, who emphasize the ideal of the married couple (Roald, 2001, p. 123). The idea is that every Muslim should be married if possible, and thus the family becomes a basic unit within Islam. Just as interpersonal bonds are important for recruitment, they are also central in preventing defection (Stark and Bainbridge, 1980, pp. 1392–1393). Pre-existing social networks may play an important role in the stability and outcome of fundamentalist movements.

Gender

When movements attempt to mobilize, they often define and target their efforts at people they would like to join their movements. Is the movement for women only, for men only, or for both? Whereas historic political movements tended to be limited to men only, historic Christian movements tended to recruit both women and men (Furseth, 2001). Contemporary fundamentalist movements vary when it comes to recruitment according to gender, and also in structure. Many movements include, for example, both women and men, but they will separate the genders into different organizations.

Some fundamentalist movements emphasize gender equality. In the fundamentalist Buddhist movement Santi Asoka in Thailand, attempts have been made to eliminate social distinctions, including gender distinctions between monks and laypersons. The ideal community for the People of Asoka is a single moral community of male and female monks and laypersons (Swearer, 1991, pp. 672–673).

Other fundamentalist movements emphasize gender difference. Protestant fundamentalist movements are eager to recruit both women and men, but they emphasize that the two genders have different roles. For example, in the 1970s and 1980s it was common among gender essentialists within this movement to view gender difference and hierarchy as divinely ordained and biologically self-evident. However, these views were contested within the movement, as biblical feminists claimed that hierarchy was a result of the Fall, and that God's intent was gender mutuality (Gallagher, 2003, p. 49). The gender hierarchy also implies that fundamentalist men have more power than women in outlining goals and ideals, although studies show that fundamentalist women have more power in the home and in the congregations than previously assumed (Brasher, 1998). Indeed, attempts have been made among Protestant fundamentalists to organize movements that only include men, such as the Promise Keepers. One aim of this movement was to develop a Godly masculinity among the participants, even if there were various versions of this ideal (Bartkowski, 2004).

The emphasis on gender difference and gender separation is common among Muslim fundamentalist movements. For example, the Dakwah community in Malaysia strictly enforces gender separation. Dakwah women are expected to perform domestic tasks and limit their activities to household chores, children, and husband. They are in strict *purdah* (isolation), so that only the males can do the shopping. The gender separation is also practiced among Dakwah students, where the basic unit, a cell-like body called the *usrah* is made up of students of the same sex (Nash, 1991, pp. 695–696, 709).

In spite of the idea of gender separation among Muslim fundamentalists, many fundamentalist Muslim organizations have male and female members. In the case where there are female and male organizations, they frequently cooperate. This is not a new phenomenon. For example, when Zaynab al-Ghazali was the head of the Muslim Women's Association in Egypt in the 1940s, she worked closely with the Muslim Brotherhood. And after the Nasser regime began to suppress the Brotherhood in the 1950s, her home became a study centre for its members. In the 1960s, the group around Zaynab al-Ghazali was connected to several other groups which had ties to the old Brotherhood organization (Voll, 1991, p. 373).

The most extreme form of gender separation among Muslim fundamentalists, one which has also led to abuse of women, was found in Afghanistan during the 1980s and 1990s. A feature of the Afghan *mujahidin* which fought against Russian occupation forces was that it excluded women. Women who became too visible or vocal were threatened or sometimes killed (Moghadam, 1992). Later, during the Taliban regime, women were not allowed to attend school or work. They were not permitted in the public sphere unless accompanied by a man, and they were forced to cover themselves. The violence against women extended to brutal public beatings, and even murder (Benard, 2002).

Most fundamentalist movements target both women and men as their potential recruits. Yet, in many ways, gender is a controversial issue in these types of movements. Although it is tempting to draw the conclusion that most fundamentalist movements emphasize gender difference, traditional gender roles, and gender hierarchy, this is not always the case. Even in movements that teach gender difference and traditional roles for women and men, internal discourses often challenge these views, and gender practices vary (Gallagher, 2003). Many movements do not practice what they teach, even if there are fundamentalist movements that teach and practice the subjugation and oppression of women.

Conclusion

This chapter has explored different theories of social and religious movements that may be useful in studies of urban fundamentalism. I have attempted to show that although the term 'fundamentalism' has its limitations when it comes to explaining why fundamentalisms grow and develop, it can still be fruitful as a sensitizing tool in distinguishing between different forms of religious movement. Furthermore, the main argument of crisis theory is that social problems are transformed into fundamentalist movements. Because these movements are understood as effects of external factors, crisis theory focuses on the preconditions for religious collective action. These conditions cannot explain the process whereby people are actually mobilized into fundamentalist movement activity. In contrast, theorists who use resource-mobilization theory and political-process theory direct their attention to the conditions under which movement activity takes place and the opportunities for movements to act. These perspectives relate the emergence of fundamentalist movements to the changing availability of resources and opportunities for collective action. The combination of perspectives taken from these theories emphasizes several external and internal factors that affect mobilization to fundamentalist movements and their outcome.

This chapter discusses what it is about the city which creates an environment conducive to fundamentalist movement growth. In the 1950s and 1960s, sociologists and historians assumed that the city was a space of irreligion; however, the 1970s witnessed the rise of seemingly new religious phenomena in urban space, such as new religious movements, the growth in Protestant evangelicalism, and political Islam. Whereas urbanization and migration to cities had previously been interpreted as factors that would produce secularization, these factors were now interpreted as challenges for traditional religion – but not for religion in general. In addition, profound changes were taking place in the cities, especially in Europe and the United States, including a new emphasis on knowledge and higher education, production of new technology, and a much stronger emphasis on cultural production and consumption. The immigration from non-Western countries also led to religiously heterogeneous cities where there was sometimes little mutual toleration and a high degree of isolation among different ethnic and religious groups. For some urban dwellers, these trends led to a secular outlook or to the adoption of more individualistic and subjective forms of religion. For others, these transformations led them to join fundamentalist movements that offered alternative ways of interpreting society, traditional values, strict moral values, and more specific social class and ethnic identities.

Although these aspects of the urban condition are important to understanding how cities become fertile grounds for fundamentalist movement growth, these factors cannot in themselves explain how movements expand. Several additional aspects of the city are useful when fundamentalist movements attempt to access a pool of potential joiners and recruit them. Movements rely on recruiting strangers in urban public and private space. They use the mass media as well as pre-existing networks such as informal networks and formal organizations. Cities provide several avenues for fundamentalist movements to mobilize and establish themselves.

By discussing additional factors that are important in the mobilization process, such as movement ideology, organization, pre-existing social networks, and gender, it is possible to see that there are many ways fundamentalist movements organize externally and internally, and many relationships between these movements and other sections of society. The growth in fundamentalist forms of faith means that urban religion can become a source of division. As exclusivist religious groups live in close proximity to one another, the level of mutual toleration is not always high. Religious belonging connected to social class and ethnic identity may increase the significance of these differences, which again may result in religious conflict.

As scholars attempt to come to terms with recent growth in urban fundamentalist movements across the world, what emerges most prominently is the complexity of the phenomenon. Whereas social scientists and historians previously assumed that modernization and urbanization would eventually lead to irreligious cities, contemporary scholars realize that urban space is a site of religious growth, decline, and conflict.

References

Ammerman, N. (1991) North American Protestant fundamentalism, in Marty, M.E. and Appleby, R.S. (eds.) *Fundamentalisms Observed*. Chicago, IL: University of Chicago Press.

An-Na'im, A.A. (1999) Political Islam in national politics and international relations, in Berger, P.L. (ed.) *The Desecularization of the World*. Grand Rapids, MI: Wm. B. Eerdmans.

Bartkowski, J.P. (2001) *Remaking the Godly Marriage*. New Brunswick, NJ: Rutgers University Press.

Bartkowski, J.P. (2004) *The Promise Keeper*. New Brunswick, NJ: Rutgers University Press.

Bell, D. (1976) *The Coming of Post-Industrial Society*. New York: Basic Books.

Benard, C. (2002) *Veiled Courage*. New York: Broadway Books.

Berger, P.L. (1967) *The Sacred Canopy*. Garden City, NJ: Doubleday.

Brasher, B. (1998) *Godly Women: Fundamentalism and Female Power*. New Brunswick, NJ: Rutgers University Press.

Brown, C.G. (1992) A revisionist approach to religious change, in Bruce, S. (ed.) *Religion and Modernization*. Oxford: Clarendon Press.

Brown, C.G. (1995) The mechanism of religious growth in urban societies, in McLeod, H. (ed.) *European Religion in the Age of Great Cities 1830–1930*. London: Routledge.

Bruce, S. (ed.) (1992) *Religion and Modernization*. Oxford: Clarendon Press.

Bruce, S. (2007) *Fundamentalism*. Cambridge: Polity Press.

Callahan, W.J. (1995) An organizational and pastoral failure: urbanization, industrialization and religion in Spain, 1850–1930, in McLeod, H. (ed.) *European Religion in the Age of Great Cities 1830–1930*. London: Routledge.

Cox, H. (1990) *The Secular City* (25th anniversary edition). New York: Collier Books.

Deiros, P.A. (1991) Protestant fundamentalism in Latin America, in Marty, M.E. and Appleby, R.S. (eds.) *Fundamentalisms Observed*. Chicago, IL: University of Chicago Press.

DiMaria, E. and Vergani, S. (2005) Culture: an engine for urban economic development, in Robinson, J. (ed.) *EUROCULT21*. Helsinki: Lasipalatsi Media Centre.

Ebaugh, H.R. and Chafetz, J.S. (2000) *Religion and the New Immigrants*. Walnut Creek, CA: AltaMira Press.

Esposito, J.L. (2002) *Unholy War*. Oxford: Oxford University Press.

Furseth, I. (1996) The impact and relevance of social movement theories in the sociology of religion, in Repstad, P. (ed.) *Religion and Modernity*. Oslo: Scandinavian University Press.

Furseth, I. (2001) Women's role in historic religious and political movements. *Sociology of Religion*, **62**(1), pp. 105–129.

Furseth, I. (2002) *A Comparative Study of Social and Religious Movements in Norway, 1780s–1905*. New York: Edwin Mellen.

Furseth, I. and Repstad, P. (2006) *An Introduction to the Sociology of Religion*. Aldershot: Ashgate.

Gallagher, S.K. (2003) *Evangelical Identity and Gendered Family Life*. New Brunswick, NJ: Rutgers University Press.

Gill, R. (2003) The future of religious participation and belief in Britain and beyond, in Fenn, R.K. (ed.) *The Blackwell Companion to Sociology of Religion*. Oxford: Blackwell.

Ginsberg, F. (1993) Saving America's souls: operation rescue's crusade against abortion, in Marty, M.E. and Appleby, R.S. (eds.) *Fundamentalisms and the State*. Chicago and London: University of Chicago Press.

Glazer, N. and Moynihan, D.P. (1963) *Beyond the Melting Pot*. Cambridge, MA: MIT Press.

Haddad, Y.Y. and Smith, J.I. (2002) *Muslim Minorities in the West*. Walnut Creek, CA: AltaMira Press.

Hammond, P.E. (ed.) (1985) *The Sacred in a Secular Age*. Berkeley, CA: University of California Press.

Heberle, R. (1951) *Social Movements*. New York: Appleton-Century-Croft.

Hertzke, A.D. (1990) Christian fundamentalists and the imperatives of American politics, in Sahliyeh, E. (ed.) *Religious Resurgence and Politics in the Contemporary World*. Albany, NY: State University of New York Press.

Husain, E. (2007) *The Islamist*. London: Penguin.

Jenkins, J.C. (1983) Resource mobilization theory and the study of social movements. *Annual Review of Sociology*, **9**, pp. 527–553.

Juergensmeyer, M. (2001) *Terror in the Mind of God*. Berkeley, CA: University of California Press.

Juergensmeyer, M. (2008) *Global Rebellion*. Berkeley, CA: University of California Press.

Kepel, G. (2002) *Jihad*. Cambridge, MA: Harvard University Press.

Kitschelt, H.P. (1986) Political opportunity structures and political protest: anti-nuclear movements in four democracies. *British Journal of Political Science*, **16**, pp. 57–85.

Klandermans, B. and Oegema, D. (1987) Potentials, networks, motivations, and barriers: steps towards participation in social movements. *American Sociological Review*, **52**, 519–531.

Marty, M.E. (1992) Fundamentals of fundamentalism, in Kaplan, L. (ed.) *Fundamentalism in Comparative Perspective*. Amherst, MA: University of Massachusetts Press.

Marty, M.E. and Appleby, R.S. (1991) Conclusion: an interim report on a hypothetical family, in Marty, M.E. and Appleby, R.S. (eds.) *Fundamentalisms Observed*. Chicago, IL: University of Chicago Press.

McAdam, D. (1982) *Political Process and the Development of Black Insurgency 1930–1970*. Chicago, IL: University of Chicago Press.

McCarthy, J.D. and Zald, M.N. (1977) Resource mobilization and social movements: a partial theory. *American Journal of Sociology*, **82**(6), pp. 1200–1241.

McLeod, H. (ed.) (1995) *European Religion in the Age of Great Cities 1830–1930*. London: Routledge.

Moghadam, V. (1992) Patriarchy and the politics of gender in modernizing societies: Iran, Pakistan and Afghanistan. *International Sociology*, **71**, pp. 35–54.

Morris, A. (1984) *The Origins of the Civil Rights Movement*. New York: The Free Press.

Nash, M. (1991) Islamic resurgence in Malaysia and Indonesia, in Marty, M.E. and Appleby, R.S. (eds.) *Fundamentalisms Observed*. Chicago, IL: University of Chicago Press.

Niebuhr, H.R. (1975) *The Social Sources of Denominationalism.* New York: Meridian.
Nielsen Jr., N.C. (1993) *Fundamentalism, Mythos, and World Religion.* Albany, NY: State University of New York Press.
Oberschall, A. (1973) *Social Conflict and Social Movements.* Englewood Cliffs, NJ: Prentice Hall.
Otte, H. (1995) 'More churches – more churchgoers': the Lutheran Church in Hanover between 1850 and 1914, in McLeod (ed.) *European Religion in the Age of Great Cities 1830–1930.* London: Routledge.
Rath, J., Penninx, R., Groenendijk, K., and Meyer, A. (2001) *Western Europe and its Islam.* Leiden: Brill.
Roald, A.S. (2001) *Women in Islam.* London: Routledge.
Sahliyeh, E. (1990) Religious resurgence and political modernization, in Sahliyeh, E. (ed.) *Religious Resurgence and Politics in the Contemporary World.* Albany, NY: State University of New York Press.
Shupe, A. and Hadden, J.K. (1989) Is there such a thing as global fundamentalism? in Hadden, J.K. and Shupe, A. (eds.) *Secularization and Fundamentalism Reconsidered,* New York: Paragon House.
Singh, K. (1990) The politics of religious resurgence and religious terrorism: the case of the Sikhs of India, in Sahliyeh (ed.) *Religious Resurgence and Politics in the Contemporary World.* Albany, NY: State University of New York Press.
Singleton, G.H. (1979) *Religion in the City of Angels: American Protestant Culture and Urbanization, Los Angeles 1850–1930.* Ann Arbor, MI: UMI Research Press.
Snow, D.A. (1987) Organization, ideology, and mobilization: the case of Nichiren Shoshu of America, in Bromley, D.G. and Hammond, P.E. (eds.) *The Future of New Religious Movements.* Macon, GA: Mercer.
Snow, D.A., Burke Rochford, E. Jr., Worden, S.K., and Benford, R.D. (1986) Frame alignment processes, micromobilization, and movement participation. *American Sociological Review,* **51**, pp. 464–481.
Snow, D.A., Zurcher Jr., L.A., and Ekland-Olson, S. (1980) Social networks and social movements: a microstructural approach to differential recruitment. *American Sociological Review,* **45**, pp. 787–801.
Stark, R. and Bainbridge, W.S. (1980) Networks of faith: interpersonal bonds and recruitment to cults and sects. *American Journal of Sociology,* **85**(6), pp. 1376–1395.
Strikwerda, C. (1995) A resurgent religion: the rise of Catholic social movements in nineteenth-century Belgian cities, in McLeod, H. (ed.) *European Religion in the Age of Great Cities 1830–1930.* London: Routledge.
Swearer, D.K. (1991) Fundamentalistic movements in Theravada Buddhism, in Marty, M.E. and Appleby, R.S. (eds.) *Fundamentalisms Observed.* Chicago, IL: University of Chicago Press.
Tarrow, S. (1988) National politics and collective action: recent theory and research in Western Europe and the United States. *Annual Review of Sociology,* **14**, pp. 421–440.
Voll, J.O. (1991) Fundamentalism in the Sunni Arab World: Egypt and the Sudan, in Marty, M.E. and Appleby, R.S. (eds.) *Fundamentalisms Observed.* Chicago, IL: University of Chicago Press.
Warner, R.S. and Wittner, J.G. (1998) *Gatherings in Diaspora.* Philadelphia, PA: Temple University Press.
Wickham, E.R. (1957) *Church and People in an Industrial City.* London: Lutterworth Press.
Williams, R.H. (2003) Religious social movements in the public sphere: organization, ideology and activism, in Dillon, M. (ed.) *The Handbook of the Sociology of Religion,* Cambridge: Cambridge University Press.

Williams, S. (1995) Urban popular religion and the rites of passage, in McLeod, H. (ed.) *European Religion in the Age of Great Cities 1830–1930*. London: Routledge.
Wilson, B. (1966) *Religion in Secular Society*. London: C.A. Watts.
Wilson, J. (1973) *Introduction to Social Movements*. New York: Basic Books.
Zald, M.N. and Ash, R. (1966) Social movement organizations: growth, decay and change. *Social Forces*, **44**, pp. 327–340.

Chapter 3

The Civility of Inegalitarian Citizenships

James Holston

If this book considers the relation between cities and fundamentalism from a number of perspectives, my concern is that of citizenship. My interest is with the civility of inegalitarian regimes of citizenship, of which fundamentalist regimes are an extreme. I begin from the premise that all citizenships have their civility, including inegalitarian ones. Indeed, all perduring citizenships must if they are to maintain themselves without a continuous and costly application of violence against their own citizens. From this point of view, I consider civility to mean conformity to standards of citizen identity and behaviour deemed necessary to create a sense of civic belonging, of commensurability among citizens for that purpose. Thus, civility constitutes a consensus, and in that sense a 'good', that a regime of citizenship must claim the right to enforce. It does so preferably through social and cultural means of disciplining citizens, using non-violent methods to develop their capacity to participate in civic life according to its standards through education, institutional design, politeness codes, body language, and so forth. However, ultimately, its right to enforce civility is based on intolerance, exclusion, and violence.

One of my objectives is thus to erode the idealization of civility, common in Western social theory, that associates its consensus with such values as civilization, liberty, trust, generosity, inclusion, secularism, and democracy; and opposes it to barbarism, dogma, fundamentalism, exclusion, violence, and religion. I do so by examining the concept of civility in light of various examples of urban civilities that generate consensus on the basis of a denial of rights. If all citizenships have their particular civilities, which share an underlying basis in exclusion, then I ask what characterizes fundamentalist regimes of citizenship and the civility that imposes their 'good'? Some fundamentalist regimes, especially of small scale, may be composed entirely of believers in a comprehensive doctrine that supplies it with pervasive values. But contemporary fundamentalist states typically include different kinds of people as citizens, many of whom are not true believers. They may therefore be described as extremes of the all-too-common type of inegalitarian national citizenship that denies some of its own citizens full citizenship.

Thus, the question I ask is what conditions sustain regimes of national citizenship that deny some of their own citizens an existence worthy of a full citizen? They are the same conditions, I would add, that maintain the nation-state as a convention of territorial borders and membership rules within a system of states that renders many people stateless, relegates them to the status of 'persons without rights', and therefore excludes them from citizenship altogether. Thus, the internally excluded and the externally excluded – the marginal national citizen and the stateless non-citizen – issue from the same font of exclusionary citizenship. This common sort of inegalitarian citizenship has the following characteristic: it incorporates vast numbers of different kinds of people as citizens and uses differences among them that are *not* the basis of national membership (differences, for example, of race, religion, class, and gender) to distribute rights, powers, and privileges differentially.

Such regimes of differentiated citizenships, as I call them, are thus mechanisms to incorporate, distribute, and manage inequality among a population. They legalize some attributes for special treatment – some religious membership, racial trait, or principle of descent – and enforce certain kinds of public practices that identify those attributes as markers of categories of people that deserve a privileged distribution of rights and powers. They are, both historically and currently, the most common type of national citizenship. Fundamentalist citizenships are a subset of these inegalitarian regimes. They give absolute value to certain differences in their management of the public and its space, assuming both the right and the duty to impose this good as the transcendental referent of civility for all citizens. In what follows, I investigate the discriminations of citizen and non-citizen that depend on these modes of incorporation and distribution to enforce the consensus of civility.

Civility and Exclusion

Rules of incorporation create various kinds of non-citizens not only as the outside limit that establishes the inside, but also as the included excluded, that is, the excluded (e.g., 'the barbarous') which indicates to the included ('the civilized') the qualities of inclusion ('civilization'). In turn, rules of distribution commonly create various kinds of internally marginalized citizens on the basis of social differences that the citizenship regime deems legitimate grounds for discrimination. On the basis of that legitimation, some citizens deny other citizens rights: they feel justified in treating other citizens as if they were not citizens because they are the wrong sort of person or exhibit the wrong sort of behaviour or belief. Such judgments of civility establish a consensus – even among the marginalized – about what is morally good and socially correct and of sufficient worth to deserve the full distribution of citizen rights. Thus, these regimes of differentiated citizenships distribute inequality among citizens and non-citizens alike by referencing assumptions about the fundamental worth of certain social differences. They equate some differences for purposes of national inclusion, creating a national society of citizens who identify with the nation, and they legalize other differences as the basis for different treatment and both internal and external exclusion.

Such regimes of citizenship depend as much on institutions of government, law, and education as on violence and repression. Yet they also rely on the cultural conventions, everyday performances, aesthetic codes, and foundation narratives which are typically understood to constitute a citizenship's civility. Admittedly a complex and ambiguous term, civility refers to the standards of behaviour and common measure – to the 'etiquettes, manners, and virtues' – that make public life coherent and hence possible under a particular regime of citizenship. Thus, Western political theory overwhelmingly presents civility as an essential good for political life and incivility as a detriment, contrasting it both to violence and to religious intransigence (the latter is often seen as inciting the former) and construing it as a means of communication rather than coercion.

But what if the regime of citizenship is fundamentally inegalitarian, foundationally intolerant of claims that would lead to redistributions of citizen powers? In such regimes, as I show with a Brazilian example, civility's idioms of inclusion and consensus create habits of the public which entrench citizenship's inequalities. Moreover, from the perspective of stateless non-citizens, all national civilities are exclusionary, if not intolerant and violent. No one calls Brazilian citizenship fundamentalist, and yet its intransigence is remarkable. Its regime of legalized privileges and legitimated inequalities has persisted for centuries under colonial, imperial, and republican rule, thriving under monarchy, dictatorship, and democracy. Indeed, its foundations have only begun to change in the last few decades. Given such persistence, incivility may in fact be an effective strategy of public interaction to puncture the civility that perpetuates the fundamentals of discrimination.

The problem of the idealization of civility is compounded in the tradition of civic humanism which inspires so much Western political theory. It is descended from the Athenian formulation of citizenship and refracted in laments for the lost civility of the *polis*, which has infused political theory from Rousseau's transformation of the *sujet* into *citoyen* to the current communitarian assault on liberalism. In this paradigm, civility is conceived and measured by the standard of the active participation of citizens in making the (re)public. It is understood as the *praxis* of citizens 'ruling and being ruled', as Aristotle defines the core characteristic of membership in the *polis*. Moreover, this ideal of civility refers not only to the capacities of citizens to be active participants in constructing the political community – in administering justice and holding office, specifically – but also to their capacities to find their honour, self-fulfilment, and mutual respect in the life of participation. This commitment to the primacy of the public, connecting active membership with personal virtue, is the conception of civility that the Western canon of political theory lionizes.

Never mind that this commitment requires specific social conditions: structures of government, scales of urban life, regimes of labour, types of families, conceptions of person, and so on. As Walzer (1989, p. 217) perceptively diagnoses, political theory has for centuries, and in the most diverse socio-historical contexts, consistently valorized this republican paradigm of civility over others. For example, Walzer contrasts the participatory with a jural paradigm which developed to accommodate the mass societies of empires and nation-states. The jural citizen is the recipient of rights and entitlements merely for having the status of citizen, without any need to participate directly in the business of rule. It is 'a status, an entitlement, a right or set of rights passively enjoyed'. In this paradigm, 'private interests' replace 'public virtues' as the primary focus of the citizens' lives. Furthermore, protecting the liberties of those private affairs becomes the primary objective of political citizenship itself. Thus, Marx (1967, p. 227) observed that 'the political liberators [in the French Revolution] reduce citizenship, the *political community*, to a mere means for preserving these so-called rights of man and the citizen is thus proclaimed to be the servant of the egoistic man'. In the same vein, the current communitarian critique claims that the welfare state and modern liberalism reduce the civility of citizenship, if it may be called that at all, to empty politeness, condemn private citizens to isolation and atomization, and doom their public interactions to oscillations of passivity, egoism, and legalism.

The argument that some citizenships foster civility and some do not is precarious. It can only be based on an idealized construction of citizen behaviour that assumes what it wants to prove. To avoid the fallacy of this *petitio principii*, it makes sense to argue instead that all regimes of citizenship reproduce themselves through citizen performances fostering standards of behaviour that confirm their specific modes of incorporation and distribution. In this sense, all regimes have their active practice, which includes a vast range of possibilities of constituting public affairs. In this sense as well, civility is not inherently incompatible with coercion, violence, or religious intolerance. All citizenships of the most diverse political types, cultural conditions, and religious persuasions construct and maintain public spheres. The questions are what kind and how.

To investigate this constitution without idealization, I propose the following conceptualization of civility, inspired by Bourdieu's (1977) work on symbolic capital and *habitus*. Civility is as a set of dispositions inscribed in routines of bodily actions and verbal expressions – and therefore in routines of perception, expectation, and thought – that permit citizens to produce the practices that confirm the categories that govern the distribution of powers consistent with a specific regime of citizenship. Such dispositions are sometimes regulated by force, and often by institutions that sometimes apply force (e.g., the police). Indeed, the threat of violence secures them. Mostly, however, civility entails practices that are regulated without explicit force. An enduring regime of citizenship must be able to produce generations of citizens from all walks of life who exhibit a practical mastery of the dispositions of civic interaction – 'tact, dexterity, or savoir-faire – presupposed by the most everyday games of sociability and accompanied by the application of a spontaneous semiology, i.e., a mass of precepts, formulae, and codified cues' (Bourdieu, 1977, p. 10).

Certainly, not all everyday sociability involves issues of citizenship. But everyday trafficking in and about public space – crossing a street, making a left-hand turn, standing in line at the post office, as well as exercising one's profession and working with neighbours on a common residential problem – is indeed the realm of modern society in which people most frequently and predictably experience the state of their citizenship. The quality of such mundane interactions may in fact be more significant to people's sense of themselves as citizens and non-citizens than the occasional heroic experience of soldiering or demonstrating, or more ordinary ones like voting or serving on a jury. Everyday citizenship entails the practical mastery of performances that turn people, however else related, into fellow citizens related by measures specific to citizenship. These may be empowering or debilitating, equalizing or differentiating; but, in that manner, they become evident.

For example, the same generalized expectation of impunity among citizens in Brazil motivates an enormous range of probabilities: that motorists will run a red light when in a hurry, and especially at night, and will probably not stop after hitting a pedestrian; that a contractor will substitute inferior materials for those required in building a public school; that a rich landowner will order the killing of a labour organizer trying to mobilize workers; that a logging company will cut down mahogany trees in an ecological preserve to sell on the black market; that a private banker will launder money for wealthy clients and supply under-the-table funds to politicians; and that Congress will institute a special committee to investigate the corruption of its own members composed in the majority of allies of one or several of the accused. These examples come from one week's worth of newspapers and conversations in Brazil during August 2008. Brazilians will agree when asked directly that impunity is 'bad'. But they will also say that they cannot get anything done without 'bending the rules', and that only 'fools' obey the letter of the law. As a result, their 'acts of exception' have become the norm, perpetuating the expectation of impunity as an essential disposition in the distributions of power that constitute Brazilian citizenship.

Fundamentally, the disposition that relies on impunity denies other citizens the fullness of their citizenship rights. Its success in structuring behaviour depends not only on assessing the likelihood of 'getting away with it', but also in knowing that those who demonstrate the capacity to break the rules, or only use them when rewarded. Doing so proves that they are 'big men', not bound by obligation to the rights of others. The civility of power and status in Brazil requires this normalization of exception and exclusion.

The Brazilian norm of impunity may appear to be a perverse case. But let me give two other historic examples of highly touted urban civilities that generate consensus on the basis of exclusion, intolerance, and special treatment. One is from fifth-century BCE Athens, and the other sixteenth-century Holland. These configurations of civility have been idealized as exemplars in the cultivation of civic life, though of different types. I choose them to reinforce the point that civility may well be based on treating some citizens as if they were not citizens.

The citizens of Athens enacted a series of constitutional reforms under the direction of Cleisthenes after their victory against a tyranny of elites backed by Sparta (510–508 BCE). These reforms created the most democratic *polis* in Greek history. One reform refashioned the ancient institution of ostracism (Kagan, 1961; Ober, 1989, pp. 73–75). It gave the citizens of Athens the right to decide each year to convene an assembly with the power to banish fellow citizens. If a quorum of 6,000 voted to do so, each citizen in the assembly would submit the name, scratched on a potsherd, of another he wished to expel from the city. No justification whatsoever needed to be given. The ballots were duly counted and the 'winner' was ostracized for ten years. The exiled were by no means necessarily misfits, conspirators, and renegades. Ober (1989, p. 75) cites the story of an ostracism, recounted in Plutarch, during which the famous politician Aristides was 'asked by an illiterate citizen to inscribe his own name on a shard; when queried, the citizen stated he was tired of hearing Aristides called "The Just" all the time'. Thus, the decision to ostracize could be irrational, arbitrary, and unjust and nevertheless be reached in the most democratic and 'civilized' manner.

The second example comes from the civic conduct book, *Vita Politica* (translated as *Civic Life*), first published in 1590 by the Dutch engineer and scientist Simon Stevin – a discussion of which may be found in Turner (1992), with subsequent comments by Burchell (1995). Like other civility books (manuals instructing citizens on the appropriate conduct of civic life) of the era, Stevin explains that the primary objectives of civility are good government, civil and religious peace, and the disciplined self-restraint of all 'errant passions' necessary to ensure both. With regard to dissent, both political and religious, Stevin's primer is clear. It is the duty of citizens to obey the legitimate authorities of the city where they reside. If they consider these authorities unjust, they have the right to rebel, but only after first having left the city (Stevin, 1966, pp. 493–500). Similarly, he (*Ibid*., p. 477) instructs that peace requires citizens to subordinate individual religious conscience to the doctrines of the state religion established by the sovereign and upheld by the civic and ecclesiastical authorities. Different religious beliefs and practices cannot, therefore, be tolerated. Those who hold such religious views are free to do so, but they must leave the city for another where they are accepted.

These examples demonstrate that the oppositions often drawn between practices of civility and those of exclusion, intolerance, and fundamentalism are false. Both the ancient and the early modern practices of excluding citizens from their city of residence served to enforce the binding nature of proper behaviour on all citizens by compelling them individually to conform to collective decisions. In these civic performances of exclusion and intolerance, the generally implicit rules of civility – which in the interactions of daily life are spontaneous performatives – are made palpable. The Greek case is especially revealing because the Athenians explicitly recognized that the decision to expel might be arbitrary and unjustified. Yet the ostracized citizen was morally and legally bound to accept the democratic will of the majority. He became the included excluded who makes publicly explicit the usually implicit force securing the everyday practices of civility that produce consensus. As Ober (1989, p. 75) suggests, ostracism in democratic Athens indicated that no member of the elite was above the law, or powerful enough to be immune to the rule of the masses. The arbitrary expulsion of any citizen for standing out in any way also demonstrated the sovereign power of the *polis* to impose conformity to *popular* conceptions of proper citizen behaviour.

That this practice of expulsion was a pedagogic performance of citizenship's civility is, moreover, clear from its conclusion: after ten years in exile, having obeyed the consensus of the Assembly, the ostracized citizen had the right to return to Athens with his full citizenship, status, and property intact. The collective judgment of ostracism had not turned him into a non-citizen. Rather, it had made him into an excluded and, in some sense, exemplary citizen, whose exclusion served as a disciplinary reminder of the binding nature of civility in everyday life.

Civility and Violence

If the sanction of exclusion ultimately secures the obligations of civility, it also raises the problem of the moral imperative it assumes: that the city (or the state) has the right and the duty to impose the good on others in managing its public. A host of questions immediately appear. What are the limits of this imposition – the city, the nation, the world; citizens, all residents, humankind? What are its means – legal due process, repression, violence? What are its institutions – the courts, the police, the church, vigilantes, the 'moral majority'? What is 'the good' – a certain aesthetic code for interpersonal relations, heterosexuality, a particular religious doctrine, reason, democracy? Merely raising these questions begins to erode contrasts typically drawn between civility and barbarism, citizenship and fundamentalism, and consensus and dogma, because it becomes clear that all of these options have legitimated the imposition of civility. Let me discuss one to investigate this legitimation: what is the foundation of right and duty in this imperative? I do so by way of a debate about an extreme version: the justification of waging war, in this case a 'just war' against the Indians of the New World by a fundamentalist regime of empire.

The debate was engaged in 1550 at Valladolid before the Spanish court by two exemplary opponents: Ginés de Sepúlveda, renowned Aristotelian scholar and Renaissance courtier, and Bartolomé de Las Casas, the Dominican bishop of Chiapas, Mexico. Before a jury of philosophers, jurists, and theologians, Sepúlveda advanced the arguments of a treatise he had written on the just causes of the Conquest, while Las Casas defended the contrary position. The debate had the form of a legal conflict as much as a theological one. After long presentations by each side (Las Casas's is reported to have lasted five days), the jurors retired to consider their verdict but could not reach a final decision. The debate dragged on in various incarnations for years without resolution, though it appears that the weight of the court leaned to Las Casas, as Sepúlveda never received authorization to publish his treatise. By then, however, the case was moot: more than 90 per cent of the Indians of Mexico had already been decimated.

Sepúlveda based his justification on two sorts of argument. He took the first from Aristotle's so-called natural theory of slavery, developed in the *Politics* (which he had translated into Latin): some people are born masters and others born slaves. 'He is by nature slave who … shares in reason to the extent of apprehending it without possessing it' (Aristotle 1978, p. 1254b). After setting out the logical extensions and implications of this natural hierarchy of inequality (e.g., the subordination of women to men, body to soul, emotion to reason, and evil to good), Sepúlveda marshalled a great deal of what can only be called ethnographic evidence about Indian beliefs and practices to prove that the Indians were inherently inferior to the Spanish. He based this evidence on descriptive accounts of the Conquest written by the scribes and priests who participated in it, which he took to be verified on the basis of observation and experience. He focused in particular on Aztec cannibalism, human sacrifice, and polytheism; and on the Indians' apparent lack of private property, clothing, writing, legal courts, and domesticated animals.

However, at one point in his argumentation, he advanced a key premise of a different sort – one that could not be subject to empirical verification. This was to posit a transcendental truth, an absolute which admitted no debate, but only acceptance or rejection. Sepúlveda cited an epistle of Saint Augustine to argue that 'the loss of a single soul dead without baptism exceeds in gravity the death of countless victims, even were they innocent' (cited in Todorov, 1984, p. 162). Here he shifted from Aristotle's genetic and analytic mode of reasoning to the authority of the Church to propose that the supreme value of baptism as a means of salvation justified the most extreme violence. His last clause suggested that although this kind of premise is often considered an 'article of faith' and attributed to religious fundamentalism, it can in fact supply the foundation of right for the imposition of any 'good' because it does not depend on the condition (belief, practice, life) of its object, its other. Like Aristides in Athens, the other may be entirely innocent. Indeed, the other's point of view is irrelevant to the certitude of this projection of the value of the good. In this case, the absolute value was the Christian religion, instituted in each individual soul through the sacrament of baptism. By comparison, the value of the biological life that temporarily bears this soul is of only incidental worth. If it does not submit to baptism, even if unaware of its truth, it may legitimately be sacrificed. This truth is so distant from the worth of an individual life, that the salvation of one soul justifies the termination of countless lives.

The core logic of this fundamentalist Christian civility is that the consensus of accepting the transcendental truth of baptism self-constitutes the right of imposition – that is, the right of those who accept its validity to impose its truth on those who reject it. Thus, the right of Sepúlveda's imperative to proselytize was not constituted in a social relation with other people who had a duty to it. It was irrelevant to the Spaniard whether or not the Indians believed in the Christian god and baptism, or believed that they had a duty to be baptized or to accept Spanish rule as inherently superior. This kind of self-constituting right engenders no correlative duty or other kind of social relation in the other. If it did, its legitimacy could be debated from *that* point of view. Moreover, as duties themselves engender reciprocal rights when constituted in social relations, the Indians would then have had rights with regard to the Spanish. But no such reciprocity was entailed in Sepúlveda's premise. Rather, the only duty constituted in this claim of right was upon the Spaniard himself, who already knew its transcendental worth – namely, the duty to impose it. The value assigned the right and the duty to the same person. Imposing it was therefore justified; not doing so was a dereliction of duty. With regard to the Indians as living people, it may be more accurate to say that Sepúlveda claimed to have neither rights nor duties. Their fate was irrelevant to him – as it was to all the Spaniards who applied this imperative.

This irrelevance may help us grasp, although with difficulty, the extraordinary nature of the violence that the Spanish inflicted upon the Indians. The application of Sepúlveda's imperative to compel the salvation of the New World natives resulted, paradoxically, not merely in their death. It also produced (or at least allowed to pass as 'uncriminal') an extreme cruelty with which the Spanish marked their civility on the bodies of the 'saved'.

Thus, among the many abuses against the natives which Todorov (1984, pp. 137–142) describes, we learn that the Spanish tested the sharpness of their swords by slicing off noses and breasts; roasted Indians alive; fed them to ravenous dogs; and branded their faces so often with the initials of their masters that they were turned into paper, yet paper so over-marked that it was illegible for all signs other than cruelty itself. For whatever happened to the lives of the Indians as a result of Sepúlveda's civilizing imperative, he and his company were immune from personal responsibility. Their responsibility was to the glory (the truth) of the imperative.

It was precisely this premise that Las Casas rejected. In doing so, he condemned its transcendental value. Yet he did not abandon religion. He was, and remained, a Christian bishop throughout his long life. Rather, his life among the Indians led him to shift the moral basis of the ecclesiastical mission from the problem of the hierarchical inequalities of natural superiority and inferiority to the acquisition of faith. Las Casas argued that because all human beings can acquire faith, they share equally an inherent capacity for being human. This premise of a universally inherent equality led him to counter Sepúlveda with the assertion of the immanent value of human life over transcendental absolutes of reason or nature: 'It would be a great disorder and a mortal sin', he replied, 'to toss a child into a well in order to baptize it and save its soul, if thereby it died' (cited in Todorov, 1984). Las Casas's premise was not only that the death of thousands of people was not justified by the salvation of one. It was more radical. The death of just one child was not worth its salvation – which is to say that the life of a single human being, however foreign, has greater worth than all the doctrines of salvation. Las Casas appears at this moment in the debate to have introduced a strikingly modern argument for the sanctity of individual human rights; indeed, his premise was that there are no persons without rights, even the most radically different.

Todorov (1984, pp. 162–167) develops an insightful assessment of this argument. On the one hand, Las Casas seemed to posit a radical equality among human beings, regardless of all other differences, not only for the capacity to acquire faith but also to possess natural rights. Hence, he argued that 'the natural laws and rules and rights of men are common to all nations, Christian and gentile, and whatever their sect, law, state, color and condition, without any difference' (cited in Todorov, 1984, p. 162). Las Casas further emphasized this equality between self and other by switching subject positions. He applied to the Indians the very same standards that he held for himself: 'All the Indians to be found here are to be held as free: for in truth so they are, by the same right as I myself am free' (*Ibid.*). Note his use of 'right' as a universal human attribute exercised individually. Also note that Las Casas's premise about universal human rights is no more verifiable than Sepúlveda's imperative. They are both theoretical postulates, though the application of Las Casas's appears to have far more just and sane consequences for the world of the living.

At the same time that Las Casas advanced this universalizing premise about human rights, he also made a set of 'ethnographic' arguments about the practices of the Indians based on his claim that they were 'innocents'. He did so to counter Sepúlveda's assessment of these same practices as barbarous, damned, and inferior. It is with these arguments that we encounter a major problem. Las Casas always interpreted native practices in terms of supposed psychologies and intentions that indicated the pre-Fall state of innocence of the Indians, and therefore their original capacity to acquire the true Christian faith. They were exemplary in being gentle, kind, humble, obedient, frugal, and so forth; in fact, he implied that the Indians exhibited the attributes (civilities) of 'good Christians' far more than the rapacious Spaniards. The very traits Sepúlveda cited as evidence of their inferiority, Las Casas consistently interpreted as evidence of their capacity to become model Christians. Their lack of private property indicated, for example, that they desired only what was necessary; their most shocking practices – human sacrifice and cannibalism – indicated their extraordinary capacity for love, for they gave what was most precious – life – to their gods.

Thus, in all examples, Las Casas never really saw the Indians as different – that is, as having different cultural and social configurations that might account for these features. Rather, he identified the Indians not only with himself but also with his own ideal. He did not realize or accept that to be equal, the Indians did not have to be identical to the Spanish. As Todorov argues, he was in this way just like Sepúlveda. Neither accepted that the Indians might be equal but different. They both observed differences but prejudged them according to their own criteria as either barbarously evil (Sepúlveda) or barbarously good (Las Casas). Neither view was one of the equal coexistence of different worlds.

Las Casas's originality in the debate over the imposition of the good through just warfare was to advance an argument for immanent individual worth and rights. However, he only disputed the means (coercion) but not the end of the Conquest. This end was the assimilation of the Indians into Christendom and with it their submission to the Spanish colonial order. As assimilation is ultimately the elimination of difference, it surely constitutes a kind of annihilation. Moreover, as Todorov (1984, p. 166) notes, Las Casas did not hesitate to condemn 'Turks and Moors, the veritable barbarian outcasts of the nations' because, obviously, he could not identify them as proto-Christians. Thus it was not that Las Casas refused to denigrate others just because they were different. He loved the Indians only because he assimilated them to himself. Like Sepúlveda, he despised those whom he perceived as truly different.

The world citizenship of natural and individual human rights that Las Casas advocated and that seems initially generous and promising thus relies on a civility not much different from those of more exclusive citizenships. It prejudges members and non-members alike by its own standards, including those it identifies as similar and excluding the rest. Its equality is one of sameness, and its norm one of a certain kind of correct status and behaviour to be acknowledged by those who define the regime of citizenship (the Church). Therefore, although Las Casas maintained that human rights are naturally immanent in each individual, the actual realization of these rights depends on the identification of people with pre-approved categories. In other words, incorporation into the world citizenship of Christendom is universally inclusive – anyone can be a Christian – but distribution is inegalitarian: Las Casas's church differentially distributes its benefits to presorted categories of people. In this differentiated citizenship, actual access to rights is the privilege of those assimilated to these categories.

Thus, we have returned to the initial question with which we began: what conditions sustain regimes of citizenship that deny some of their own citizens an existence worthy of a citizen? If we are all citizens of world Christendom, how can some of its citizens deny other citizens ('Turks and Moors') rights?

Rights as Privilege

The conceptualization of rights as special treatment for certain kinds of citizens grounds the civilities of all the systems of differentiated citizenship I have examined thus far. Surely, some measures of special treatment are necessary in every citizenship, including those that seek fairness. Hence, exceptional rights for seniors, veterans, and the disabled are widely seen as promoting fairness. In differentiated citizenships, however, such exceptions become the norm. As long as special treatment prevails as the general conception of rights, citizenship remains overwhelmingly a means for distributing and legitimating inequality and therefore for creating both marginal (national) citizens and (stateless) non-citizens. Let me return to the case of the urban peripheries of contemporary Brazil to consider the conditions of impunity that perpetuate special treatment citizenship.[1] For it is Sepúlveda's sense of impunity in the absence of any reciprocity with the Indians that secured his assumption of the right to impose a civilizing good.

In the peripheries of São Paulo where I have done fieldwork for many years, a resident described to me the regime of citizenship against which he has struggled his whole life:

> In the time of my parents, a citizen was a guy loaded with money. It's true. The citizen was the chic, the rich, the owner of a business... The worker was not a citizen, no. That didn't exist. The worker was a peon (*peão*). Peon, peon, peon, his whole life. My father came to São Paulo a simple share cropper and died a simple construction worker. But he met all his obligations, all his duties. And when he went somewhere and needed some right, no one treated him as a citizen. They made him into a marginal, as if he were trash. I saw that, and I experienced that too. The injustice made me furious.

The paradigm of citizenship this man describes produces the paradox he denounces: Brazilians can consider some Brazilians as 'citizens' who have rights, and other Brazilians as 'marginals' who lack rights. This distinction only makes sense from within the system of differentiated citizenship that treats some Brazilians as if they were not citizens because, for reasons that have nothing to do with their national membership, they are denied rights.

In the urban peripheries where this man lives, residents use the word 'right' in three modalities.[2] In the singular, it denotes a specific right (*direito de*), usually a political or civil right, like the right to vote or strike. In the plural, it means a condition of having rights (*ter direitos*). The scope of these rights has expanded considerably over the last forty years with the successes of urban citizenship movements. If the plural tended in the past to refer only to labour and welfare rights, today it is just as likely to include political and civil rights and a host of rights to the city all grounded in the 1988 Citizen Constitution. Much has changed in this regard. Nevertheless, in both the singular and the plural, residents continue to think that having rights depends on a third modality of right – namely, that of being right (*ser direito*) or walking right (*andar direito*). This mode refers to a moral condition of correctness. Having rights depends on being right and being right is a matter of achieving certain statuses, basically those of 'a good worker, family provider, and honest person', as residents say. Those who have citizen rights deserve them because they are morally good and socially correct in these publicly recognized terms. Similarly, those who fail to be morally right – criminals, squatters, deviants … an expandable category – deserve to be denied rights. By extension, the logic of this special-treatment citizenship also produces the *a priori* judgment that those who lack rights – the poor, for example – must be assumed – even by the poor themselves – to have failed morally. Both negative judgments allow Brazilians to assume that other Brazilians lack rights in relation to themselves, and therefore that they have no duty to them if they consider them marginals in one way or another.[3]

Access to rights in this conceptualization of special treatment depends on two conditions. On the one hand, people think they have rights because they hold statuses recognized and legalized by the state. On the other, the state only bestows these rights on the right people. Laws establish both conditions. For example, the 1937 Constitution created a perduring construct of social marginality and exclusion with regard to unemployment and informal work. However, having or not having rights is not only a determination of law. Rather, legal rights may be available to all citizens in theory (as populism proposes), but they can only be acquired and realized by those who deserve them in terms of specific personal attributes (e.g., whether they are good workers or literate or registered in a profession). For most Brazilians, therefore, the exclusions of differentiated citizenship often appear to result less from legal and political causes than from personal failings. This depoliticization perpetuates the legitimacy of exclusionary citizenship rights by blaming the excluded for not having them, and it establishes a consensus about who has rights and according to what standards – a civility, in short.

It is significant to note that because these rights can only be acquired by the right citizens, people who need to use them 'have to chase after their rights'. In the context of special-treatment citizenship, the ubiquitous phrase 'look for your rights' means not only knowing what rights adhere to a particular status. Above all, it means having to prove to the proper authorities that you possess the right status and deserve its rights. This proof is not only a matter of having the right paperwork to show that the petitioner is an 'honest person' with a clean record, and not a marginal of any sort. To be rewarded with a right, the correct status and behaviour of the petitioner must be acknowledged by the provider, typically a bureaucrat, official, or employer. This personal acknowledgment is required not only because special-treatment rights always depend on the identification of subsets of statuses within the general status of citizen. More significantly, it is necessary because the application of law in Brazil is rarely routine or certain. Rather, it must be made to apply through the personal intervention of someone in a position to acknowledge the good standing and just deserve of the petitioner. The need for such special pleading exacerbates the struggle of the poor to run after their rights. It always puts them on the defensive, forces them to find the right person to intercede on their behalf, renders uncertain their dignity and respect, and makes them acknowledge their inferiority. Consequently, proving one's worth to find one's rights is always frustrating and often impossible. It is therefore not surprising that being 'treated like trash' is a reason I frequently hear to explain why people desist from pursuing their rights.

The personalization of rights means that their exercise depends on the discretion, not the duty, of someone in a position of power to recognize the personal merit of the petitioner and grant access to the right. This discretionary power converts rights into privileges, in the sense that it becomes a privilege to obtain what is by law a right. A right creates a duty when it makes someone vulnerable to a claimant's legal powers. In that sense it empowers the claimant. When these relations depend on personal intervention, discretion, and mediation, they become legally subverted. In Hohfeldian terms, the acknowledger now has the power to decide when rights apply, and yet no duty to make them available. He is not liable to the claimant's legal power and has thus gained an immunity. In turn, the claimant is vulnerable to the exercise of that power, having no right to determine its course. He therefore suffers a disability that can only be overcome by personal intercession. When the latter occurs, the claimant exercises his right only as the favour of the person who grants it. In a system of citizenship rights thus based on the immunity/impunity of some and the disability of others, rights become relations of privilege between some who act with an absence of duty to others who, in turn, have no power to enforce claims. The disprivileged lack rights and are vulnerable to the power of others. The privileged experience citizenship as a power that frees them from the claims of others, leaving them unconstrained by legal duty and exempt from legal responsibility. These relations of privilege and disprivilege are the central features of the civility that characterizes citizenships in which special treatment dominates.

The 'search for rights' in such citizenships therefore engages the poor in a perverse exercise which those with immunity and privilege bypass: it not only perpetuates but also legitimates the distribution of inequality, because it gets the disprivileged to seek and defend special treatment for themselves and disqualification for others as the means to confirm their particular worthiness and attain their hard-won recognition, respect, and recompense. In this exchange, it induces them to accept the legitimacy of citizenship's distribution of unequal treatment as a just means to compensate for, if not reward, pre-existing inequalities. In this way, the lived experience of differentiated citizenship – its civility – gets the disprivileged to approve compensating inequalities of privilege by legalizing more privilege.

Brazilian civility thus turns the poor into advocates of the maxim of justice that the republican Rui Barbosa is credited with coining at the end of the nineteenth century, and that has become a fixture of legal education in Brazil ever since: 'Justice consists in treating the equal equally and the unequal unequally according to the measure of their inequality'. Barbosa was in fact reiterating a view usually traced to Aristotle's *Nicomachean Ethics*: a just distribution, Aristotle argued, allocates the right share to the right person, such that 'the ratio between the shares will be the same as that between the persons. If the persons are not equal, their just shares will not be equal' (1962, p. 118).

When I have asked residents of the peripheries to explain Barbosa's maximum, those who do not get twisted up in its play of words often give the very same example that lawyers and judges I have asked also give: the law permits women to retire five years earlier than men. This discrimination is just because over the course of a normal life, working women 'have more service' than working men; in addition to work outside the home, they have to do the housework and childcare in which they are little aided by their husbands. 'Thus', one renowned law professor and scholar concluded, 'she has an overload of services that is just to compensate by allowing her to retire with less time of service and less age'. This is the same reasoning that a textile worker in the urban peripheries gave me:

> I think it is just. Because if you think about it, a housewife who has a job outside works double. When I arrive home from work, what do I do? I take a shower, watch television, sit on the sofa doing whatever; or I go to the bar and have a beer. What does the woman do? When she arrives from work, she makes dinner, takes care of the children, cleans the house, arranges the kitchen, washes and irons clothes. She works about double my work, if you analyze the question. Therefore, I think that she should have even more time [than five years] to retire before a man, because there still exists a lot of discrimination in the work of women.

In both explanations, the solution for the social facts of inequality – that working women are unequal because they work more – is not to propose to change the social relations of gender and work. Rather, it is to produce more inequality, in the form of the compensatory legal privilege of earlier retirement.

We have thus come full circle to find Aristotle in the urban peripheries of Brazil. Most Brazilians, including the urban poor, understand this justice to mean that unequal treatment is a just way to offset pre-existing inequalities, especially among the poor. It may be a means, therefore, to compensate an inequality of disprivilege by legalizing privilege. However, it may also compensate an inequality of privilege by legalizing more privilege. In either case, the civility of Brazilian citizenship lets few grasp that it reproduces privilege – and therefore impunity – throughout the social and legal system. It is, moreover, a static concept of justice. It does not contest inequality. Rather, it accepts that social inequalities exist as prior conditions of either disprivilege or privilege and treats them differently by distributing resources accordingly. Thus the civility in which Barbosa's maxim is a taken-for-granted standard produces consensus for a justice system that perpetuates differentiated citizenship. Through countless dispositions of everyday social relations, it maintains a society of social differences organized according to legalized privileges, disprivileges, and impunities.

Conclusion

My arguments here have focused on the foundations of civility that establish its conformity to standards of citizen behaviour. I have been concerned to show that even inegalitarian citizenships have their civility, as a means to erode its idealization. Hence, I analyzed examples in which civility is based explicitly on exclusion, denial of rights, and impunity. If it is the case, as I suggest, that all citizenships have their civility, that all claim the right and the duty to impose their conceptions of the good on citizens, and that they do so ultimately and often explicitly on the basis of exclusion, then what distinguishes a fundamentalist civility from other kinds?

Civilities do not exist without a point of reference which establishes the standards of conformity. This reference may be considered a good that legitimates the consensus of conformity and justifies the sacrifices, judgments, and exclusions required to achieve it. Such goods are of different sorts – for example, a notion of honour or merit that infuses a hierarchical structure of classes, of purity and pollution in a system of racial categories, or of the immanent worth of an individual life. Citizenships use such goods to justify their modes of incorporating members and distributing rights among them. What distinguishes the civility of a fundamentalist citizenship is that it is based on a comprehensive doctrine rather than a political conception of the good. The former constitutes itself as a transcendental truth whose validity admits of no rational debate and whose value, as the case of Sepúlveda illustrates, assigns both the right and the duty of realization to the believer regardless of the interests of non-believers. The latter are irrelevant to this right and have no recourse other than capitulation or resistance to its exercise. A political conceptualization of goods also legitimates imposition and, more often than not in the history of citizenships, inegalitarian distributions of resources. But it can be challenged, debated, undermined, overthrown, and displaced by political contestation, however difficult that process turns out to be in particular cases. Political contestation can dismantle privilege, establish responsibility, and end impunity.

A judgment in favour of political over comprehensive and transcendental conceptions of the goods that legitimate civility may seem obvious to many readers – as easy as the condemnation of Sepúlveda for inciting crimes against humanity. Yet a lauded book on civility by a prominent contemporary American jurist and declared Christian suggests that transcendental conceptions endure. Stephen L. Carter (1998, p. 23) defines civility as 'the set of sacrifices we make for the sake of our common journey with others, and out of love and respect for the very idea that there *are* others'. The apparent generosity of this conceptualization turns out to be grounded in two religious precepts. Carter derives the sacrifices we have to make to ensure civility from 'the duty to love our neighbours', even if we despise them, which 'is a precept of both the Christian and Jewish traditions' (*Ibid*., p. 23). He bases the commonality civility aims to create and the respect for the other on which it is grounded on an ultimate transcendent: faith in God, 'for there is no truer and more profound vision of equality than equality before God... In the absence of that language of loving sacrifice, that connection to the transcendent, civility, like any other moral principle, has no firm rock on which to stand' (*Ibid*., 31).

Carter's distrust of 'the shifting sands of secular morality' and 'political wind' (*Ibid*., 31) to provide firm foundations for civility hinges on what he means by firm. Surely religious conflicts are no less frequent or deadly than secular ones. But the transcendental truths that collide in religious conflict are to each side of believers as unimpeachable and immovable as secular politics appears mutable. To be firm for Carter, therefore, the foundation of civility must be transcendental. The problem of such an unimpeachable basis is the one we encountered in Sepúlveda's justification of a just war against the Indians: it assumes a right of imposition that is self-constituting and irrefutable (and always pre-empts any Las Casaian claim to 'love the other'). Hence, in a discussion of civility and religious proselytizing, Carter argues:

> But once you [who?] decide that it is time to get the message out ... you must preach to those you want to reach. The process is the same whether you are electioneering, collecting for charity, or gathering signatures on a petition ... or

> trying to persuade others that you possess a religious truth that they lack... [For those] who believe that the Gospel message is universal in the sense that nobody who does not accept it will be saved, it would plainly be unloving – uncivil – not to propagate it. Integrity would permit nothing less. (*Ibid.*, pp. 255–257)

The problem with this argument is that it is a mistake to think that gathering signatures on a petition and trying to convert others to a religious truth they lack are 'the same'. They are not. The civilities entailed are entirely different. The one involves a political conception of goods that is subject to debate and that focuses on a specific issue, about which agreement or disagreement to lend support entails no other commitments or at least none of a comprehensive and pervasive nature. The other involves precisely that kind of comprehensive commitment beyond the reach of reason. For believers, proselytizing entails nothing less than salvation or damnation, a transcendental dilemma for fundamentalist civility that can, as with Sepúlveda, be used to justify 'the death of countless victims, even were they innocent'. Political conceptions of the goods that legitimate civility are also used to justify its imposition, often with the denial of rights, the assumption of impunity, and the loss of life. However, the political is both imposable and impeachable by rational means. Its values can be defined and delimited by deliberation among citizens and between citizens and non-citizens. These are, perhaps, small differences given the barbarisms that have been committed in the name of civility, both secular and religious. But for human life, they seem to me rather more than less significant.

Notes

1 A full discussion of this case may be found in Holston, 2008.

2 In the late 1970s, Teresa Caldeira (1984, pp. 224–235) identified these modalities in her study of one neighbourhood in these peripheries. My own fieldwork in the same neighbourhood beginning a decade later confirmed that these modalities persisted, albeit with additional meanings and different foundations.

3 The social and citizenship movements of the urban peripheries created two new foundations for rights based on the condition of being a stake holder in the city (derived from the autoconstruction of houses and neighbourhoods) and on the 1988 Citizen Constitution. I analyze the former as contributor rights, and the later as text-based rights, in Holston, 2008. Nevertheless, the moral basis for having rights continues to be prominent as well. Now, however, it is evoked with the other two, often mixed in the same discussion, creating a hybrid and at times contradictory foundation for rights.

References

Aristotle (1962) *Nicomachean Ethics* (translated by M. Ostwald). Indianapolis, IN: Bobbs-Merrill.

Aristotle (1978) *The Politics* (translated by E. Barker). Oxford: Oxford University Press.

Bourdieu, P. (1977) *The Outline of a Theory of Practice* (translated by Richard Nice). Cambridge: Cambridge University Press.

Burchell, D. (1995) The attributes of citizens: virtue, manners, and the activity of citizenship. *Economy and Society*, **24**(4), pp. 540–558.

Caldeira, T.P.R. (1984) *A Política dos Outros: O Cotidiano dos Moradores da Periferia e o que Pensam do Poder e dos Poderosos* [The policy of the other: the daily life of residents of the periphery and what they think of power and the powerful]. Sao Paulo: Brasiliense.

Carter, S.L. (1998) *Civility: Manners, Morals, and the Etiquette of Democracy*. New York: HarperPerennial.

Holston, J. (2008) *Insurgent Citizenship: Disjunctions of Democracy and Modernity in Brazil*. Princeton, NJ: Princeton University Press.

Kagan, D. (1961) The origin and purposes of ostracism. *Hesperia*, **30**, pp. 393–401.

Marx, K. (1967) On the Jewish question, in Easton, L.D. and Guddat, K.H. (eds.) *Writings of the Young Marx on Philosophy and Society*. New York: Anchor Books.

Ober, J. (1989) *Mass and Elite in Democratic Athens: Rhetoric, Ideology, and the Power of the People*. Princeton, NJ: Princeton University Press.

Roberts, J.T. (1982) *Accountability in Athenian Government*. Madison, WI: University of Wisconsin Press.

Stevin, S. (1966) *Civic Life* (A. Romein-Verschoor edition). Amsterdam: CV Swets & Zeitlinger (original edition, *Vita Politica*, 1590).

Todorov, T. (1984) *The Conquest of America*. New York: Harper & Row.

Turner, B. (1992) Outline of a theory of citizenship, in Mouffe, C. (ed.) *Dimensions of Radical Democracy: Pluralism, Citizenship, Community*. London: Verso.

Walzer, M. (1989) Citizenship, in Ball, T., Farr, J. and Hanson, R.L. (eds.) *Political Innovation and Conceptual Change*. Cambridge: Cambridge University Press.

Part II: Fundamentalisms and Urbanism

Chapter 4

American National Identity, the Rise of the Modern City, and the Birth of Protestant Fundamentalism

Rhys H. Williams

The early-twentieth-century birth of Protestant fundamentalism in the United States, as a reaction to 'modernism' in religion and culture, is an oft-told story, and more or less conventional academic wisdom. Modernist developments in culture and religion, especially among Protestants, spurred a forceful reassertion of theological orthodoxy that grew into a significant religious, and eventually social, movement. Less appreciated are other aspects of the societal milieu during the turn-of-the-century formative period to which fundamentalism was also responding – the transformation of American cities due to large-scale immigration, the development of an urban, industrialized economy, and an attendant increase in socio-geographic mobility. These developments can be termed 'modernity' – a set of social, structural, and socio-cultural changes that began to define industrial society in the mid-nineteenth century. Because fundamentalism was so self-consciously theological – that is, it was originally an 'intellectual movement', even as it was rejecting much of the intellectual world of its day – the standard focus is on its relationship to modernism as a set of ideas. In this chapter I focus more on its reaction to modernity as a set of social arrangements. And in turn, I argue that the turn-of-the-century cultural and social milieu – the ascension of modernism and modernity – shaped a crisis in popular conceptions of American national identity.

These last factors were linked to the rise of Protestant fundamentalism more significantly than is often recognized, in part, I argue, because fundamentalism's urban profile is underappreciated. After the trial of John Scopes for teaching evolution in Tennessee in 1925, fundamentalist religion became characterized – primarily by Northeast urban intellectuals – as the desperate reaction of the uneducated, the rural, and those in the South (where, indeed, levels of formal education were lower, and rates of poverty and rural population were higher). This was partly a rhetorical victory for fundamentalism's modernist opponents (such as H.L. Mencken), eager to portray it as backward and ignorant. Not coincidently, these opponents were also committed to a new, more cosmopolitan definition of American national identity. Further, the typification of fundamentalism as backward also aligned with the reigning social-science theory of the day – a narrative of 'modernization' that predicted that rationality, industry, and institutional differentiation would secularize society in the name of science and social progress.

These developments, whatever truth there is in them, should not obscure either the founding impulses that produced fundamentalist religion or fundamentalism's significant urban presence in the early twentieth century – at the exact time when immigration and industry were transforming American cities. This period of social change produced a number of innovations in American religion, and fundamentalism was but one such product. Thus, fundamentalism's relationship to urban society is both foundational and deeply ingrained. It should not be ignored. While fundamentalist Protestantism did diffuse throughout the rural South and Midwest, it also maintained a significant urban presence.

I pursue this argument through a historically grounded analysis of the socio-cultural setting in which fundamentalism arose, but one that also has obvious implications for the contemporary United States, particularly as American cities are again being transformed by immigration and incorporation into global markets. The chapter proceeds as follows. First, I consider 'fundamentalism' as a religious and historical concept, attempting some analytic precision, or at the least consistency, in its use. Contemporary usage of the concept is often quite expansive and ahistorical. At the same time, the potential advantages of a comparative and diachronic perspective should neither be dismissed nor accepted uncritically. Second, I chart the transformation of urban America in the early twentieth century, with a particular focus on immigration, and examine the religious responses to this transformation. Third, I show how both of these developments – fundamentalism and urban transformation – were connected, sometimes explicitly and sometimes only implicitly, to contests over and a crisis in American national identity.

Fundamentalism as Religious Innovation

A basic hurdle in establishing this argument is recognizing that the concept of fundamentalism is often treated too expansively in popular media and everyday usage. Indeed, it is often conflated with any form of religious traditionalism, or used as a pejorative attached to any belief system thought to be inflexible and politically conservative. In the current U.S. scene, fundamentalism is often used synonymously with the term 'evangelicalism'. Particularly since the publication of the five-volume 'Fundamentalisms Project' (see Marty and Appleby, 1991, 1995) there has been considerable scholarly debate about the usefulness of the term fundamentalism. There is no disputing the origins of the religious movement in North American Protestantism in the early twentieth century (Ammerman, 1991; Marsden, 1980). In that regard, many scholars have taken what is essentially a historicist position, and believe that the term should remain dedicated to that referent. They maintain that the theological characteristics that marked Protestant fundamentalism as a religious movement are culturally and religiously specific, and that stretching the term dilutes its accuracy. There is concern that such dilution is a form of reductionism in that it dismisses Protestant fundamentalism's specific religious tenets and beliefs. The original 'fundamentals' were a self-consciously theological reaction to developments in Protestantism.[1]

Other scholars, of course, have argued that there is an analytic usefulness in using the term to examine what seem to be similar religious movements in other faith traditions. For example, if we foreground the importance placed on particular scriptural interpretations by some conservative Jewish and Islamic groups, 'fundamentalism' could be applied across the three Abrahamic monotheisms. This is probably the most common use of the term, and it focuses on fundamentalism as a religious orientation marked by scriptural literalism or legalism in authority. On the other hand, if we emphasize fundamentalism's tendency to construct an idealized past and then employ a totalizing religious worldview in an attempt to re-create that past (a feature that figures in Marty and Appleby, 1991), we can see that tendency among some Hindu and Buddhist religious movements. These Eastern faiths have also had to respond to the modern industrial world and its use of scientific rationality – both modernity and modernism – and thus they share some characteristics with similarly situated religious movements in other faiths.[2]

Marty and Appleby's comparative fundamentalisms project was built on the assumption of the analytic usefulness of comparisons across traditions, and it focused on the encounter with the modern world within each faith. They recognize the controversy in their usage, but defend it by noting that there are commonalities across a 'hypothetical family', even as they analytically preserve the religious character of the movements themselves: '… we have begun by emptying the term of its culture-specific and tradition-specific content and context before examining cases across the board to see if there are in fact 'family resemblances' among movements commonly perceived as "fundamentalist" ' (Marty and Appleby, 1991, p. 816).

Significantly, Marty and Appleby point to the importance of the basic conditions for fundamentalism's emergence, what they call the 'contemporaneity of their enterprise' (*Ibid.*, p. 814). They see it as a twentieth-century phenomenon which draws on an idealized past for standards and values, but is without direct ideological precedents.[3] Thus, they argue, fundamentalism shows a closer affinity to modernism – as an ideological project, as an organized religious expression, and in using modernism as a self-conscious foil – than it does to traditionalism. Fundamentalism was created in the encounter with modernism and modernity, and shares some modernist assumptions about the world. But it uses its modernist features to oppose certain societal developments.

As one of the contributors to the Marty and Appleby project (Williams, 1994), and as a sociologist, I generally side with the comparative analytic approach as opposed to the historicist critique. However, I am mindful of the dangers of stretching the term too thinly; specifically within the American case, I reject the conflation of 'fundamentalism' with 'conservative Protestantism' or 'evangelicalism', as is too often done in popular media and political writing. That approach violates the phenomenology of the religious believers, as well as masking analytic differences in the beliefs and the social consequences of these different forms of religious expression. Thus, for the purposes of clarity regarding this project, I introduce some distinctions here.

Fundamentalism is, sociologically, distinctly different from the religious 'traditionalism' which preceded it, even though that traditionalism may well have objected to some of the same developments in religious and social 'modernism' that sparked fundamentalism. However, fundamentalism's self-consciousness about its opposition produced what could be considered religious innovation. For example, the elevation of a particularly rigid interpretive form of understanding Scripture – and the emphasis on that textual authority in preference to other potential sources of authority (e.g., church doctrine, past practices, human reason) – is not entirely consistent with 'traditional' practices.

As sociologist Max Weber noted, 'traditional authority' – the justificatory cultural apparatus that is used to legitimate traditional social arrangements – is typified by the phrase 'we have always done it this way'. In settings with relatively slow paces of social change and population development (such as rural areas), cultures often become highly internally integrated, with various aspects of the culture reinforcing belief and practice in other dimensions. Religious beliefs, practices, and the interpretation of Scripture may be quite traditional, but they need not be theorized directly; nor need the Scripture be necessarily seen as inerrant or to be interpreted literally, precisely because other aspects of culture and social life are structured in ways that reinforce traditional meanings and practices.

With the advent of modern urban society, life was so utterly transformed that the integrated cultural forms of traditional society no longer fit together so smoothly. Further, the spread of literacy and the availability of Bibles meant that more people had access to sacred writing than ever before. Traditional authority could not as easily guide religious and social relationships. Some form of authority needed to be elevated and systematized to replace the taken-for-grantedness of traditionalism, and for fundamentalist religion it was the inerrantist and often literal interpretation of Scripture. Thus, it is only with the development of the modern world and its accompanying socio-cultural changes that a religious reaction can be termed 'fundamentalism'. Fundamentalism is reactive to threats to what is understood to be 'traditional' culture and its sets of meanings, and it often claims to be restorative of this lost or threatened golden age. But fundamentalism relies on a particular interpretive literalism in textual analysis, and a rigid boundary-construction in creating cultural meanings, because its other cultural underpinnings – homogeneous populations, slower paces of social mobility and social change – are less reliable. Things that were taken for granted must be justified directly. In Clifford Geertz's (1973) terms, 'culture' – a set of meanings that hold people in place and reproduce the existing social world – must give way to 'ideology' – a set of precepts or meanings that must be decided upon and held fervently after taken-for-grantedness is no longer possible (see Williams, 1996).

Just as there is an important distinction between fundamentalism and traditionalism there are important differences, particularly in contemporary North America, between fundamentalism and 'evangelicalism', a term with even more historical overlays in its definitions. Currently, what is called evangelicalism is a Protestant religious orientation found among a quarter to a third (depending upon how one measures) of the American people. But what is currently called evangelicalism, while sharing some precepts with fundamentalism, was also a mid-twentieth-century reformist reaction among conservative Protestants against aspects of fundamentalism.

The two orientations share an emphasis on Scripture as the dominant – even sole – source of religious authority. And they share a prescriptive desire to convert others to their faith (i.e., to 'evangelize'). However, historically, evangelicalism has emphasized experiential approaches to faith, often expressed emotionally. Evangelicalism spread through religious 'revivals', which increased the emphasis on subjectively realized, emotionally driven faith. Often the people involved were barely literate, making the reading and comprehension of Scripture problematic, and occasionally leading into individualistic interpretations (built on experience) while making a conversion or 'born again' experience a premier qualifying mark of faithfulness. Carpenter distinguishes this from fundamentalism:

> While these groups [Holiness and Pentecostal forms of evangelical Protestantism] emphasized moral and experiential answers to modern secularity's challenge, fundamentalism favored the cognitive and ideological battleground. Although fundamentalists shared the other two movements' concerns for right living and the power of the Holy Spirit, they cared more about fighting for right doctrine. (Carpenter, 1997, p. 5)

Another important distinction between evangelism and fundamentalism is the relative pre-eminence of religious 'separatism', in which believers are called to separate themselves from 'the world' – including not just the secular world but also those mainstream religious institutions which have made their peace with modernity. Fundamentalism was as much a response to liberal, accommodating religion as it was to secular society. Ammerman (1991, pp. 7–8) takes this separatism seriously enough to make it a 'central feature' of fundamentalism and part of her definition. Carpenter (1997, chapters 2–4) makes it less definitional, but shows how theological, social, and institutional separatism developed within North American fundamentalism, as individuals and groups were urged to 'come out' of the fallen world. Carpenter (1997), Marsden (1987), and Smith *et al.* (1998) credit this separatism with making many conservative Protestants extremely uncomfortable by the mid-twentieth century, leading to the founding of a religious movement that self-consciously called itself 'evangelicalism', and that sought to maintain its faith while engaging the world more fully.

For my own working definition of fundamentalism, I focus on what might be called 'structural' features, rather than the specific content, of fundamentalist doctrine. That is, I emphasize certain features of the fundamentalist worldview because of the ways they shape fundamentalism's stance towards social, cultural, and religious worlds. The specific content of religious doctrine is less important in this case, as different beliefs could take what I think of as the fundamentalist form. This approach keeps a focus on the importance of 'right belief' that so animates fundamentalism, but also allows comparisons across religious traditions, and across historical and social settings. Thus, while focusing this chapter on the birth of Protestant fundamentalism in the U.S., I use a definition that does not limit the term to this case. I draw attention to three features as emblematic of fundamentalism.

First, fundamentalism is 'restorationist', in the sense that it draws upon an idealized past for standards and values, with a call to return to that social time and setting. As a practical matter, this often necessitates some historical and theological reconstruction, but that is usually done within an inerrantist, often literalist, interpretive framework. This idealized time, it is important to note, is an image of the religious collective – a 'public good' and a societal order that fundamentalists are often committed to attaining for the wider society.

Second, fundamentalisms produce totalizing, boundary-creating cultural fields, which characteristically separate the world into binary sets of categories. The differences demarcated by these boundaries are theologized by fundamentalists, and divide the social world not just into separate groups, but into the religious and the secular (or the fallen), the moral and the immoral – or in stark terms, good and evil. A binary classification scheme is not unique to fundamentalist thought, but it totalizes the world and infuses it with religious significance (Marty and Appleby, 1991, pp. 814–842). The boundaries creating these binary distinctions pull together in powerful and easily comprehended terms a religio-moral world which can be superimposed on the social world. They align moral identity with hierarchy and legitimacy.

Third, and attendant to the concern with boundaries, has been the 'separatism' noted above in which believers are called to separate themselves from 'the world' – including both the secular world and those religions that have not policed their boundaries with significant rigor. It follows that if boundaries become identified as 'sacred', their importance and need for protection is that much greater. For religions that claim all divine truth for themselves rival faiths can form as much a threat as the secular world. Because most religions offer clear prescriptions for organizing human relationships within society (as well as dictating how humans should relate to the divine) sacralized boundaries often translate into a suspicion of religious pluralism in society. Similarly, the secular world and its pleasures represent a danger to religious purity. And yet, in modern societies, fundamentalist religious people must live surrounded by other faiths and those who are not religious. 'Defensive' fundamentalism (Williams, 1994) becomes a religious impulse in which groups become primarily interested in keeping the world out, and they maintain a view of themselves as a faithful, holy remnant.

One element of fundamentalist separatism has been a decided tendency towards the establishment of parallel institutions, such as schools (see Peshkin, 1988), to keep the community of believers away from the temptations and threats represented by those who do not share their commitments. Fundamentalism's separatist worldview maintains the purity of its faith by refusing to 'fellowship' with non-believers or even other conservative Christians who are not sufficiently fundamentalist (Carpenter, 1997, p. 7; Marsden, 1987, p. 5). When the separatist taboo becomes paramount, there is a disengagement from the world. Fundamentalists first expressed this separatism through denominational schisms and the founding of 'parachurch' organizations. Separatism was then reinforced in the U.S. by the South's (and other rural areas) regional marginalization, its agricultural economy, and the relative lack of immigration to and social diversity in those regions through much of the twentieth century.

However, fundamentalism's concerns with boundaries and separatism have always existed in tension with the basic Protestant Christian charge to spread religious truth through evangelism and conversion. Fundamentalists see part of their religious mission as changing this world to better accord with Divine Will; and to the extent that they have viewed themselves as heirs of the quintessential American religion, they see themselves as needing to restore their religion to its rightful public and institutional position. To do that, however, they have had to interact with other religious groups or the secular world. The tension has been consistent in fundamentalism – the threat of boundary 'violations' to separatist purity has had to be balanced with the capacity to allow the faithful to reach out and convert the world. Thus, the boundaries separating fundamentalists from others are both important, and potentially problematic. If one pursues an evangelical and universalistic faith, one is committed to the idea that anyone can be, potentially, a convert. Conversion and evangelism are key to, and definitional of, fundamentalist identity. With the concomitant duty to help create a godly society, or at least to attempt to do so, fundamentalists must reach across boundaries even as they police them. All the while the godly must be careful not to fall themselves.

In sum, fundamentalism as a broad analytic category is best understood as a religious innovation born of the encounter with a modernizing, secularizing, and diversifying world, simultaneously dependent upon but in opposition to key features of modern secular society. Fundamentalisms are marked by particular doctrinal features (such as, for Christians, premillennial theology), religious teachings that can be seen to exist across religious traditions (a codification and literal interpretation of Scripture and scriptural authority), a number of shared cultural practices (such as the development and policing of bright moral boundaries to order the religious and secular world), and the need to be simultaneously universalist and exclusionist. This constellation of characteristics can make fundamentalist groups both defensive and proactive, depending on context. Such groups have often also relied on a number of modernist organizational and technological forms to achieve their ends, even as their concerns with boundaries and religious purity lead them towards separatist withdrawal. I explore these features for North American Protestant fundamentalism through an examination of the social milieu of its origins.

The City in U.S. Religious History

Since the mid-nineteenth century, urban America has been the site of social, religious, language, and cultural diversity, provided mainly by the arrival of immigrants.[4] Certainly, not all immigrants moved into cities – especially in the nineteenth century, hundreds of thousands of European immigrants merely passed through on their way west to exploit the opportunity offered by new areas of cultivatable land that were being opened to settlement there. But cities have been the places where the United States met the world. Moreover, the city, as the site of religious diversity and innovation, went hand-in-hand with the development of the city as the site for industrial economic production and the accompanying social forms of 'modernity'.

This transformation of the city, and its social organization, represented a change in the organization of religion in the United States. Much of the religious innovation and diversity in the colonial and early national periods was not in cities so much as on the frontier. Established religious authorities often controlled institutions and policed doctrinal orthodoxy in the communities along the seaboard. This was especially true in colonial New England, but the established Anglican churches in the southern colonies were also stronger in the coastal areas. However, on the western frontier population density lessened, clergy were harder to come by, stable congregations were rarer, and various types of religious 'entrepreneurs' could try to sell their doctrinal wares without fear of establishment persecution (Finke and Stark, 1992). Further, populations were often less literate, less bound to particular places, and thus less bound by traditional religious authority. As a result, many new religious movements in American history began in less densely populated areas. What is often known as both the First (roughly 1730–1770) and Second (about 1790–1840) 'Great Awakenings' of evangelical religious fervour were primarily frontier phenomena – from western New England, through up-state New York, to Appalachia (see McLoughlin, 1978; Hatch, 1989).

This began to change in the mid-nineteenth century. Legal disestablishment of the federal government, and eventually of the state governments, meant that former state religions (such as those in Massachusetts and Connecticut) could no longer control public religious expressions. New immigrant populations, and the beginnings of a serious factory system in the Northeast, meant that cities began to grow. Irish Catholics began to pour into New York, Boston, and Philadelphia. German Catholics and Jews, increasingly common in the Ohio River Valley, began to swell the populations of smaller cities such as Cincinnati, St. Louis, and Milwaukee. After 1870, as the U.S. economy began to recover from the Civil War and the industrial factory system took off in earnest, Southern and Eastern European Catholics and Jews formed the large immigrant populations that led to the opening of Ellis Island in 1892. Most emphatically perhaps, the Statue of Liberty was erected in New York Harbor in 1886, and a plaque with the sonnet by Emma Lazarus which bids welcome to 'huddled masses yearning to breathe free' was attached to it in 1903. These icons mark the fabled period of European immigration that is so idealized in contemporary American culture. The city became the place where America met the world, and urban America became a great source of religious and ethnic diversity (Christiano, 1987). In contrast to the first two Great Awakenings, McLoughlin (1978) portrays the innovations of the 'Third Great Awakening' (1890–1920) as a primarily urban phenomenon.

Religious and ethnic 'diversity' did not automatically translate into 'pluralism' as a normative value. That is, the fact of social, economic, and religious difference – which could not be avoided when one walked the streets of Boston's North End, New York's Lower East Side, the South Side of Philadelphia, or St. Louis' Hill District – was not always celebrated as a positive social development. Nativist reactions sometimes resulted in violence, including street battles between native Protestant and immigrant Catholic groups. Hosts of anti-immigrant organizations – from the 'Know-Nothing' political party, to the American Protective Association, to the Ku Klux Klan – used political, legal, and sometimes violent measures to preserve native-born white Protestant domination. Discrimination, segregation, and cultural suspicion kept new arrivals in subordinated social and economic positions, legitimated religious prejudice, and made cities a place of supposed danger and foreignness.

The emergence of urban industrial society thus saw a cultural demonization of cities and a valorization of small-town, or rural, agricultural life. Sometimes this was romantic nostalgia, but it also played a role in the ethno-religious politics and the cultural construction of 'American-ness'. Working the land came to be seen as a more honest, more 'authentic' way to make a living. Farms were thought to promote families, a myth that still fills out the image of the 'family farm', and the Jeffersonian 'yeoman farmer' was constructed as the lynchpin of American democracy. The Populist movement, and political party, swept through the Midwest and South, often fired up by anger at the 'plutocrats', bankers, and elites of the cities of the Northeast, often using evangelical religious rhetoric (Williams and Alexander, 1994). Thomas (1988) shows how revivalist religion was both part of and a reaction to the expansion of a market economy and the incorporation of small towns and rural areas into national markets. This change in the social landscape – this 'transformation of America' as Daniel Walker Howe (2007) calls it – was a time of territorial growth, religious revival, booming industrialization, a recalibrating of American democracy, and the rise of nationalist sentiment.

In popular culture and media, in political rhetoric, and in national mythology, the agrarian and the small-town were increasingly celebrated as the basis for true American values, with a corresponding confluence of national, racial, and religious identity (Williams, 2004). Historian Robert Orsi notes a cultural template that emerged out of the ante-bellum period of urbanization and immigration and the corresponding growth of cities:

> ... the city was cast as the necessary mirror of American civilization, and fundamental categories of American reality – whiteness, heterosexuality, domestic virtue, feminine purity, middle-class respectability – were constituted in opposition to what was said to exist in cities. (Orsi, 1999, p. 5)

Orsi points out that the discourse about cities was at once titillating and censorious. The city as a land of desire made it a bourgeois warning of temptation and moral failing, even as it also made it alluring and a symbol of unrestrained freedom. The industrial city seemed 'out of control' to most Americans (*Ibid.*, p. 16), and they have wondered ever since whether it endangered the health – whether physically from disease or materially and morally – of their families, or of the nation.

> The beginning of urbanization in the United State coincided with and partially occasioned a crisis of national identity, furthermore, and from this point on, what it meant to be an American – morally, religiously, politically, even physically – was defined in opposition to the cities, even as the nation became increasingly and inexorably urban ... cities became and have remained central locations on the map of American paranoia. (*Ibid.*, pp. 16–17)

While Orsi finds the roots of these cultural and social developments in ante-bellum urbanization and immigration, the trends of industrial expansion and urban growth accelerated in the late nineteenth century, as immigration rates rose and cities expanded (see Cronon, 1991, pp. 357–359). Even after the cessation of immigration in the 1920s in the wake of the national quota system, the 'Great Migration' of thousands of African Americans, moving from the Jim Crow South to northern factory jobs acted as *de facto* immigrants, filling the bottom rungs of the economic and social ladders as well as residential ghettos. This did not make cities any less foreign or dangerous to white middle- and working-class America, particularly those from rural areas and small towns.

However, there was also a significant internal migration of the native white population during this time period. Along with the Great Migration of African Americans, other Americans were finding the lure of city jobs pulling them out of small towns and off farms. Social and geographical mobility was relocating the white and black native-born population, bringing them to new places, new jobs, and transfiguring the roles of religion and family in their lives. Cities became places that seemed foreign, not just in terms of who lived there (ethno-religious others), but also in terms of how life, work, and family life were organized. Neighbourhoods were less 'neighbourly'; religious congregations organized less of a social life and had to compete for members; and the visibility of many traditional 'vices' or social problems (such as the consumption of alcohol) was enhanced. As Giggie and Winston (2002) argue, urban commercial culture became an enormous force by the turn of the twentieth century. It transformed life along every dimension, including the religious. It was a different world for those moving into the industrial urban economy.

Life in urban settings necessarily involves encounter with other social groups. The two defining characteristics of cities are population density and heterogeneity. People who do not know each other interact directly; but more importantly, they may not know how to place each other in a cultural category. How to classify others? How to know who others are and what they are like? And, perhaps most importantly, are they 'like us', or are they 'other'? The city has long been a place for such encounters, and cities have often resulted in creative, innovative, and synthetic religious outcomes (see Orsi, 1999; Williams, 2002). New religious visions, hybrid religious practices, and shared symbolic worlds can often be found in what we might call 'urban religion'. New groups of believers observe and often borrow elements from other groups, and at a certain level of diversity one is likely to find some example of almost every type of religious expression (see, for example, Warner and Wittner, 1998).

This syncretism is consistent with many theoretical expectations for what happens in situations of contact between different social groups; diversifying societies are often sites of assimilative and accommodating social and cultural processes.[5] From things as everyday as the street language that incorporates words or phrases from a variety of different groups, to the development of syncretic religious movements, urban settings promote cultural change and innovation. Early-twentieth-century social theory expected this cultural change to secularize society and make religion less important. Both urbanization and secularization were thought to be a part of modernization.

But it is important to note that encounters with social others sometimes result in retreat, distrust, and higher social boundaries. In these cases the results are polarizing, solidifying distinctions among people as they find a more intense sense of internal community through focusing on differences between themselves and others. In the immediacy of everyday contact on urban streets, it often does not take much imagination to see religious, social, and cultural differences; they easily take on the aura of being self-apparent. As Lowell Livezey (2000) notes, diversity produces both 'community' and 'enclaves'. These can be complements or antagonists of each other, depending upon how community becomes defined by those monitoring its external boundaries.

Emmanuel Sivan has thus written about fundamentalisms as 'enclave cultures' that protect their boundaries with a 'wall of virtue':

> The enclave is usually the response to a community's problems with its boundary. Its future seems to be at the mercy of members likely to slip away. For some reason, usually the appeal of the neighboring central community, it cannot stop its members from deserting. Devoid of coercive powers over the members, it cannot punish them; lacking sufficient resources, it cannot reward them. The only control to be deployed in order to shore up the boundary is moral persuasion. The interpretation developed by this type of community thus stands in opposition to outside society. (Sivan, 1995, p. 17)

When one considers the concern of Protestant fundamentalism at its founding with the threat posed by liberal religion (even more than with secular society), and the consistent tendency towards separatism and boundary maintenance, the concept of the enclave culture seems quite fitting.

Protestantism, the City, and the Origins of Fundamentalism

It is not coincidental that the social, cultural, and religious milieu of the early-twentieth-century American city was the site for the genesis of Protestant fundamentalism. As clearly implied by the previous section, the early-twentieth-century city was a site of significant religious innovation; fundamentalism was not the only religious development of the period. The type of religious expression that came to be known as 'Pentecostalism', while not new to the U.S., became a full-blown religious movement after the Azuza Street revivals in Los Angeles in 1906 (Wacker, 2001). Pentecostalism was marked by ecstatic religious experiences and practices revealing the 'gifts of the spirit', especially healing and speaking in tongues.

Also, progressive religious reformers such as Washington Gladden and Walter Rauschenbusch developed what came to be known as the 'Social Gospel' and worked to infuse Christianity with a worldly reformism, and vice versa (Luker, 1998; McLoughlin, 1978). Social gospellers found the new city emblematic of a society that had lost its moral moorings. But rather than cajoling believers into renewed piety, they worked to apply Christianity – combined with insights from the newly emerging social sciences – to solve social problems. They emphasized the need for the Gospel to be relevant, by shaping it to the issues of industrial urban society.

As another example, Winston (1999) documents how the Salvation Army, which arrived in the U.S. in 1891, developed as a particular type of urban religion. The Army brought conservative theology and a strict evangelical doctrine into the 'cathedral of the open air', working the streets with missionaries, brass bands, and altar calls. Their use of the techniques of popular culture, such as music, publicity, and entertaining services, in order to spread their message scandalized many established churchgoers. But the Army was unapologetic, arguing that it needed a contemporary package for its message to compete in the cultural marketplace that was the new industrial city.

Nonetheless, a standard narrative of American Protestantism is that the growth of the turn-of-the-century city was the beginning of the end of

Protestantism's dominance of American society. Not only were many more non-Protestants coming to, and prospering in, the U.S., but the industrializing and urbanizing economy was reshaping life away from the homogeneous small towns and faithful culture that had become emblematically 'American'.

Significant historical revisionism is now questioning that easy narrative (e.g., Butler, 1997). Christiano (1987), using national quantitative data; Lewis (1992), doing a single city history; and Winston (1999), examining the Salvation Army, show that Protestantism was adapting both its messages and its organizational forms to the new reality of urban life. While social and religious change was a given, different forms of Protestantism were proving to be 'at home in the city' (Lewis, 1992). While the large, downtown Presbyterian, Episcopal, and Congregational churches that dominated so many American towns were to eventually lose the culture-defining influence they once had, Protestantism continued to change and adapt – a trend that included the rise of fundamentalism.

Along with internal societal changes, the turn of the twentieth century also saw the United States move into a position to compete for global economic and political dominance. The country's economic base was expanding; its population was beginning to take control of a continent full of natural resources; and it won a war against a fading European military power (Spanish-American War, 1898), and thereby assumed control of colonies in several parts of the world (e.g., the Philippines). The U.S. society of the ante-bellum period was either destroyed or was being remade by the Gilded Age (Menand, 2001). A distinctive national culture was beginning to emerge, but it was not the culture that marked earlier periods. While Thomas (1988), Howe (2007), and others rightly point to the importance of the ante-bellum decades in shaping American life, the twentieth century presented new challenges, new ideas, and new relationships with the rest of the world. By the 1920 U.S. census, for the first time a majority of Americans lived in urban settings.

Pertinent to the story here, the institutions of higher education in the northeastern U.S. were beginning to struggle with identity – both in national and intellectual terms. European intellectual, literary, and artistic currents continued to make important impressions in the U.S., but there were also urges – some nationalist and xenophobic, some organically home grown in the U.S.'s particular social history – to develop distinctly 'American' schools of thought. The 'German higher criticism' in literary theory and the spread of Darwinian-influenced thought in the natural and social sciences were two European-inspired developments that brought some of these tensions into open conflict. Ammerman (1991), in particular, notes the extent to which German influences, in culture, in intellectual ideas, and in politics, were beginning to worry many American nationalists.

Conflicts over the interpretation of literary texts were multifaceted, but the German higher criticism basically applied rationalized literary theory to all texts, including the Bible. The variations in the contexts of production, the analysis of variations in interpretations, the sense that many scriptural imperatives were historically specific rather than timeless – these views threatened common-sense notions of what the Bible said and what it meant. The implicit rationalism also threatened such traditional orthodox Christian beliefs as the virgin birth and the physical resurrection. For many, this seemed to shake the foundations of the faith.

As these theories and critical interpretive practices found their way into Protestant seminaries, one reaction was what has come to be known as fundamentalism – originally articulated in a series of twelve books, published between 1910 and 1915, called *The Fundamentals: A Testimony to the Truth*. The books were written by conservative clergy at Princeton Theological Seminary (where this controversy resulted in an organizational schism) and were intended for distribution to parish clergy, missionaries, and interested laity as well as seminary professors. The volumes reasserted Christian orthodoxy via a stance of 'inerrancy' towards Scripture. About 1920 the term 'fundamentalism' began to be used to describe the religious impulse grounded in the beliefs expounded in these books. But soon the term was also used to refer to a religious movement grounded in inter- and intra-denominational organizations and networks, committed first to reforming what they considered wayward Protestant churches and denominations, and second to withdrawing into righteous communities if such attempts at reform failed. Thus, from the beginning, fundamentalism was a normatively loaded term, implying not just an analytic stance but also attitudes towards a particular religious approach. Also present from the beginning were the twin themes of reform and separatism – of convincing others of a restorative truth and yet advocating withdrawal from contact with the world in order to maintain doctrinal authority and purity.

The Social and Political Development of Fundamentalism in the U.S.

The argument above credits the origins of Protestant fundamentalism in the U.S. with the social changes accompanying urbanization and modernization. And yet, of course, the conventional wisdom is that fundamentalism has been the faith of rural and small-town America, particularly in the South and

Midwest, with the urban Northeast as its symbolic 'other'. These two arguments need not contradict each other. As alluded to earlier, some of the problem resides in confusing post-1920s fundamentalism with the context of its founding. Reconciling these two perspectives requires some qualification.

First, after its emergence, fundamentalism went through a diffusion process much like any cultural or religious innovation. Many intellectual or religious movements originate in one place, among a particular group of intellectual leaders, but then spread to other populations who find the messages resonant. Thus it is not surprising that fundamentalism could emerge in intellectual and social centres, which were in or near cities undergoing significant change (such as Princeton, New Jersey – near New York City), but then take off among people in another social location. As cities increasingly became sites of social and cultural 'foreignness', fundamentalist religion could present itself as a counter identity. At the same time, communication and media technologies were increasingly nationalizing the country into a single market, spreading awareness of the extent to which urban culture challenged traditional Protestant values (e.g., through movies, mass culture and entertainment, the spread of higher education, etc.). And those living in small towns and the non-Northeast became aware of a sense of cultural marginalization in ways which were new in American life. They had once imagined themselves the centre of America, but that seemed to be changing. That fundamentalism found receptive audiences in places other than those of its founding is thus not unusual.

Second, fundamentalism did indeed take hold in America's cities, even if the proportions of followers there were not as predominant as in rural and small-town populations. The history of revivalism in the United States began on the frontier, but later moved into cities, and included numerous urban campaigns. Fundamentalists did not originate the revival form, but they were quick to adopt urban revivals as a way to reach large numbers of people and get into local congregations (Ammerman, 1991, pp. 18–20). This was particularly true in the late nineteenth and early twentieth centuries behind such figures as Dwight Moody (whose Moody Bible Institute in Chicago is still engaged with a mission to apply conservative Protestant theology to urban realities) and Billy Sunday. While often successful, fundamentalism was not a religious orientation that could dominate growing cities (no single religious orientation could), and it often developed over time into doctrines and practices distinct from their original forms. As Marsden notes (1987, p. 10), the term fundamentalism was initially applied to a broad coalition of anti-modernists; only with time were *The Fundamentals* used to codify the movement and narrow the term. Nonetheless, it is an exaggeration to think that fundamentalism was only rural, and simply not true that it did not find some home in the city.

It is true that, in general terms, fundamentalism 'withdrew from the world' following the Scopes trial in Tennessee in 1926. Fundamentalists did not disappear, of course; but the trial and the public ridicule heaped on religious conservatives by the Eastern media and cultural elites did convince many fundamentalists to forego attempts at civic engagement and political change and instead reinforce their boundaries. An 'other-worldly' orientation became paramount, in which fundamentalists paid particular attention to their own salvation and to staying right with God, rather than trying to reform the fallen world. Parallel social institutions such as schools kept members of fundamentalist churches distinct from a modernizing culture and provided a safe setting in which to rear children in the faith.

Fundamentalist Protestant Christianity began to emerge from a half-century of withdrawal in the mid to late 1970s. Beginning with a threat to the tax-exempt status of Christian schools run by fundamentalist congregations and denominations (usually over their racial exclusion or segregationist rules) in the Carter administration, fundamentalist leaders began to organize to defend their terrain. However, led by key organizers such as the Reverend Jerry Falwell, and advised by conservative political consultants and activists, several groups began to promote a political vision that went beyond a defensive stance. By 1978 Falwell had organized a group he called the 'Moral Majority', and by the 1980 presidential election season they, and similar groups, became well known and somewhat influential, especially within the Republican party.[6] This period represents the re-emergence of the 'reform' impulse in fundamentalism. After a half-century of relative withdrawal from politics and other worldly concerns, fundamentalists were again asserting themselves in public life, and articulating a duty to impose the truth on that public.[7]

Fundamentalism as a Response to the Crisis in American National Identity

To this point, I have argued that Protestant fundamentalism was an essentially religious reaction to a changing socio-cultural context within the turn-of-the-century American city. While the secular character of much of modern culture was a threat, what prompted the fundamentalist reaction was the perception that secular modernity was moving into the churches – that liberal churches were being tainted rather than maintaining a 'wall of virtue'.

One other part of the cultural context is also worth considering – the extent to which the changes described here were producing a crisis in American national identity. While fundamentalists were not necessarily or intentionally working to restore a particular but fading concept of American national identity, the challenges to that identity in the early twentieth century helped produce the sense that such a response was needed.

New England Calvinism and its progeny in many ways set out the cultural terms in which the United States became a nation. While there was religious diversity across the colonies, New Englanders first articulated a sense of nationhood and how that nationhood could be connected to religious identity (McKenna, 2007). These Calvinist orientations were marked by a cultural tendency to divide the social and moral world into dualisms (see Williams, 1999) and to see the American nation as bound in a 'covenanted' relationship to God. For Calvinists, the U.S. was God's chosen nation, charged with bringing God's Kingdom to realization here on Earth, and blessed with a certain prosperity as long as it remained faithful to that charge (*Ibid.*). There is a clear 'corporatist' aspect here: the nation as a body politic is responsible for this mission, and suffers as a whole for transgressions or wandering from the way, just as Jews were covenanted with Yahweh after the Exodus. This notion of corporate sin is not as widespread as it once was, but it has not disappeared. It is still common for fundamentalists to argue that their nation's collective forsaking of the Christian path is the cause of many of its national social problems. Further, many fundamentalists see themselves as uniquely positioned to bring the nation back to the path of righteousness. They have a duty to keep the United States as God's chosen nation. Combining the covenanted relationship with God to a cultural worldview of binary categories of good and evil produced a thorough identification of Protestant religious affiliation and American national identity. Their missions in the world are aligned – America to fulfil God's purpose, and fundamentalism to keep the nation in pursuit of this destiny.

As noted, many things that had been taken for granted in the U.S. were challenged by the arrival of large numbers of non-Protestant immigrants beginning in the 1840s. A consistent basis for the demonization of immigrants was the charge that they were not capable of assimilating American national identity. Their differences were often articulated as involving a lack of commitment to 'democracy'. Catholics were charged with having a first allegiance to a foreign 'potentate', the Pope, rather than to the U.S., and to worshipping an ecclesiastical hierarchy rather than practicing a congregational style of democracy. Further, they accepted many social practices, such as alcoholic drink, that many Protestants abhorred.

Anti-Semitism joined anti-Catholicism for many Protestants when significant numbers of Eastern European Jews immigrated to the U.S. in the 1880s and 1990s. Anti-Jewish stories often paralleled anti-Catholic narratives, such as their responsibility for sexually corrupting naïve Protestant youth – especially those who migrated to the city. Importantly, in the nineteenth and early twentieth centuries, much of the Protestant criticism of Judaism, like nativist suspicions about Catholicism, concerned the fact that it was part of, and was seen as essentially, an international religion. It was a 'foreign' religion, and the integration of its adherents into American culture seemed questionable.

R. Laurence Moore (1986, p. 208) notes that Protestants also considered a series of religious 'outsiders' – such as Mormons, Christian Scientists, and Adventists (along with Catholics and Jews) – to be both 'aberrational or not-yet-American'. Given the important role played by many Protestants in the abolition, temperance, and other moral reform movements, evangelical Protestants for a significant period of American history assumed themselves to be the moral guardians of the nation (Marsden, 1980; Smith *et al.*, 1998, p. 4). Further, they assumed that the alignment of their religious commitments with national identity was the clear cornerstone of American democracy. Non-Protestant immigrants – and the 'Old World' cultures that both European Catholics and Jews brought with them – highlighted and potentially questioned these assumed connections. As Geertz (1973) or McLoughlin (1978) might have argued, when received cultural meanings seem inadequate for making sense of a new social situation, one response is a forceful reassertion of social, cultural, or religious orthodoxy, with the goal of re-establishing the previous associations. Fundamentalism offered that reassertion in religious terms.

Thus, connections between conservative Protestantism and American identity and loyalty became linked in many strands of American political and religious thought. For example, Kenneth Jackson (1967) demonstrates how these concerns came together in the twentieth-century re-emergence of the Ku Klux Klan. Undermining the conventional view of the Klan as Southern, small-town, and consumed with the racial domination of African Americans, Jackson shows that urban centres in the Northeast and Midwest had considerable Klan memberships. Moreover, Klan membership was fuelled in these cities by anti-immigrant, anti-Catholic, anti-Jewish nationalism. In the very places where more non-Protestant immigrants were arriving, the Klan's profession of a nativist, white Protestant, American identity found adherents. As with fundamentalism, the Klan was more visible in the South and in rural and small-town areas. But it was also an urban phenomenon – one that explicitly connected Protestant religious and native-born American national identity.

Conclusion

Although it could be inferred from this historical narrative, it is incorrect to think of Protestant fundamentalism as an overtly political response to threats or perceived threats to American nationhood and well-being. Fundamentalism was foremost a religious reaction to a sense that Christianity had lost its way in dealing with the changes of the modern industrial world. However, it is also true that several social developments at the turn of the century cut the long-assumed knot that tied together Protestant religious identity, Anglo-Saxon Western-European ethnic identity, and American political identity. Protestant fundamentalism helped retie those connections, implicitly and often explicitly. And fundamentalism had within its cultural worldview the assumption that it had a duty to revive an American public sphere that would have an orthodox Protestant religious institution and doctrine at its core. The growth of the modern industrial city was a challenge to the world American Protestantism had known, and fundamentalism urged a restoration.

In sum, in many 'modernization-theory' paradigms, the changes that are associated with urbanization are thought to entail enough education, worldly experience, and social diversity so that the religious particularism known as 'fundamentalism' would be little more than a rear-guard action by a dying subculture. In this chapter, I have argued the reverse. Without the development of the modern city, and the density and diversity involved in that development, fundamentalism would not have emerged as the religious reaction it has. Further, the combination of the social contexts of its founding – 'in' the emerging industrial city but not quite 'of' it – and the inherent tension between its impulse to withdraw into a religiously pure community of the right-minded and attempts to reach out and shape the public sphere give fundamentalism the particular urban presence and character it has today.

Notes

1 Both Ammerman (1991) and Carpenter (1997) agree on three basic theological principles, often translated into religious practice, as the definitive core of North American Protestant fundamentalism: (*a*) the requirement for active *evangelism* of non-believers toward the goal of voluntary conversions; (*b*) a strict attitude of *inerrancy* towards Scripture, arising from literalist interpretive approaches; and (*c*) a set of *premillennialist* beliefs, i.e., the expectation of the immanent Second

Coming of Jesus, ushering in a thousand-year reign of the Kingdom of God and separating the saved from the unregenerate. A definitional requirement for premillennial theology makes fundamentalism specifically Christian, while the other two factors could apply to other religions.

2 David Smith (2003) devotes a book to Hinduism and modernity, but does not engage the fundamentalism debate. He analyzes 'Hindu nationalism' as a particular religio-political hybrid that is one product of the encounter with modernity. However, Smith does use in passing the phrase 'Hindu fundamentalist' (e.g., 2003, p. 185) without definitional concern. Kurien (2007), while writing for a different purpose, demonstrates that there are some specific ways in which Hindutva nationalists are moving towards positions more typical of Abrahamic fundamentalists – e.g., as they attempt to standardize Hindu practice and belief, they increasingly emphasize particular Hindu sacred writings as a type of 'scripture' that has uniquely inspired and divine status. One might infer that Hindutva activists are 'becoming' more 'fundamentalist' as they move toward a Scriptural inerrancy position regarding religious texts (even, of course, as they work to distance themselves from Muslims and Christians). Kurien notes that Hindu nationalists both valorize and question Hinduism's historic tolerance and pluralism – in part creating the totalizing worldview usually seen as 'fundamentalist'.

3 Marty and Appleby note that they share this feature with Lawrence (1989).

4 For most of the nation's history, this applied to the cities of the Northeast and upper Midwest. The cities in the American South were different – plantation agriculture required fewer cities, and without the employment opportunities in industrial production they attracted fewer immigrants. In turn, ethnic distinctions among whites were masked by the enormity of the white-black racial divide. That has changed in the post-World War II era, particularly with the population shift to the 'sunbelt' and the expansion of immigrant populations since the 1970s.

5 I note that the textbook definition of 'assimilation' is the lowering of boundaries between different socio-cultural groups. This need not entail, at least by definition, imbalance in the power and social status of the groups involved. But real life has not worked that way: smaller, minority groups have usually assimilated towards the dominant majority. Certainly in American history, assimilation has been used to denote the ways in which immigrant groups were expected to engage in 'Anglo-conformity' – becoming more like the Anglo-Saxon cultural majority. But that is not the only dynamic involved. Dominant groups often have accommodationist processes of their own, and smaller social populations often have to adjust to each other, not just to the societal majority.

6 Debates abound in the scholarly literature over how much so-called 'Christian Right' organizations, such as the Moral Majority, contributed to the election of Ronald Reagan in 1980. Those groups were eager to take credit, and many in the media, and some scholars, agree. Others, however, believe the main factors were the economy and dissatisfaction with the Carter Administration and established Washington. The Christian Right's influence in subsequent elections, and particularly in the administrations of George W. Bush, is more agreed upon. Nonetheless, whether it was 'values voters' or 'security moms' that gave G.W. Bush his second term in 2004 is still a matter of contention.

7 While some aspects of this political mobilization have been successful, many fundamentalists fear that involvement in the 'dirty' world of politics will pollute religious truth. Cal Thomas and Ed Dobson (2000), once Moral Majority insiders, have written that it is time for fundamentalist Christians to move closer toward the 'withdrawal' mode.

References

Ammerman, N.T. (1991) North American Protestant fundamentalism, in Marty, M.E. and Appleby, R.S. (eds.) *Fundamentalisms Observed.* Chicago, IL: University of Chicago Press.

Butler, J. (1997) Protestant success in the new American city, 1870–1920, in Stout, H. and Hart, D.G. (eds.) *New Directions in American Religious History.* New York: Oxford University Press.

Carpenter, J.A. (1997) *Revive Us Again: The Reawakening of American Fundamentalism.* New York: Oxford University Press.

Christiano, K. (1987) *Religious Diversity and Social Change: American Cities, 1890–1906.* Cambridge: Cambridge University Press.

Cronon, W. (1991) *Nature's Metropolis: Chicago and the Great West.* New York: W.W. Norton.

Finke, R. and Stark, R. (1992) *The Churching of America: Winners and Losers in the American Religious Economy, 1776–1990.* New Brunswick, NJ: Rutgers University Press.

Geertz, C. (1973) *The Interpretation of Culture.* New York: Basic Books.

Giggie, J.M. and Winston, D. (eds.) (2002) *Faith in the Market: Religion and the Rise of Urban Commercial Culture.* New Brunswick, NJ: Rutgers University Press.

Hatch, N.O. (1989) *The Democratization of American Christianity.* New Haven, CT: Yale University Press.

Howe, D.W. (2007) *What Hath God Wrought: The Transformation of America, 1815–1848.* New York: Oxford University Press.

Jackson, K.T. (1967) *Ku Klux Klan in the City, 1915–1930.* New York: Oxford University Press.

Kurien, P.A. (2007) *A Place at the Multicultural Table: The Development of an American Hinduism.* New Brunswick, NJ: Rutgers University Press.

Lawrence, B.B. (1989) *Defenders of God: The Fundamentalist Revolt Against the Modern Age.* San Francisco, CA: Harper and Row.

Lewis, J.W. (1992) *The Protestant Experience in Gary, Indiana, 1906–1975: At Home in the City.* Knoxville, TN: University of Tennessee Press.

Livezey, L.W. (2000) Communities and enclaves: where Jews, Christians, Hindus and Muslims share the neighborhood, in Lowell, L.W. (ed.) *Public Religion and Urban Transformation: Faith in the City.* New York: New York University Press.

Luker, Ralph E. (1998) *The Social Gospel in Black and White: American Racial Reform, 1885–1912.* Chapel Hill, NC: The University of North Carolina Press.

Marsden, G.M. (1980) *Fundamentalism and American Culture: The Shaping of Twentieth-Century Evangelicalism, 1870–1925.* New York: Oxford University Press.

Marsden, G.M. (1987) *Reforming Fundamentalism: Fuller Seminary and the New Evangelicalism.* Grand Rapids, MI: William B. Eerdmans Publishing.

Marty, M. and Appleby, R.S. (eds.) (1991) *Fundamentalisms Observed.* Chicago, IL: University of Chicago Press.

Marty, M. and Appleby, R.S. (eds.) (1995) *Fundamentalisms Comprehended.* Chicago, IL: University of Chicago Press.

McKenna, G. (2007) *The Puritan Origins of American Patriotism.* New Haven, CT: Yale University Press.

McLoughlin, W.G. (1978) *Revivals, Awakenings, and Reform: An Essay on Religion and Social Change in America, 1607–1977.* Chicago, IL: University of Chicago Press.

Menand, L. (2001) *The Metaphysical Club: A Story of Ideas in America.* New York: Farrar, Straus and Giroux.

Moore, R.L. (1986) *Religious Outsiders and the Making of Americans*. New York: Oxford University Press.
Orsi, R.A. (ed.) (1999) *Gods of the City*. Bloomington, IN: Indiana University Press.
Peshkin, A. (1988) *God's Choice: The Total World of a Fundamentalist School*. Chicago, IL: University of Chicago Press.
Sivan, E. (1995) The enclave culture, in Marty, M.E. and Appleby, R.S. (eds.) *Fundamentalisms Comprehended*. Chicago, IL: University of Chicago Press.
Smith, C. (ed.) (2003) *The Secular Revolution: Power, Interests, and Conflict in the Secularization of American Public Life*. Berkeley, IL: University of California Press.
Smith, C., with Emerson, M., Gallagher, S., Kennedy, P., Sikkink, D., (1998) *American Evangelicalism: Embattled and Thriving*. Chicago, IL: University of Chicago Press.
Smith, D. (2003) *Hinduism and Modernity*. Oxford: Blackwell.
Thomas, C. and Dobson, E. (2000) *Blinded by Might: Why the Religious Right Can't Save America*. Grand Rapids, MI.: Zondervan Publishing Company.
Thomas, G.M. (1988) *Revivalism and Cultural Change: Christianity, Nation-Building and the Market in the 19th Century United States*. Chicago, IL: University of Chicago Press.
Wacker, G. (2001) *Heaven Below: Early Pentecostals and American Culture*. Cambridge, MA: Harvard University Press.
Warner, R.S. and Wittner, J.C. (1998) *Gatherings in Diaspora: Religious Communities and the New Migration*. Philadelphia, PA: Temple University Press.
Williams, R.H. (1994) Movement dynamics and social change: transforming fundamentalist ideology and organizations, in Marty, M.E. and Appleby, R.S. (eds.) *Accounting for Fundamentalisms: The Dynamic Character of Movements*. Chicago, IL: University of Chicago Press.
Williams, R.H. (1996) Religion as political resource: culture or ideology? *Journal for the Scientific Study of Religion*, **35**, pp. 368–378.
Williams, R.H. (1999) Visions of the good society and the religious roots of American political culture. *Sociology of Religion*, **60**, pp. 1–34.
Williams, R.H. (2002) Review essay: religion, community, and place: locating the transcendent. *Religion and American Culture: A Journal of Interpretation*, **12**(2), pp. 249–263.
Williams, R.H. (2004) Religion and place in the Midwest: urban, suburban, and rural forms of religious expression, in Barlow, P. and Silk. M. (eds.) *Religion and Public Life in the Midwest: America's Common Denominator?* Walnut Creek, CA: Altamira Press.
Williams, R.H. and Alexander, S.M. (1994) Religious rhetoric in American populism: civil religion as movement ideology. *Journal for the Scientific Study of Religion*, **33**, pp. 1–15.
Winston, D. (1999) *Red-Hot and Righteous: The Urban Religion of the Salvation Army*. Cambridge, MA: Harvard University Press.

Chapter 5

Producing and Contesting the 'Communalized City': Hindutva Politics and Urban Space in Ahmedabad

Renu Desai

The rise of Hindutva politics in India over the recent decades has been one of the most significant developments in India's postcolonial history. A coalition led by the Bharatiya Janata Party (BJP) – the political arm of the Sangh Parivar, the family of right-wing Hindu organizations which promotes the ideology of Hindutva (the vision of India as a Hindu nation) – first came to power at the national level in the late 1990s. Although the coalition was voted out of power in the 2004 national elections, and failed to make a comeback in 2009, the divisive and exclusionary politics of Hindutva remain a powerful force. Hindutva politics has had, and continues to have, profound implications at the local level, and the relationship between it and urban conditions is strong. This chapter explores this relationship in the context of Ahmedabad, India's seventh largest city and the commercial capital of the western state of Gujarat. It examines the practices through which Ahmedabad's urban landscape has been reconfigured by Hindutva politics and the ways in which the ideology has been embedded in the city and in the lives of its residents.

The first section of the chapter draws on the extensive scholarship on the Hindu Right to discuss Hindutva ideology and the reasons for the emergence of Hindutva politics since the 1980s in India – and more specifically, in Gujarat. It also considers the problematic use of the rubric of religious fundamentalism to describe the ideology and politics of Hindutva. The second section interrogates the production of the 'communalized city' by examining how the nature of urban segregation in Ahmedabad has been transformed since the late 1960s by Hindu-Muslim violence – also called 'communal violence' in India – and other practices born out of fear, suspicion, and exclusion of the Muslim Other. The notion of the 'communalized city' is then used to conceptualize the specific forms and meanings of urban segregation that have emerged in Ahmedabad through the politics of Hindutva, and to investigate their implications for the entrenchment of the Hindutva ideology and the reproduction of power by the Hindu Right. The third section examines how the contemporary globalizing city reproduces the 'communalized city' as well as creates the conditions for challenging and contesting it.

The Rise of Hindutva Politics

The Sangh Parivar's Hindutva ideology has its roots in the early-twentieth-century writings of D.V. Savarkar, who argued that the Aryans who settled in India had formed a nation, and that this is today embodied in Hindu culture. Savarkar coined the term Hindutva to describe their Hindu-ness, which he conceived as resting on three pillars: geographical unity, racial features, and a common culture (Jaffrelot, 1996, p. 26). Christophe Jaffrelot explains that although the importance of religious criteria was minimized by Savarkar in his conception of Hindu-ness, his notion of a common Hindu culture included the idea that India was the fatherland of the Hindu nation. India's Christians and Muslims could not be part of this nation because they did not look upon India as their holy land, thus making their loyalty to it suspect (*Ibid.*, pp. 28–31).[1] Hindutva's territorial imagining of India as a Hindu nation is thus founded upon the exclusion and threat of the religious Other – particularly Muslims, whose threat is magnified by their construction of a precolonial Indian history that emphasizes the Muslim (Mughal) invasion (Bhatt, 2001, pp. 92–93), as well as by the postcolonial identification of Indian Muslims with Pakistan – and more recently, with a militant transnational Islam.

The rise of the Hindutva movement in the 1980s can be seen as part of a broader resurgence of religious movements across the world since the 1970s. Hindutva remained a marginal ideology throughout most of the twentieth century. However, scholars see its recent rise as a reaction to modernity, emerging from a search for meaning in a fast-changing world and from a rejection of secularism. The Fundamentalism Project, sponsored by the American Academy of Arts and Sciences in the early 1990s, studied many of these religious movements, including Hindutva, under the larger rubric of religious fundamentalism, and it identified various family resemblances and common ideological and organizational traits. Among these were their reaction to the marginalization of religion in public life, selectivity in reshaping religious traditions, adoption of a dualistic world-view, the belief in the inerrancy of religious texts, adoption of some form of millennialism, and creation of sharp group boundaries (Almond *et al.*, 1995). Critics have pointed out, however, that while many of these characteristics were shared across the movements studied, other characteristics were not (Emerson and Hartman, 2006, p. 135).

Scholars have since continued to grapple with both the historical specificity of the term 'fundamentalism' (first used to refer to a particular conservative strain of Protestantism that developed in the United States) and the immense differences in resurgent religious movements across the world. Most relevant to the discussion of Hindutva, however, is the critique that fundamentalism, as a concept, may conflate conservative religious movements with postcolonial national religious movements (*Ibid.*, p. 130). In the Indian context, therefore, some scholars reject the term because they see Hindutva as a symptom of social, economic, and political factors rather than being essentially about, or motivated by, religious concerns (Juergensmeyer, 1996, p. 130). Scholars studying the rise of the Hindu Right in India have thus explained it in the context of colonial history and postcolonial formations. And instead of using the term 'fundamentalist', they have preferred the terms 'Hindutva', 'Hindu nationalism', or 'communalism' – the last of which is a specifically Indian usage describing conflict and dissension between religious communities, mainly Hindus and Muslims (Needham and Rajan, 2007, p. 12).

Gyan Pandey (1990), for example, has situated the rise of communalism in the context of British colonial rule, which employed a divide-and-rule policy to create dissension and competition between Hindus and Muslims. For Pandey, the rise of communalism took place in tandem with the rise of modern nationalism. Nonetheless, after the partition of the subcontinent into India and Pakistan on the eve of independence from Britain, India was founded as a secular nation-state, in the sense of constitutionally guaranteeing equality of all religions, promoting religious tolerance, and preventing the state from favouring or discriminating against any religion. Sumantra Bose (1997) has thus argued that the rise of the Hindu Right from the early 1980s is an expression of the crisis of the Indian state, both in secular and developmental terms.

The social, economic, and political factors that constitute this crisis have been explained in a number of ways. Some attribute it to the failure of the Indian state to address competing demands from different socioeconomic groups, which has created various underlying class and caste tensions (*Ibid.*). Others point to a crisis in the hegemony of the secular Congress Party (the dominant political party in postcolonial India until the 1970s), which contributed to a general erosion of secularism since the 1980s as the party has sought to retain power by deploying religious and caste categories rather than campaigning on substantive issues (Jaffrelot, 1996, pp. 330–337). Others have cited anxieties among privileged socioeconomic groups and the urban petty bourgeois (comprising mainly middle- and upper-caste Hindus) over the political mobilization of lower-caste Hindus, and the subsequent displacement of class-and-caste tensions into communal ones (*Ibid.*, pp. 432–446). In this context, Hindutva, by arguing for an overarching Hindu unity, has been interpreted as an attempt by the middle- and upper-castes to co-opt the lower castes. Crucial to the success of this effort has been the growth of Sangh Parivar organizations. The VHP (Vishwa Hindu Parishad or World Hindu Council), for instance, has more than ten thousand branches in the state of Gujarat alone (Bunsha, 2006, p. 240).

Gujarat is, in fact, one of the regions where Hindutva has become most deeply rooted. The BJP has been in power here since 1995, and after the violence of 2002, the state even came to be known as the 'laboratory of Hindutva'. Scholars and other commentators have attempted to understand Hindutva's entrenchment in Gujarat by analyzing the region's distinctive political economy and socio-political formations. In this regard, many have written about the formation of a winning electoral alliance (the KHAM alliance), which brought together Muslims and certain lower-caste Hindu groups (including Dalits – who are considered 'untouchable') under the Congress Party in the early 1980s. This, in turn, led to a profound reaction by Gujarat's dominant castes, who managed to break this alliance by mobilizing the lower castes around Hindutva (Shah, 1998; Shani, 2007; Yagnik and Seth, 2005).

Among other commentators, Radhika Desai has argued that Gujarat represents the 'Hindutva of development', as it is one of India's most prosperous states and a model of successful capitalism. Its economic and political elite include both upper and middle castes, for whom Hindutva is useful as a way to control labour and gain a sense of identity as a ruling class (Desai, 2004, pp. 119–120). Dionne Bunsha has also referred to Gujarat's prosperity, observing that the state has a large middle class as well as a diaspora of foreign workers who are largely conservative. She notes that the links between commerce, religion and politics are more clearly etched in Gujarat than in other regions. This, she argues, has provided fertile soil for cultivating a militant Hindu identity and consciousness (Bunsha, 2006, pp. 238–242). Jan

Breman (2002) has furthermore argued that Gujarat's economic growth pattern is one of 'lumpen capitalism', with the new liberalizing economy creating large numbers of underemployed and unemployed people who have been actively recruited by Sangh Parivar organizations.

Within Gujarat, Ahmedabad is one of the urban strongholds of the BJP. The city has a reputation for violence based on a history of recurring communal riots since the late 1960s. Ashutosh Varshney (2002) has argued that local incidents have been able to trigger larger communal riots in Ahmedabad because of a decline in the city's civic institutions, including the secular Congress Party and the Textile Labour Association. Breman (2002), on the other hand, has argued that these riots must be seen in the context of the changing political economy of the city; particularly important has been the closure of the city's textile mills, which resulted in the informalization of labour, an increase in social and economic insecurity among marginalized groups, and the demise of working-class solidarities. Darshini Mahadevia (2002) has further argued that the exclusion which began in the employment sector has expanded and deepened under economic liberalization to include overall development processes in the city. This situation has been successfully exploited by the Hindu Right, which has incited anti-Muslim sentiment among marginalized groups and orchestrated violence which systematically targets Muslims.

As the above discussion shows, Hindutva needs to be understood in its specific historical and geographical context. Though it might be framed within the rubric of religious fundamentalism, the usefulness of the fundamentalist concept in this case lies in its ability to provoke consideration of the many ways in which a militant rejection of secularism may be articulated. Secularism separates church and state, and according to some understandings, relegates religion to the private sphere. According to other understandings, it repositions religious organizations as simply one voice among many in the public sphere, rather than as special keepers of the truth. A militant rejection of secularism entails exclusions based on religion; and while this rejection by Hindutva politics is profoundly linked to, and even emerges from, various social, political, and economic historical processes in which class, caste, and religious tensions are intertwined, it has created in Ahmedabad an urban landscape that embodies these exclusions.

Producing the 'Communalized City'

In recent years, scholars have begun to turn their attention to the spatialities of Hindutva. Following Satish Deshpande (2000), Rupal Oza has examined various 'spatial strategies' through which, she argues, public spaces have been converted into Hindu spaces. This includes nationwide pilgrimages organized by the Hindu Right since the 1980s which have attempted to integrate Indian territory with Hindu nationalism (Oza, 2007, p. 160). It also includes the movement spearheaded by the Hindu Right to build a Ram temple on the disputed site of the Babri Mosque in Ayodhya, which was forcefully occupied in 1992 and transformed into a 'symbol for occupying national space' (*Ibid*., p. 161). Oza also describes the 2002 violence in Gujarat as a spatial strategy through which Muslims were 'permanently removed and signs of their presence erased from the landscape' (*Ibid*., p. 163).

A key spatial strategy in Ahmedabad has been the orchestration of communal violence, which has fostered fear and suspicion between Hindus and Muslims, and through which residential and commercial property belonging to Muslims has been systematically destroyed in certain urban localities. This has led to a profound transformation in the nature of urban segregation in the city. Urban segregation by itself is not new in Ahmedabad, and it is therefore important to bring a historical perspective to the changes to its forms and meanings over recent decades. I thus begin with a brief discussion of the patterns of urban segregation in the city through the 1970s. Following this, I trace the geography of the communal violence which occurred in the late 1960s, mid-1980s, early 1990s, and in 2002; the consequent reconfigurations in the city's housing landscape; and the new forms and meanings of urban segregation that emerged. I also examine other practices born out of the fear, suspicion, and exclusion of the Muslim Other, which continue to reproduce this new urban segregation.

To conceptualize the specific forms and meanings of urban segregation produced in Ahmedabad through Hindutva politics, I propose the notion of the 'communalized city'. In doing so, I am choosing not to borrow the concept of the divided city or the segregated city (Low, 1999). While urban segregation and division are features of the communalized city, my attempt is to situate them in the context of Hindutva politics, which constructs Muslims as the religious Other according to a larger imaginary of India as a Hindu nation.

As mentioned earlier, in its Indian usage, communalism refers to conflict and dissension between religious communities, mainly Hindus and Muslims. Rather than paying attention to the construction of and instability of religious identities, some scholars have argued that the notion of communalism presupposes that Hindu and Muslim are fixed identities (Shani, 2007, pp. 10, 136). My attempt here is to show how the very idea of the 'communalized city' is produced – that is, how Ahmedabad's urban landscape is reshaped through orchestrated violence and other practices in ways that seek to construct and fix Hindu and Muslim identities – and, furthermore, to construct them as antagonistic. The 'communalized city' also points to a space where this separateness, and even the notion of antagonism, is increasingly naturalized. It is in these ways that the production of the 'communalized city' is crucial to the entrenchment of the Hindutva ideology and the reproduction of power for the Hindu Right.

Patterns of Urban Segregation in Ahmedabad up to the 1970s

Ahmedabad was founded in 1411 by Sultan Ahmed Shah on the eastern banks of the Sabarmati river. The city's social organization was based on strict caste, religious, and occupational groupings. The generic morphology which made up the original city (the area known today as the Walled City – see figure 5.1) was the *pol*, a house cluster organized along a narrow lane(s) that was linked to a particular caste and religion. Wealthy members of a caste lived among their poorer caste fellows in the *pol* and looked after them. Under the sultans of Gujarat and then the Mughal viceroys, the city's administrative and bureaucratic apparatus was dominated by Muslims, while the majority of the financiers and merchants were Hindus and Jains. Although the city's organization was thus rigid and hierarchical, Gillion (1968, pp. 20–26) argues that it involved cooperation between different classes and communities.

With the decline of the Mughal Empire, the city witnessed almost two centuries of political instability and economic decline. Its re-emergence is commonly attributed to the political stability brought by colonial rule, which began in 1817. The city began to industrialize in the late nineteenth century with the emergence of the cotton textile industry on the outskirts of the Walled City (the area known today as eastern Ahmedabad).[2] Large numbers of people migrated from surrounding rural areas, and later from other regions of India, to work in these mills. Work in the mills was organized along caste and religious lines (Gillion, 1968; Breman, 2004). Thus, Dalits were employed only in the spinning departments since they were considered an 'untouchable' caste, while Muslims and higher-caste Hindus worked in the weaving departments (Breman, 2004, p. 17). Caste- and religion-based segregation was duplicated outside the mill walls in the *chawls*, or working-class neighbourhoods. Dalits, due to their social exclusion and stigma as 'untouchables', lived in separate neighbourhoods

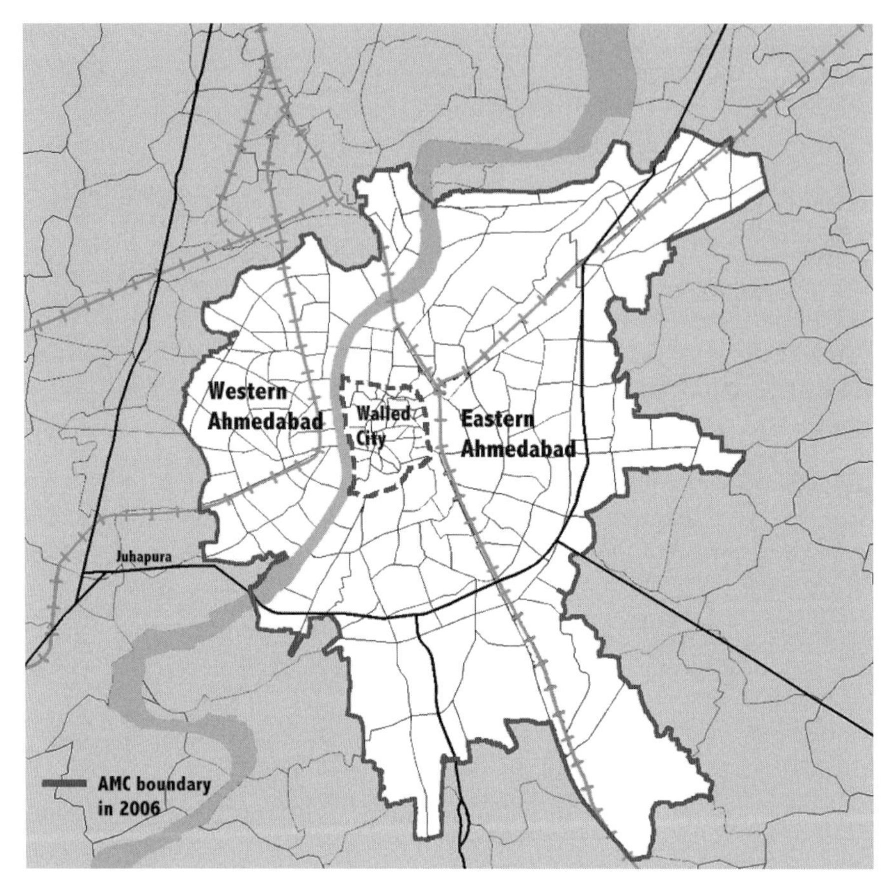

Figure 5.1. Map of Ahmedabad. (*Source*: drawn by author based on Map 2.3 in Ahmedabad Municipal Corporation, *Ahmedabad City Development Plan 2006–2012*, Prepared for the Jawaharlal Nehru National Urban Renewal Mission, 2006: 12)

from middle- and upper-caste Hindus. And middle- and upper-caste Hindus, especially from Gujarat, followed strict vegetarian diets due to their religious practices, and thus rarely lived in the same neighbourhoods as Muslims. But as Dalits did not follow vegetarian diets, Dalit and Muslim neighbourhoods were often adjacent to each other, and the two communities sometimes occupied the same neighbourhood (*Ibid*., pp. 35–36). Thus, the primary lines of residential segregation in the industrial localities of eastern Ahmedabad were between middle- and upper-caste Hindus, and Dalits and Muslims.

The early twentieth century saw the development of western Ahmedabad (see figure 5.1). This began as affluent families from upper-caste backgrounds began to move out of the Walled City and build lavish bungalows in private compounds there. However, much of the later development in western Ahmedabad which accommodated the upper-middle class (who also began to move out of the Walled City), took the form of cooperative housing societies. These were generally developed on a community basis with a group of people from the same caste, geographical origin, or occupational background coming together to form a cooperative (Mehta and Mehta, 1989, p. 75). This pattern of *de facto* segregation was assisted by existing landowners who were selective in selling their land only to particular class- or caste-based cooperatives (*Ibid.*, p. 44). Around the 1970s, the middle classes, particularly Hindus, also began to move out of the Walled City to western Ahmedabad in large numbers. And at the other end of the economic spectrum, the 1970s saw the proliferation of slums, particularly in eastern Ahmedabad and its periphery (*Ibid.*, p. 184). Western Ahmedabad thus developed as a predominantly upper-middle- and middle-class and upper- and middle-caste Hindu area with some upper-middle- and middle-class Muslim neighbourhoods. Muslims were often able to buy commercial properties in western Ahmedabad, and some upper-middle-class Muslims even bought residential properties in Hindu neighbourhoods.

Achyut Yagnik and Suchitra Seth thus note that by the late 1960s there were three Ahmedabads. The first was the Walled City 'where the upper castes, Dalits and Muslims lived cheek by jowl, each community in its own pol'. The second grew in the early twentieth century around the textile mills, in which 'Dalit and Muslim textile workers who together formed two-thirds of the labouring population' lived. The third was the new Ahmedabad, which began to grow after Indian independence on the western side of the river, and which was 'an elite area of upper- and middle-caste Hindus, a few well-to-do Muslims and some Dalit slums' (Yagnik and Seth, 2005, pp. 230–231). With the development of Ahmedabad as a modern city, class increasingly became a defining factor in its urban landscape. Caste and religion did not, however, disappear as social criteria, and class, caste, and religion overlapped to shape the urban landscape. The result, therefore, was a complex pattern of urban segregation which featured three distinct areas – the Walled City, industrial eastern Ahmedabad, and western Ahmedabad – each with its own particular class, caste, and religious composition, and within which Hindu and Muslim residential space overlapped in different ways. However, important aspects of this urban segregation would be transformed over the next four decades, as a result of recurring episodes of communal violence in which Hindu right-wing organizations played an increasingly central role.

Hindutva, Violence, and New Patterns of Urban Segregation

In 1969 Ahmedabad witnessed large-scale communal riots for the first time since India's independence. Ironically, just the previous year an urban history of Ahmedabad had been published, and its author, Kenneth Gillion, had written that 'the most tragic problem of modern Indian political history – the communal problem – has fortunately played only a small role in the history of Ahmedabad' (Gillion, 1968, p. 171). During the 1969 riots, Muslims were systematically targeted, and within a week, approximately six hundred people had been killed; more than six thousand houses, huts, and shops had been looted and destroyed; and many women were physically abused and raped (Shah, 1970, p. 195; Shani, 2007, p. 162).

The riots began following an incident outside one of the city's temples, involving a clash between ascetics from the temple and a crowd of Muslims who had gathered to celebrate the anniversary of the death of a Sufi saint. Soon after the incident, a large meeting was organized by the Hindu Dharma Suraksha Samiti (Committee to Defend Hindu Religion), which had been formed some months earlier by religious leaders and leaders of the Jana Sangh (the precursor to the BJP), during which anti-Muslim slogans were shouted (Shah, 1970, p. 193). Accounts of the subsequent riots reveal that the violence was planned. Handbills giving exaggerated and provocative news about the incident were distributed; transport was arranged to take rioters from one place to another; organized attacks were made against Muslims based on lists of Muslim properties; and Muslim shops in Hindu-owned buildings were merely looted, whereas Muslim shops in Muslim-owned buildings were set on fire (Shah, 1970, pp. 193–199; Shani, 2007, p. 162). One scholar thus wrote that prejudices against Muslims, which had been building among Hindus, were 'actively nursed and incited' to transform the local incident near the temple into large-scale and brutal violence (Shah, 1970, p. 193).[3] A government-appointed commission found that the state government had not acted appropriately to contain the violence, and that in many cases the police had been biased in their actions against Muslims – observations that were made in later communal riots in the city as well. It concluded, however, that 'Hindu communal elements were on the aggressive and the Muslim violence was a reaction' (Reddy Commission, quoted in Shani, 2007, p. 149–151).

The widespread violence originated in scattered parts of the Walled City as a series of arson attacks, but soon spread to the industrial areas of eastern Ahmedabad. In certain mixed localities, every house and shop belonging to Muslims was gutted (Shah, 1970, p. 187). Western Ahmedabad also saw some violence, and a Muslim hostel for university students there (where some Hindu students also lived) was completely destroyed (*Ibid.*, p. 195). More importantly, through spatialized violence which involved the destruction of Muslim property in certain urban areas, the 1969 riots began to create a new pattern of segregation linked to concerns for the safety of both lives and property. It is difficult, however, to construct a more detailed picture of the segregation emerging at that time.

The next round of large-scale rioting in Ahmedabad, which took place in 1985, disrupted life in the city for six months. These riots began as caste-related agitations, with the middle and upper castes targeting lower castes who were beneficiaries of the Congress Party government's reservation policy (a form of positive discrimination policy towards Dalits and other socially and economically backward castes). Scholarly research has since revealed that the transformation of these caste-related agitations into communal riots involved complex and multilayered events and a range of actors, including bootleggers, local politicians, and the police (Shah, 1998; Yagnik and Seth, 2005; Shani, 2007). More significantly perhaps, it has pointed to the ways Hindutva politics and communal violence in Gujarat have been tied to caste politics.

Based on an ethnography of people's experiences of the 1985–1986 violence, Ornit Shani (2007, p. 132) argues that 'the communal antagonism that emerged in 1985 was related to Hindus' changing experiences of caste and the relations between caste and class in politics and in their social and economic lives'. From a different perspective and referring to the KHAM alliance mentioned earlier, Atul Kohli (1990, p. 239) suggests that this violence sought to 'undermine governmental legitimacy and destroy the electoral alliance on which the Congress' power rested'. Thus, 'in spite of an inter-Hindu caste reservation conflict and prevailing class tensions among them, an all-Hindu consolidation against Muslims emerged' (Shani, 2007, p. 132). The Sangh Parivar organizations played a central role in creating this all-Hindu consolidation through the distribution of provocative pamphlets about the slaughtering of a cow belonging to a local temple. These then urged Hindus to awaken and take revenge against Muslims by 'beating them to death'; by locally organizing the violence against Muslims; and by providing aid to Hindus and even Dalits during the riots, and thus increasingly winning the sympathy of Dalits (*Ibid.*, pp. 113–118).

The communal violence began in Dariapur, a mixed area in the Walled City where upper-caste Hindus, Dalits, and Muslims lived in separate but proximate neighbourhoods (*Ibid.*, pp. 108–109). But it spread to many other areas of the Walled City, particularly localities with mixed populations of lower-middle-class and poor Hindus, Muslims, and Dalits (Engineer, 1985*a*, p. 628). Shani reports that after the riots, many economically better-off, upper-caste Hindu families moved from one of the Dariapur neighbourhoods to the western side of the river, and some lower-caste families also moved into this neighbourhood from Muslim-dominated areas (Shani, 2007, pp. 121–122). The close proximity of Hindu and Muslim neighbourhoods in many localities of the Walled City had thus become an opportunity for the Hindu Right to incite communal violence more easily. For their inhabitants, living in these localities created both heightened concerns about safety as well as daily inconvenience during the riots. Communal riots were therefore one of the factors that led to the movement of the Hindu middle classes out of the Walled City, mostly to the western side of the river, which, as mentioned earlier, had been developed as a predominantly Hindu area. Many well-off Muslims also moved out to the western side of the river, but they came to be increasingly confined to certain localities on its periphery, such as Juhapura, which I shall discuss in further detail later.

In 1985 the violence spread beyond the Walled City when Hindus attacked Garibnagar, a Muslim slum neighbourhood in Bapunagar, a locality in eastern Ahmedabad. Narratives of this violence reveal that this attack was planned in advance (Engineer, 1985*b*; Shani, 2007). The violence subsequently spread to other industrial localities. In Shani's (2007, p. 126) interviews in the late 1990s, residents in Bapunagar argued that the main problem in the locality in the mid-1980s had been growing interest in slum dwellers' lands, and that riots were a means of driving people off these lands. Shani provides this glimpse of the population shifts that took place after the 1985 riots, and that increased after the riots in the early 1990s:

> After the 1985 riots the Bapunagar area was partitioned. The local residents call one part 'Pakistan' and the other 'Hindustan' or 'diamond-nagar'... By the end of the 1990s only Muslims lived in Morarji Chowk, which until 1985 used to be a mixed locality of both Hindus and Muslims. Small Indira Garibnagar became a lower-caste Hindu locality. All the Muslims who had lived there had moved, most of them to Juni Chowl near General Hospital and Bapunagar Post Office. (*Ibid.*, p. 127)

Since Garibnagar was reoccupied by lower-caste Hindus, it is unclear whether the communal riots there, like the 1992 riots in Calcutta (Das, 2000), were originally fomented as a pretext for vacating and grabbing slum lands for commercial purposes. But it is clear that this 'eviction through riots' of Muslims from Garibnagar was an instance of what Kothari and Contractor (1996) – in a study of the 1992–1993 communal riots and slum evictions in Mumbai – called 'planned segregation'. Hindus also moved out of neighbourhoods located near Muslim ones, though this was linked less to evictions through riots than to a sense of fear and suspicion of the Muslim Other increasingly fostered by the Hindu Right. An elderly Aslambhai who lives in one of eastern Ahmedabad's working-class neighbourhoods explained to me:

> My father had rented this house in 1930. Since then we have lived here. This has always been a Muslim chali [neighbourhood]. There were never any Hindus living here. But Parmanand Patel ni *chali* [the name of a nearby neighbourhood] was entirely Hindu in 1930. After 1985–1986 things changed. Until 1986, we used to go there, and they used to come here. But then most of them moved away out of fear, and now almost all are Muslim households there. (Personal interview, 7 November 2005)

Many working-class neighbourhoods of eastern Ahmedabad had also been mixed, and these too started to become mono-religious. The breaking up of the Congress Party's KHAM alliance was, in fact, poignantly reflected in these reconfigurations of working-class localities and neighbourhoods through communal violence that prised apart Dalits and Muslims.

In 1986 the Gujarat government passed the Disturbed Areas Act, a legislation preventing a member of one religious community from selling his property to a member of another religious community if that property lay within what was considered a 'disturbed area' (one where violence would be likely to take place during communal riots). The purpose of the law was to prevent the transformation of mixed localities, particularly in the Walled City and eastern Ahmedabad, into mono-religious localities. Not surprisingly, the fear of violence, which came to be associated with close proximity to the religious Other, prompted residents of such mixed localities to find other, informal ways to sell their property to members of the other community (Shani, 2007, pp. 127–128). And after 1986 the reconfigurations of the urban landscape started to create a patchwork of Hindu and Muslim localities, with the meeting points between these referred to as 'borders'.

The demolition of the Babri Mosque in Ayodhya in northern India in 1992 was followed by large-scale riots in many parts of the country, including Ahmedabad. This third round of rioting led to even further segregation between the two communities in the city. This time, however, the communal violence was not only restricted to the Walled City and eastern Ahmedabad. Although western Ahmedabad, which had developed as a predominantly Hindu area, had witnessed sporadic violence during earlier outbreaks of rioting, this time the peripheral, rapidly growing areas of Vejalpur and Juhapura (see figure 5.1) in southwestern Ahmedabad were deeply affected. Vejalpur had developed as a predominantly Hindu locality, while adjacent Juhapura had seen an influx of Muslims over time, many of whom had moved there from the Walled City and eastern Ahmedabad after the 1980s riots.

In 1990, after the first violent conflict in these two localities (incidentally, between a Hindu and a Muslim group over some vacant land), a high wall was erected between a Hindu and a Muslim neighbourhood by their more peace-loving inhabitants as a means of preventing future clashes. This came to be known as the 'border' between Vejalpur and Juhapura. However, there were still some lower-caste Hindu families living in Juhapura, and some Muslims living in Vejalpur. This changed after the 1992–1993 riots. During the first round of disturbances, Hindus stormed the houses of their Muslim neighbours and threatened to kill them if they did not leave immediately. Some two hundred Muslim families fled to Sankalitnagar, a neighbourhood in Juhapura, leaving all their possessions behind; these were then looted, after which their houses were smashed and burned (Breman, 1999, p. 266). The lower-caste Hindu families who lived in Sankalitanagar were also threatened by Muslims, and they fled to Vejalpur, after which their houses, too, were looted and destroyed. Several Hindus still remained in Sankalitnagar with the help of their Muslim neighbours, but when the second round of rioting started, they too left (*Ibid*., pp. 276–278).

Juhapura had already started to become a Muslim ghetto. But in the years that followed Muslims found it nearly impossible to develop new neighbourhoods or buy property elsewhere in western Ahmedabad. As a result, by the mid-1990s, Juhapura was home to more than 200,000 Muslims. An investigative journalist noted that in Juhapura, 'judges and businessmen live alongside carpenters and hawkers. A common tag binds them all – their religion. Well-off Muslims who would like to live in elite areas of the city cannot. People will not sell or rent houses to them' (Bunsha, 2006, p. 243). The area also came to be referred to as 'mini-Pakistan' by many Hindus.

The fourth outburst of rioting in Gujarat, in 2002, has come to be referred to as the Gujarat carnage or Gujarat genocide. It led to the death of more than two thousand Muslims, the rape of Muslim women, the damage or complete destruction of more than 100,000 houses and 15,000 businesses belonging largely to Muslims (Concerned Citizens Tribunal, 2002, p. 27; Human Rights Watch, 2002, p. 4). The riots were triggered by the torching, allegedly by a Muslim mob, of two train cars carrying right-wing Hindu activists returning from Ayodhya, where the VHP had sought to begin construction of a Hindu temple on the disputed site of the Babri Mosque. The train cars were torched in the town of Godhra, after which violence broke out in different parts of Gujarat.[4] In Ahmedabad, Muslim-owned residential and commercial properties situated in the midst of predominantly Hindu areas were systematically targeted, and were looted, destroyed, and often burned to the ground by mobs. Lists of Muslim-owned shops had been prepared by the VHP, and voter lists were also reported to have been used. In some Muslim neighbourhoods situated in Hindu localities, mobs brandishing swords, daggers, axes, and iron rods massacred groups of Muslims and committed the most horrifying acts of sexual violence against Muslim women. Bodies were indiscriminately buried in mass graves in various parts of the city. Many mosques were also destroyed, and makeshift temples were erected on their ruins. At a few places, retaliatory attacks on Hindus also took place.

The brutality and scale of the violence, as well as the BJP-led state's complicity in it, led to further segregation and ghettoization across the city. Yet some mixed localities and neighbourhoods within the Walled City and eastern Ahmedabad remained peaceful, and they remain mixed even today. How they have remained so is an important research question. But the fact is that these are exceptions to the rule.

While class, caste, and religious segregation existed in varying forms in different areas of the city before the 1970s, communal violence and concerns about safety and the fear of the religious Other that it has produced, reconfigured this segregation. Spatial divides between the two communities in the Walled City have sharpened, as the well-off have moved to western Ahmedabad, and those who remained have moved across localities. Where Muslims and Hindus still live in close proximity, walls have often been erected between neighbourhoods. In eastern Ahmedabad, communal violence has transformed most formerly mixed Dalit-Muslim working-class neighbourhoods into mono-religious ones. In both the Walled City and eastern Ahmedabad, the language of 'borders' defines the new separation. Western Ahmedabad has become a rigidly Hindu space, leaving only a few older Muslim neighbourhoods and a large and growing Muslim ghetto there for Muslims to inhabit. This new segregation between Hindus and Muslims is, furthermore, not only a matter of separation but also antagonism, with Muslim spaces such as Juhapura discursively equated with Pakistan, India's 'enemy'. In contrast, many urban localities are declared as part of the 'Hindu Rashtra' (the 'Hindu Nation') on street signboards (see figure 5.2).

Figure 5.2. The street signboard on the top says: 'Welcome to the Hindu Rashtra's Naranpura [the name of a locality in Ahmedabad] prakhand [referring to a unit]'. (*Photo*: Renu Desai, 2005)

For Satish Deshpande, who has written about the spatial strategies of Hindutva, the significance of the locality lies not only in terms of what it is able to achieve within its own spatial limits, but 'rather in the possibilities it creates for [insertion into] ... the larger grid of ideological dissemination and political action' (Deshpande, 2000, p. 202). The socio-spatial language of 'borders', 'mini-Pakistan', and 'Hindu Rashtra' within the city points to how urban localities in Ahmedabad have been powerfully inserted into Hindutva's ideological grid. Traumatic experiences of communal violence, but also the discursive and material reshaping of the city have begun to recast the urban experience in communalized terms, and narrowed the possibilities for quotidian interaction between Hindus and Muslims. Many anecdotes now reveal how people experience the city in communalized terms. Rickshaw drivers of different religious backgrounds often take different routes through the city, avoiding localities and neighbourhoods dominated by the religious Other. Common perceptions of Juhapura include the idea that its Muslim residents burst firecrackers to celebrate when Pakistan wins a cricket match against India. These perceptions and urban experiences have serious implications for the creation of a society based on religious tolerance and mutual respect.

Rupal Oza (2007, p. 154) has argued that the 2002 Gujarat riots sought to erase the landscape of Muslim presence altogether. However, as she goes on to argue (*Ibid.*, p. 164), this erasure is a frustrated and unrealized goal. I would suggest that Hindutva, in fact, thrives through the presence of the Muslim Other in spaces which are constructed as the Other, and where separation is constructed through 'borders'. Breman (1999, p. 281) notes that during the 1992–1993 violence, stones and fireworks were let off over the wall dividing Juhapura and Vejalpur. The border, then, not only separates but also becomes a zone of engagement through violence, a zone where communal hostility is displayed to reinforce separateness, antagonism, and irreconcilability. All of these elements constitute the communalized city, which is crucial to the entrenchment of Hindutva ideology.

Various practices reproduce this new segregation and the exclusion and ghettoization of Muslims. Middle- and upper-caste Hindus from Gujarat, who follow vegetarian diets linked to religious beliefs, have long preferred to live in residential developments where non-vegetarianism is unwelcome. However, there are many residential developments where these unwritten rules are no longer followed, or which cater for those who follow non-vegetarian diets, including the many non-Gujaratis who live in Ahmedabad. Muslims are, however, rigorously excluded from such developments. Likewise, allowing a Muslim to buy a flat in a building is even perceived by some Hindus as inviting the 'capture' of the entire building by Muslims (Bunsha, 2006, p. 260). In one instance, the attempt by a Muslim to buy five houses belonging to Hindus just beyond a 'border' was met with protest by one thousand Hindu families living nearby, who argued that the sale would lead to a drop in the value of their properties (*Gujarat Samachar*, 2006).

Since there is limited space for them, Muslims also have to contend with paying higher real estate prices (Bunsha, 2006, p. 259). New developments by and for Muslims in western Ahmedabad are confined to Juhapura and parts of its adjoining areas. As a result, Juhapura is not only overcrowded, but having been outside the city's limits until recently, it lacks proper urban services and amenities. Its residents also often face discrimination by banks and insurance companies (Times News Network, 2004). Thus, access to land, real estate, and urban services in the most spacious and well-serviced areas of the city has increasingly become the exclusive right of Hindus.

Reproducing and Contesting the Communalized City in the Globalizing City

Urban redevelopment projects shaped by imaginaries of globally successful cities and the desire to attract investment further threaten to deepen the divides produced in Ahmedabad by communal violence and the fear and suspicion of the Other. One such project is the ambitious Sabarmati River Front Development, which involves the resettlement of more than 14,000 families living in informal settlements (known as slums) on the Sabarmati river, which runs through the centre of the city. Many of these slums are mixed even today, with Hindus and Muslims living in adjacent neighbourhoods. In many cases, this condition is the result of a lack of alternative affordable housing in the central areas. But, since 2002, many NGOs have also organized programmes in some of these slums to rebuild trust between the two communities.

Such programmes face an uphill battle, however. Although a comprehensive slum resettlement and rehabilitation policy has not been finalized for the project, the Ahmedabad Municipal Corporation (AMC) has begun to resettle groups of slum dwellers from the riverfront when they become obstacles to ongoing construction. Such resettlement frequently brings fears to the surface. In 2008, a group of approximately twenty lower-caste Hindu families had agreed to be resettled from the riverfront to housing in eastern Ahmedabad. They agreed to move to this site until they heard that it was a 'Mohammedan area' (referring to a Muslim locality), when the government had to resettle them in a Hindu locality on the outskirts of western Ahmedabad. Urban redevelopment in the globalizing city is thus likely to reproduce the segregation created by Hindutva politics.

But the globalizing city is not just a site for such convergences; it has also created frictions for Hindutva politics. Close on the heels of the 2002 Gujarat riots, concerns began to emerge that the violence had delivered a severe blow to Gujarat's economy and affected its image as an investor-friendly destination. The Gujarat Chamber of Commerce and Industry (GCCI) also made a presentation to the Prime Minister of India requesting help in restoring peace, on the grounds that it was crucial for Gujarat's economy. In its presentation, the GCCI stated that 'International experience shows that industrialists do not invest where the social fabric is weak or reported instability risks are high. This is bound to hinder Gujarat's march towards globalisation and further affect economic growth' (Gujarat Chamber of Commerce and Industry, 2002). In a bid to promote normalcy in the state, the GCCI also organized a peace march through the streets of

Ahmedabad's Walled City. According to the GCCI president, one reason for the march was to 'allay the growing concern of the international community about the situation in the state' (*The Asian Age*, 2002).

These anxieties around global investment, though centred on regional investment in Gujarat, have been articulated by an urban business class seeking to benefit from the global economy under liberalization. Their anxieties reveal the frictions between communalized space and globalizing space – particularly the possibility that the violent orchestrations of the Hindu Right might affect Gujarat's image, thus undermining the ability of its urban business class to participate in the global economy. However, such anxieties do not necessarily demonstrate concern about the human costs of the riots or the state's complicity in them. In fact, when the Confederation of Indian Industry (CII) raised questions regarding the responsibility of the Gujarat government, it not only chilled relations between CII and Narendra Modi (the Chief Minister of Gujarat and a key figure of the Hindu Right), but it also caused Gujarat's business elite to meet under the banner of the 'Resurgent Group of Gujarat' and throw its weight behind Modi and criticize groups like the CII for embarrassing the state globally (Express News Service, 2003). Thus, while there are frictions between communalized space and globalizing space, this has not created a challenge to communalized space; rather, these frictions have been managed by the Gujarat government in ways which silently reproduce the communalized city.

In 2003, instead of its earlier business summits, the Gujarat government began to organize 'Vibrant Gujarat' events annually. These extravagant government-sponsored cultural events and exhibitions that seek to portray Gujarat as an economically and culturally dynamic region have attempted to rehabilitate the image of the region and restore the confidence of investors. Ahmedabad has been crucial to these image-building events, not only serving as a venue several times, but also providing a public stage by means of cultural events and exhibitions along its riverfront. Some of these urban spectacles have been organized on land being reclaimed from the river through the riverfront project. Indeed, by allowing visitors to walk along the river for the first time, the project (still under construction) is explicitly intended to be part of the spectacle (see figure 5.3). As David Harvey (1989, p. 9) has observed, urban spectacle and display have become 'symbols of a dynamic community'. Here, the city, by showcasing both 'culture' and 'development', is crucial to the government's attempts to rehabilitate Gujarat's image and manage the frictions between communalized and globalizing space.

Figure 5.3. The Vibrant Gujarat exhibitions on the Sabarmati riverfront creating an urban spectacle in the centre of the city. (*Photo*: Renu Desai, 2007)

But to what extent has this management of frictions actually challenged communalized space by creating spaces of interaction between Hindus and Muslims, and spaces of inclusion for the latter? At the 2004 Vibrant Gujarat exhibitions, the Gujarat government invited three religious minority communities – two Gujarati Muslim sects (the Khojas and the Bohras) and the Parsis – to set up theme pavilions. Critics pointed out that while the Gujarat government was attempting to woo religious minorities, it had rolled out the red carpet only for moneyed minority communities, while various other Gujarati Muslim sects were not invited to exhibit their culture (*Times of India*, 2004).

At the 2006 event, Ahmedabad was likewise represented by a poster which was a collage of pictures of the city and its people and words such as 'vibrant', 'scintillating', 'multi-cultured', and 'historical'. This represented Ahmedabad as a culturally thriving city and as conflict-free. On the other hand, at the same event, a large display of sites of tourist interest in Gujarat showcased mostly Hindu and Jain temples and places of pilgrimage. For a region whose cultural heritage includes a vast array of architecture built under the patronage of Muslim rulers and communities, the almost complete absence of such sites constituted a symbolic erasure of Islamic history and culture.

The Vibrant Gujarat events of a globalizing Gujarat and a globalizing Ahmedabad have thus not only denied the deep divides created by Hindutva politics in the city and region, but also reproduced communalized space through symbolic exclusion. The events, by making invisible the communalization of space and society by Hindutva politics, and by constructing legitimacy for the BJP and Modi in the eyes of an urban and global elite on a developmental platform, have been crucial for the reproduction of power for the BJP and Modi in Gujarat.

Meanwhile, this reproduction of power has made it possible for the BJP and Modi to continue to produce the communalized city on other terrains. A case in point is the municipal elections in Ahmedabad in 2005, during which Modi campaigned in Ahmedabad's streets using inflammatory anti-Muslim slogans such as 'Bring an end to Congress's Mughal rule. Vote for the BJP', which attacked the Muslim woman who was mayor of the city during 2003–2005. The BJP's victory in Ahmedabad's municipal elections was largely attributed to Modi's immense presence in the campaigns. It is thus that globalizing space has served to reproduce communalized space even though there have been frictions between them.

But if the globalizing city reproduces the communalized city, then it is also important to note that the globalizing city has created the conditions for building progressive grassroots alliances which contest it. The pursuit of urban development projects in the globalizing city, which threaten to displace the lower classes and urban poor regardless of their religious backgrounds, also holds the potential to build solidarities based on shared experiences of marginalization and struggle. Since 2004 a grassroots movement known as the Sabarmati Nagrik Adhikar Manch (Sabarmati Citizens' Rights Forum, hereafter referred to as the Manch) has emerged in response to the impending displacement and resettlement of riverfront slum dwellers. Central to this housing-rights movement has been an emphasis on Hindu-Muslim unity, or communal unity. This is most clearly seen in the posters and banners of the Manch, which include the religious symbols of Hindus, Muslims, and Christians (see figure 5.4). One of the riverfront slums, Ram-Rahim Nagar, a mixed area, with 70 per cent Muslims and the rest Dalits, has always been promoted in the Manch's meetings as an exemplary case of Hindu-Muslim unity and peaceful coexistence. In the meetings the Manch organized to mobilize slum dwellers, community leaders gave passionate and moving speeches on communal unity at the same time as they argued for housing rights.

Figure 5.4. Banner of the Sabarmati Nagrik Adhikar Manch. (*Photo*: Renu Desai, 2005)

In a city where religious identities offer fertile ground for dividing the poor and weakening organizations that seek to create class-based alliances, this emphasis on communal unity not only served as a necessary strategy for collective mobilization, but also sincerely constructed non-communal collective identities. The construction of the identity of the residents as 'slum dwellers' or '*jhupdawasio*' by community leaders challenged, or rather attempted to supersede, other collective identities. In one of the Manch's meetings, a Muslim leader urged, 'We have to forget that we are Hindus, that we are Muslims. We must think of ourselves only as *jhupdawasio* [slum dwellers]'.[5] Other leaders evoked religious identities as non-antagonistic identities, and they attempted to mobilize caste and religious identities to emphasize the bonds of brotherhood between Dalits and Muslims, as they both suffered marginalization. A grassroots housing-rights movement that emerged in response to the exclusionary practices of urban development in the globalizing city thus contests the communalized city.

Conclusions

In this chapter, I have proposed the notion of the 'communalized city' to interrogate the relationship of the city to the divisive politics of Hindutva, which constructs Muslims as the religious Other as well as the primary obstacle to its larger national imaginary of India as a Hindu nation. Hindutva politics might have various spatial implications in Indian cities (see, for example, the chapter by Rajagopalan in this collection), but one of the crucial spatial implications of Hindutva politics in Ahmedabad has been the production of a 'communalized city' – that is, a city which constructs and fixes Hindu and Muslim identities and spaces as separate and antagonistic. The 'communalized city' thus embodies a militant rejection of secular modernity by the Hindu Right and substitutes a set of exclusions based on religion as a focus of identity formation. As this chapter shows, it has been produced first and foremost through communal violence orchestrated by the Hindu Right, as well as through other practices which might be seen as the long-term implications of this violence. This has reconfigured the nature of urban segregation in Ahmedabad by reshaping the urban landscape discursively and materially.

For Muslims, attacks on both their bodies and their property in mixed neighbourhoods and in predominantly Hindu localities – which at times has involved 'evictions through riots' – have forced them to search for housing elsewhere, thus creating new patterns of segregation based on concerns for the safety of both lives and property. Hindus, meanwhile, have moved out of mixed neighbourhoods and localities largely because of fear and suspicion of the Muslim Other created by sometimes real, but often imagined communal violence in which Muslims are cast as perpetrators. At times they have also moved to get away from the inconvenience of living in areas which might become sites for the Hindu Right's violent orchestrations. This has increasingly led to the exclusion of Muslims from easy access to housing, urban services, and amenities, forcing them to live in ghettos.

Furthermore, where Hindus and Muslims could not move entirely away from each other, such as in the Walled City and eastern Ahmedabad, a patchwork of Hindu and Muslim neighbourhoods has emerged. Along with the language of walls and 'borders', this too constitutes a new urban segregation. In some cases, Muslim urban space has been discursively equated with that which is considered anti-Indian, while other urban spaces in the city have been claimed as part of the Hindu nation. Here, to paraphrase Satish Deshpande (2000, p. 17), concrete (physical) spaces such as neighbourhoods, localities, and streets or walls have been linked in a durable way to abstract (imagined) spaces such as the nation, the space of an alleged national enemy, and national borders.

In Ahmedabad, fear of communal violence has created a variety of instruments of social separation. In some cases, these are concrete walls between neighbourhoods of different religious communities. But in others they are existing streets and walls infused with the meaning of a border. Through the inherent risk of crossing physical borders, language mapped onto urban space thus becomes an instrument of social separation itself. A discriminatory real estate market, which has been almost naturalized and does not even seem to need justification, is also an instrument of social separation in Ahmedabad.

The city, then, is not simply the container in which Hindutva politics has played out. Rather, the discursive and material production of urban space is crucial to it, because it recasts the urban experience in communalized terms. Over time, as quotidian interactions decrease, the separateness produced between the two communities will begin to be naturalized. It is in these ways that the 'communalized city' is crucial for the perpetuation of the politics of the Hindu Right, the entrenchment of the Hindutva ideology, and the reproduction of the Hindu Right's power.

This chapter has also briefly examined the ways in which the globalizing city reproduces and becomes a site of contestation to the 'communalized city'. Satish Deshpande (2000, p. 211) has proposed that 'globalisation and Hindutva impact on each other in contradictory as well as complementary ways (not to speak of the ways that may go beyond this dichotomy), making it difficult to hold on to any unidimensional conception of their reciprocal involvement'. This chapter, through an interrogation of the spatial implications of Hindutva politics and globalization, shows how these converge in complex ways, even though there are at times divergences and frictions.

However, the reconfiguration of space in the globalizing city has also created the conditions for evoking shared marginalization and struggles among the urban poor from different religious backgrounds, especially where they still live in close proximity. Through this, contestations have emerged against not only the exclusions of the globalizing city but also the communalized city. The 'communalized city', then, remains an incomplete project, and one that can be, and is being, challenged even in Ahmedabad. However, the relationship of Hindutva politics and urban space must be taken seriously if it is to be challenged in more powerful ways.

Notes

1 Since the sacred sites of Sikhs, Jains, and Buddhists lie within India, they are considered to be part of the Hindu nation.
2 Notably, Ahmedabad's transformation into a modern industrial city was largely carried out through the initiative of indigenous and not British or European capital and enterprise (Gillion, 1968).
3 Shah (1970) writes that prejudices against Muslims among Hindus had been building in the mid to late 1960s due to various events. These included the India-Pakistan War of 1965, large meetings organized in Ahmedabad by both orthodox Muslim and Hindu organizations, and the formation of the Hindu Dharma Suraksha Samiti.
4 The media coverage of the violence was extensive, and here I rely on the reports of the Concerned Citizens Tribunal (2002) and Human Rights Watch (2002).
5 Meeting organized by the Manch on 27 March 2005, at a community hall in the city.

References

Almond, G.A., Sivan, E., and Appleby, R.S. (1995) Fundamentalism: genus and species, in Marty, M.E. and Appleby, R.S. (eds.) *Fundamentalisms Comprehended*. Chicago, IL: University of Chicago Press.

Bhatt, C. (2001) *Hindu Nationalism: Origins, Ideologies and Modern Myths*. Oxford: Berg.

Bose, S. (1997) Hindu nationalism and the crisis of the Indian state: a theoretical perspective, in Bose, S. and Jalal, A. (eds.) *Nationalism, Democracy and Development: State and Politics in India*. New Delhi: Oxford University Press.

Breman, J. (1999) Ghettoization and communal politics: the dynamics of inclusion and exclusion in the Hindutva landscape, in Parry, J. and Guha, R. (eds.) *Institutions and Inequalities: Essays in Honour of Andre Beteille*. New Delhi: Oxford University Press.

Breman, J. (2002) Communal upheaval as resurgence of social Darwinism. *Economic and Political Weekly*, 20 April.

Breman, J. (2004) *The Making and Unmaking of an Industrial Working Class: Sliding Down the Labour Hierarchy in Ahmedabad, India*. New Delhi: Oxford University Press.

Bunsha, D. (2006) *Scarred: Experiments with Violence in Gujarat*. New Delhi: Penguin.

Concerned Citizens Tribunal (2002) *Crime Against Humanity: An Inquiry into the Carnage in Gujarat*, Volume 3. Mumbai: Citizens for Justice and Peace.

Das, S. (2000) The 1992 Calcutta riot in historical continuum: a relapse into 'communal fury'? *Modern Asian Studies*, **34**(2), pp. 281–306.

Desai, R. (2004) *Slouching Towards Ayodhya: From Congress to Hindutva in Indian Politics*. New Delhi: Three Essays Collective.

Deshpande, S. (2000) Hegemonic spatial strategies: the nation-space and Hindu communalism in twentieth-century India, in Chatterjee, P. and Jeganathan, P. (eds.) *Subaltern Studies XI: Community, Gender and Violence: Essays on the Subaltern Condition*. Delhi: Permanent Black and Ravi Dayal Publishers.

Emerson, M.O. and Hartman, D. (2006) The rise of religious fundamentalism. *Annual Review of Sociology*, **32**, pp. 127–144.

Engineer, A.A. (1985*a*) From caste to communal violence. *Economic and Political Weekly*, 13 April, pp. 628–630.

Engineer, A.A. (1985*b*) Communal fire engulfs Ahmedabad once again. *Economic and Political Weekly*, 6 July, pp. 1116–1120.

Express News Service (2003) Gujarat Inc throws weight behind Modi, takes on CII. 21 February.
Gillion, K.L. (1968) *Ahmedabad: A Study in Indian Urban History.* Berkeley, CA: University of California Press.
Gujarat Chamber of Commerce and Industry (2002) Shri Kalyan J. Shah, President Gujarat Chamber of Commerce & Industry Represented the Impact of the Godhra Carnage on the Trade and Industry. Presentation made to the Prime Minister on 4 April 2002. Available online at http://www.gujaratchamber.org/memorandum-PM1.htm. Accessed 12 March 2006)
Gujarat Samachar (2006) 1000 families oppose the sale of five bungalows in Paldi. 28 October.
Harvey, D. (1989) From managerialism to entrepreneurialism: the transformation in urban governance in late capitalism. *Geogratiska Annaler*, **71**(1), pp. 3–17.
Human Rights Watch (2002) We Have No Orders to Save You: State Participation and Complicity in Communal Violence in Gujarat. A report of the Asia Division.
Jaffrelot, C. (1996) *The Hindu Nationalist Movement in India.* New York: Columbia University Press.
Juergensmeyer, M. (1996) The debate over Hindutva. *Religion*, **26**, pp. 129–136.
Kohli, A. (1990) *Democracy and Discontent: India's Growing Crisis of Governability.* Cambridge: Cambridge University Press.
Kothari, M. and Contractor, N. (1996) *Planned Segregation: Riots, Evictions and Dispossession in Jogeshwari East, Mumbai/Bombay, India.* Mumbai: Youth for Unity and Voluntary Action.
Low, S. (1999) *Theorizing the City: The New Urban Anthropology Reader.* New Brunswick, NJ: Rutgers University Press.
Mahadevia, D. (2002) Communal space over life space: saga of increasing vulnerability in Ahmedabad. *Economic and Political Weekly*, 30 November.
Mehta, M. and Mehta, D. (1989) *Metropolitan Housing Market: A Study of Ahmedabad.* New Delhi: Sage Publications.
Needham, A.D. and Rajan, R.S. (eds.) (2007) *The Crisis of Secularism in India.* Durham, NC: Duke University Press.
Oza, R. (2007) The geography of Hindu right-wing violence in India, in Gregory, D. and Pred, A. (eds.) *Violent Geographies: Fear, Terror and Political Violence.* London: Routledge.
Pandey, G. (1990) *The Construction of Communalism in Colonial North India.* Delhi: Oxford University Press.
Shah, G. (1970) Communal riots in Gujarat: report of a preliminary investigation. *Economic and Political Weekly*, No. 5, pp. 3–5.
Shah, G. (1998) The BJP's riddle in Gujarat: caste, factionalism and Hindutva, in Hansen, T.B. and Jaffrelot, C. (eds.) *The BJP and the Compulsions of Politics in India.* New Delhi: Oxford University Press.
Shani, O. (2007) *Communalism, Caste and Hindu Nationalism: The Violence in Gujarat.* Cambridge: Cambridge University Press.
The Asian Age (2002) GCCI to organize peace process. 26 April.
Times News Network (2004) Bankers say they can't risk their lives, money in Juhapura. *Times of India*, 11 January.
Times of India (2004) Modi rolls out red carpet for moneyed minorities. 9 January.
Varshney, A. (2002) *Ethnic Conflict and Civic Life: Hindus and Muslims in India.* New Haven, CT: Yale University Press.
Yagnik, A. and Seth, S. (2005) *The Shaping of Modern Gujarat: Plurality, Hindutva and Beyond.* New Delhi: Penguin.

Chapter 6

On Religiosity and Spatiality: Lessons from Hezbollah in Beirut

Mona Harb

Religious fundamentalism remains poorly positioned *vis-à-vis* secularization theories and matters of modernity and, as a concept, it still lacks rigorous definition and methodology (Emerson and Hartman, 2006). Meanwhile, superficial and generalized references to fundamentalism by political leaders, who use it to describe highly complex geopolitical situations, have endowed the term with normative, reductionist dimensions. Scholars have recently attempted to provide a framework through which more rigorous academic investigation of the concept may take place. Most importantly, in their seminal work, Marty and Appleby (1993) presented specific characteristics by which to arrive at a more solid understanding of religious fundamentalist groups across the world. Other authors have examined the relationships of fundamentalism and social life, looking at a variety of sociological aspects, such as family, gender, ethnicity, and class (Emerson and Hartman, 2006).

In Lebanon, several groups could be categorized as fundamentalist – perhaps the most notorious being Hezbollah, the Shi'i political party (also known as the Party of God).[1] This chapter will examine the case of Hezbollah as a supposedly fundamentalist religious group and investigate its impact on the city of Beirut, given its interventions in the built environment there since its emergence in 1984, i.e., for more than twenty-five years.

My acceptance of Hezbollah as a fundamentalist organization follows Marty and Appleby's model, which specifies two sets of characteristics by which to evaluate religious fundamentalism (as mentioned in Emerson and Hartman, 2006, p. 134). First are its ideological attributes, which include the following: (*i*) the Hezbollah ideology is reactive to the marginalization of an 'authentic' application of religion and attributes this marginalization to the negative effects of Western modernity and secularization; (*ii*) although Hezbollah asserts and uses aspects of modernity (such as science and technology), its ideology relates to a specific interpretation of Shi'i religious power as developed by Imam Khomeini (*wilayat al-faqih* – the guidance of the theologian-jurist);[2] (*iii*) the worldview of the Party of God is dualistic, opposing the forces of evil (U.S., Israel, Zionism) to the forces of good (Islam); (*iv*) the sacred texts of reference are the Quran and the *hadith* (the Prophet's sayings);: and (*v*) the course of history is miraculous and holy, predicated on the reappearance of the Hidden Imam (*al-mahdi al-muntazar*).

Second are organizational attributes, which include the following: (*i*) members of Hezbollah are perceived as chosen, given their kinship ties to the Prophet or their intensive religious training in prominent Shi'i sites such as Najaf and Qom; (*ii*) the boundaries defining adherence to Hezbollah's political agenda are sharp, although some flexibility in moral and socio-cultural practices is allowed based on an individual's choice to emulate a particular religious authority (*marja'*);[3] (*iii*) the group's organizational structure is managed by charismatic leaders who mobilize followers – its secretary general, Hassan Nasrallah, epitomizing this trait; and (*iv*) behavioural requirements for Hezbollah's members are holistic and cater to a wide range of everyday concerns, as will be developed later on.

Fundamentalism can be a useful conceptual tool to compare political groups who use religious ideology to define and mobilize society. It can also be used to understand more accurately their decision-making processes, strategies of action, and impacts on economic, social, cultural, and urban transformations. However, to date, the analysis of religious fundamentalism has been divorced from an exploration of the spatial dimension and its role in shaping and furthering religious identities. If defined as an ideological reaction against political and socioeconomic inequalities, and the resulting 'cultural crisis' provoked by globalization (as developed by Karner and Aldridge, 2004), fundamentalism can provide a useful umbrella under which to investigate how religious identities (re)shape and are (re)shaped by urban conditions. It is based on this understanding that I seek here to examine the spatial conditions or urban features under which a religious-fundamentalist movement like Hezbollah has successfully managed to thrive and expand its power in the city of Beirut. But, given that the current framework for understanding fundamentalism does not provide sufficient tools to investigate the relation between spatiality and religiosity, I will rely on other, more general analytical concepts to engage in this study.

In her historical and sociological analysis of the relationship between space and religion, Hervieu-Légier (2002) identifies three registers through which this investigation can occur. The first is the 'territorial modalities of the communalization of religions' – namely, 'the study of the relations that each religious community maintains with the space in which it has become established', which are determined by its legal and political contexts. The second is the 'geopolitics of the religious', which allows the study of relationships and interactions with other groups, the distribution of local power, and the dynamics of 'exile, emigration, refuge and dispersal into diaspora'. The third register deals with 'religious symbolizations of space', and evokes dimensions of memory and utopia in the production of space. Evidently, these registers are interdependent, and need to be simultaneously mobilized to understand how space and religion affect each other. In this chapter I will use Hervieu-Légier's framework alongside a number of socio-spatial concepts related to the understanding of cities through the politics of space, embodied spatiality, and systems of signs (Tonkiss, 2005), with the goal of explaining the urban conditions under which Hezbollah has consolidated its hegemony over Beirut's southern suburbs.

The chapter will build on Hervieu-Léger's three registers, as mentioned above. The first section contextualizes the discussion and explains how the southern suburbs of Beirut have come to be labelled 'al-Dahiye'. This involves understanding the role of Hezbollah's holistic network of institutions in organizing the delivery of social services, amidst both conflicting and complementary relationships with the Lebanese political and legal systems. In particular, I highlight two interrelated processes that explain the success and durability of this network over the last quarter of the century: the building of institutions; and the production of meanings related to Shi'i faith and politics as embodied in the concept of *al-hala al-islamiyya* or the 'Islamic milieu'.

In the second section, I discuss how Hezbollah's religiosity varies according to geopolitics. I show how the Party of God uses religiosity not only to consolidate its strategic ties with its transnational Shi'i peers, but also to forge professional ties with local, national, and international stakeholders through its developmental and social work. These latter efforts deliberately build on narratives of modernity and progress, justified by Islamic rhetoric, to legitimize its policy interventions.

In the third section, I analyze the process of territorialization of the southern suburbs, showing how Hezbollah has elaborated specific spatial strategies with the aim of controlling, securing, and ordering public space and livelihoods. To this end, Hezbollah has institutionalized its mechanisms of religious symbolization in ways that appropriate and mark territories with specific meanings – rewriting histories, memories, and imaginaries in a process of spatial (re)production. I conclude with a discussion of the dialectics linking religiosity and urban space, arguing that their interrelationship varies with spatial organization and urban governance, as well as with geopolitics, and that the spatial tactics of fundamentalist groups do not always generate additional religiosity and territoriality.

Hezbollah and the Making of al-Dahiye

Hezbollah has been largely described as a terrorist group without enough attention being paid to the way it has inscribed itself into the social and spatial settings in which it has developed over the past three decades. The southern suburbs of Beirut are notoriously considered the fiefdom of the fundamentalist Party of God. However, the area's population of 500,000 is not systematically affiliated with Hezbollah's political and religious ideology. Altogether, the agglomeration of metropolitan Beirut houses about 1.5 million inhabitants, of which only one-third live within the city's official municipal boundaries. The rest live in suburbs that stretch north, northeast, and southwards, and provide a home for people with a strong relationship to the city through work and leisure. Neighbourhoods are, however, well-marked in terms of sectarian affiliation, as a consequence of the fifteen-year-long Lebanese civil war (1975–1990) that displaced much of the population and reconfigured the city into sectarian enclaves. This polarization is regularly reawakened by violent local and regional conflicts, as occurred following the assassination of Prime Minister Rafic Hariri in February 2005 and during the Israeli war on Lebanon in July 2006.

Labels, Identities and Space

Since the early 1980s, al-Dahiye – 'the suburb' in Arabic – has been the charged term used to refer to a specific section of southern Beirut, evoking specific connotations in the minds of Lebanese. Such connotations relate to the social, cultural, economic, urban, and political characteristics of the space: al-Dahiye is Shi'a, poor, backward, rural, peripheral, anarchic, illegal, and Islamist. But the al-Dahiye label also has a complex history relating to the social and political mobilization, as well as rapid urbanization, of Shi'a groups in Beirut (Nasr, 1985; Norton, 1987, 1998).

During the 1960s, in line with modernist and hygienist urban analysis, which sought to build well-organized spaces conducive to the emergence of the 'good' city, geographers and sociologists pointed to a 'misery belt' around

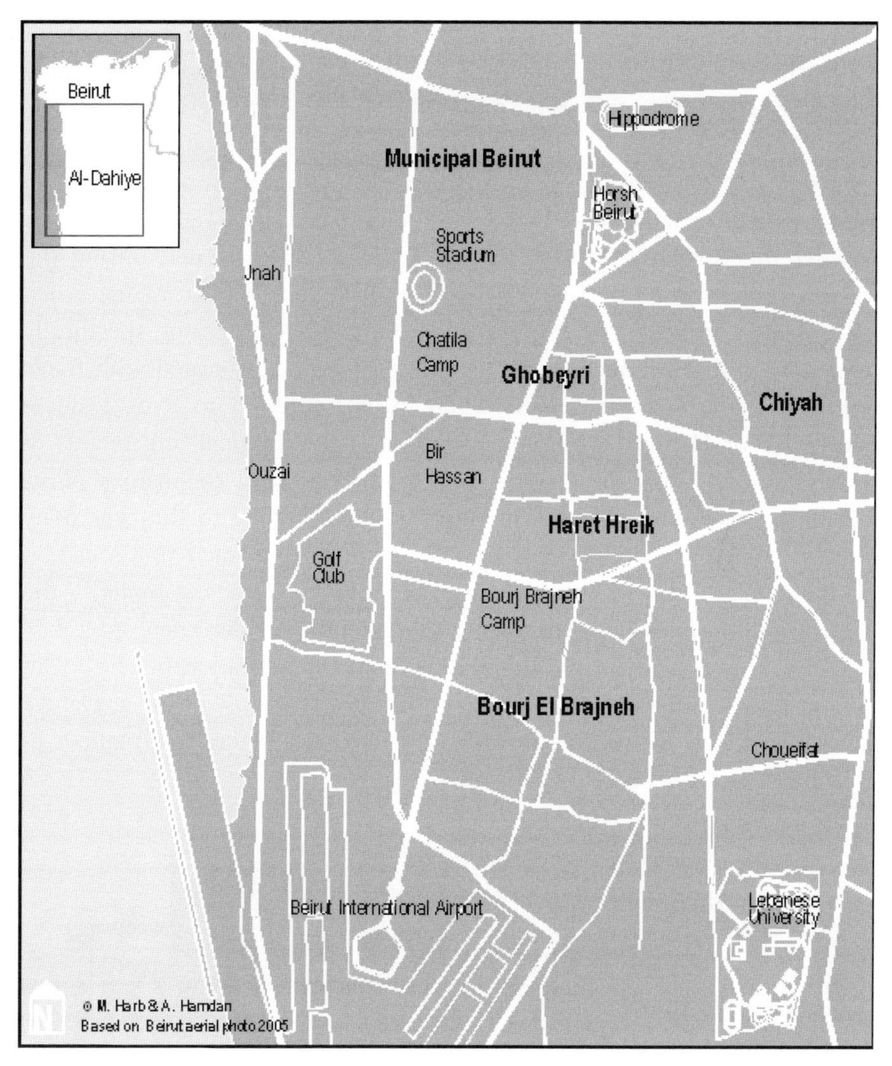

Figure 6.1. Al-Dahiye in Beirut: main neighbourhoods and selected landmarks. (*Source*: Mona Harb, 2008)

Beirut, composed of its poor suburbs (*dawahi*), which they claimed were the source of the capital's urban problems (DGU, 1973). The term *dawahi* thus entered the urban lexicon as a stigmatizing label, disseminated by the print media as well as by planners, sociologists, government officials, and political groups (Harb, 2003). Following the civil war, population displacement, and the division of the city into homogeneous sectarian enclaves, the plural *dawahi* was reformulated into two separate areas: the Christian-dominated northeastern suburbs (*al-dawahi al-shamaliyye al-sharqiyye*), and the Shi'a-dominated southern suburb (*al-dahiye al-janubiyye*).

The southern suburbs, which had formerly housed a significant minority of Christian Maronites in the towns of Haret Hreik and Mrayjeh, were now largely Shi'a.[4] And, in time, the suffix *janubiyye*/southern was slowly abandoned, to be replaced with 'Dahiye' or 'al-Dahiye', to signify a rebellious section of the city whose residents defied state authority, contested existing power structures, and claimed their own political and social rights.

After the Amal movement, the Shi'a militia that operated during the Lebanese civil war, was defeated by Hezbollah in 1989. At a time when today's Shi'a brothers were fierce enemies, al-Dahiye became specifically associated with Hezbollah. Even more so, it became associated with Iran's supreme leader – to the point where people referred to it as Dahiyet Hezbollah or Dahiyet al-Khomeyni. Some of the 'original' inhabitants of Dahiye rejected this designation and tried to reclaim the names of its neighbourhoods and streets – linking the new labelling to the hegemony that Hezbollah was now exercising over space. They formed an association by the name of the 'southern coast of Metn' (*sahel al-matn al-janubi*) – the legal name of Dahiye – to denounce the 'ghetto' (*sic*) that their space had become. However, since the mid-1990s, the Lebanonization of Hezbollah (Hamzeh, 1993) has helped normalize and regulate the use of the al-Dahiye label in everyday life. People now use the term in a neutralized manner to

Figure 6.2. Al-Dahiye's as administrative label, Airport Highway (right-hand panel says: *Banlieue* in French, *Dahiye* in Arabic). (*Photo*: Mona Harb, 2005)

Figure 6.3. Al-Dahiye as resistance label, Haret Hreik (banner says: 'Al-Dahiye is a Living Testimony of the New American Middle East'). (*Photo*: Mona Harb, 2006)

indicate their geographic destination to cab drivers, who recognize well its boundaries. Public administration officials have also co-opted the term and included it as a geographic destination on highway signs. And the five municipalities of al-Dahiye are now grouped in a union of municipalities officially registered with the Ministry of Interior as Dahiye Janubiyye.

Of course, the Dahiye label is still strongly associated with Hezbollah – an association which gained international attention during the July 2006 Israeli war on Lebanon. Indeed, Hezbollah proudly proclaims the political dimension of this linkage, and its leaders regularly describe Dahiye as the capital of the Resistance – contesting the title which was once specific to South Lebanon (cf. Bint Jbeil), and which might today be transferred to the Bekaa (ICG, 2007). The Party of God is also constructing a historicized legitimacy to the resistance dimension of al-Dahiye. Thus, its affiliated Arts NGO produced a short documentary film in 2007 about the reconstruction of Haret Hreik, whose story line was that 'Dahiye was always a resistance site in the modern history of Lebanon'. By claiming that area residents had previously contested the authority of Ottoman and French rulers and sided with pro-Arab nationalists and the Palestinian cause, the film ascribed an invented history of political and national resistance to the southern suburbs.

Antagonistic Cooperation with the Lebanese 'State'

Before going further and explaining how the Party of God in Dahiye has organized its control over urban space, I will briefly describe the national political context in which Hezbollah operates. The Lebanese political system encourages its seventeen sectarian groups to manage and govern the affairs of their own communities according to a rationale of self-protection and identity preservation. Within this structure of *communautaurisme politique*, or 'sectarian politics' (Picard, 1997), the political system cannot determine the growth and reinforcement of sectarian (and religious) identities, and their ensuing *'assabiyyat*.[5] Thus, conceiving a Lebanese state that could effectively rule over society and exercise its authority appears unrealistic.

Within this power-sharing formula, and specifically in urban contexts, sectarian groups are the primary medium through which citizens access services and jobs. However, they do so in exchange for their loyalty on election day or when politics warrant a mobilization. This historic clientelistic system has been analyzed by many authors (Denoeux, 1993; Khalaf, 1968; Binder, 1966; Johnson, 1986), who have emphasized the roles of the *za'im* (patron; pl. *zu'ama*) and its *qabadayat* (middle men).

To cut a long story further short, it is also important to understand that Shi'i groups, through most of Lebanon's modern history, have been the country's political underdogs. Persecuted by the Ottomans, marginalized by the French, stigmatized by the Christians and Sunnis, their mobilization and struggle to be recognized as a full political partner is still violently underway. Predominantly located in the south of Lebanon, these Shi'i groups were once organized around *zu'ama* and other elites (*wujaha*) (Mervin, 2000; Charara, 1996), in a context where resources were locally produced and redistributed. But the rural-urban migration of Shi'a to Beirut's suburbs in the 1960s transformed the terms of the clientelistic exchange, especially as it coincided with the loss of power of many landlords due to regional economic changes. Local urban elites (such as Rachid Beydoun, see Early, 1971) and religious leaders (such as Mohammad Hussein Fadllallah) emerged to fill the leadership gap, attempting to address the dire social and urban needs of these new migrants to the city. Simultaneously, the attempts of President Fuad Shehab to build a welfare society provided Shi'a with access to public services – especially education.

Authors have shown how mobility and rapid urbanization, together with access to modern education and services, may lead to politicization, mobilization, and radicalization of identities (Gole, 2000, pp. 93–95). In the early 1970s, Shi'i groups started demanding public benefits through their new sectarian leaders, as other communities had been doing for several decades. In particular, the political mobilization led by Sayyed Moussa al-Sadr allowed them to acquire institutionalized access to sectarian representation within the Lebanese political system (with the establishment of the Supreme Shi'i Council in 1969), and thereafter improved social services. The founding of the Movement of the Disinherited, followed closely by its armed counterpart Afwaj al-Muqawama al-Lubnaniyya (AMAL) in 1974, paved the way for a progressive transformation of these arrangements during the years of the Lebanese civil war – and after the disappearance of al-Sadr in 1978 – into a fully-fledged militia economy (Picard, 1999).

Emerging in 1982, Hezbollah was initially more inclined to follow the ideals of social work developed by al-Sadr, and it adopted a different strategy to provide services to its constituency. Clientelistic strategies in Lebanon vary between two extremes (Harb, forthcoming): in the first (direct redistribution), sectarian leaders claim state resources and redistribute them to their constituency; in the second (self-sufficiency or alternative systems), sectarian leaders generate their own resources and distribute them to their clients. Of course, many sectarian leaders combine the two strategies in ways that depend on specific variables (time, space, nature of the stake, etc.). But, in general, Amal is an example of the first system, while Hezbollah is the success story of the second. Indeed, since its foundation Hezbollah has established more than ten service-based institutions governed by a strict executive hierarchy, servicing thousands of beneficiaries over the past twenty-five years.

I will turn shortly to discussing this service network. But first it is important to explain Hezbollah's relationship to the Lebanese political system. Hezbollah joined the ranks of organized political parties in 1992 by participating in legislative elections. Since then, the Party of God/Hezbollah has consistently elected a bloc of eight to ten parliamentary representatives (out of 128). In 1998, the Party participated in municipal elections, where its slates won a significant proportion of local councils; and by 2004, Hezbollah controlled 120 municipalities out of about 900 (Harb, 2009). In 2005, it joined the government through two ministries. Through these different political forms of participation, Hezbollah now contributes to the power-sharing game that is peculiar to the Lebanese system of governance. It also negotiates with other sectarian groups – as well as with its Shi'a partner, the Amal movement – for a share of public resources, according to a subtle game of exchange which follows rules of antagonistic cooperation (Marin, 1990). Indeed, while the Party uses public governance structures to advance the interests of its constituency as well as gain access to power, it also presents itself as the sole legitimate representative of Shi'i rights and needs. But it further uses the state's weakness to justify its autonomy and independent decision-making – especially in relation to military resistance against Israel – thus challenging the state's political and social legitimacy. Hezbollah has maintained this ambiguous position over the past two decades, adapting its strategies to different situations, times, and spaces in an effort systematically to maximize its power and advance its interests (Harb, forthcoming).

Building Institutions, Disseminating Meaning

Hezbollah's institutions are nongovernmental organizations (NGOs), registered with the Ministry of the Interior. Their administrative and financial relationships to the Party vary: some are local branches of Iranian organizations and do not report directly to the Party, while others have been established by Hezbollah and report directly to its ruling bodies. Others are managed by Party cadres but are financially and administratively autonomous. Overall, these organizations fall into two main groups: those providing services to people directly or indirectly involved in military resistance; and those managing social, religious, financial, and urban services oriented to a larger audience.

The first group includes the Martyr Organization (al-Chahid, established in 1982) and the Wounded Organization (al-Jarih, established in 1990). The former is the local arm of an Iranian organization of the same name; the latter was founded by Hezbollah. Within the second group, two organizations, both founded by Hezbollah and dependent on its structure, provide education and microcredit services. The Educational Institution (al-Muassasa al-Tarbawiyya, founded in 1991) manages the education sector. The Good Loan (al-Qard al-Hassan, established in 1984) is specialized in microcredit following Islamic rules of borrowing. Four other satellite organizations are administratively autonomous from Hezbollah, though they are managed by Party cadres, and their employees are affiliated with it. Three are local branches of Iranian institutions. The Support (al-Imdâd, established in 1987) distributes social services to the needy and to the poor. The Islamic Health Society (al-Hay'a al-Suhiyya, founded in 1984) supervises forty-six health centres and a hospital. Jihad for Building (Jihad al-Binaa', founded in 1988) operates in the construction and agriculture sectors, especially in the South Lebanon and the Bekaa. The fourth organization, the Consultation Centre for Studies and Documentation (CCSD, established in 1988), is a think tank that prepares policy reports and studies.

Hezbollah's institutions operate as a holistic network that provides a comprehensive set of services to users, addressing diverse components of their daily needs. Moreover, since Hezbollah was chosen by voters to take charge of local governance in Dahiye in 1998, these institutions have been partnering with its municipal councils, which significantly rely on Hezbollah's expertise for elaborating and implementing local development strategies as well as social policies.

What are the organizational features behind the success and the durability of these institutions? First, Hezbollah's institutions are responsive and accountable thanks to a decentralized system of operations. This system maximizes the reach of Hezbollah's networks; ensures their responsiveness to local needs while disseminating participatory values; and guarantees regular feedback to the service providers, enabling them to adapt their services to local needs. Socio-spatial 'embeddedness' also helps to develop trust between service users and Hezbollah's institutions, enhancing their accountability. Second, the Party's institutions have a strong leadership structure and maintain professional standards. Each institution's leader is typically a charismatic male who is a professional and/or is well educated, who often has past experience in militant work, who is calm and composed, and who can therefore act as a role model for his employees and project a professional image for the institution. The institutions' directors also rotate on a regular basis, ensuring steady administrative turnover and building individual managerial capacities across policy sectors. Third, Hezbollah's institutions perform well and are considered effective, especially in a context known for corruption and bureaucratic laxity. The institutions are keen on scientific knowledge, rational thinking, and technical expertise, as these are believed to produce effective and 'modern' outputs. In addition, Hezbollah's institutions trust learning from experience: they evaluate their work regularly, and revise their policy agendas accordingly (Harb, forthcoming).

Such organizational features are not the only explanation for Hezbollah's institutional success, however. On closer examination, these features are systematically related to religious rhetoric which emphasizes social justice (*'adala*) and participation (*ishrak*), and which relates professionalism to excellence (*itqan*), while grounding effectiveness in modern values (*hadara*). Indeed, Hezbollah's organizational structure incorporates a pious morality that defines codes, norms, and values referred to by the terms of *hala*, *bi'a,* or *jaw* (from which the adjective *islamiyya* is deleted, as it is internalized), and which I translate as the 'Islamic milieu' or 'sphere'. Sometimes this milieu is also described as *mujtamaa' al-muqawama* – the society of the Resistance.

The Islamic milieu aggregates different pious groups around Hezbollah's institutions, irrespective of the religious authority they emulate.[6] It proposes one common denominator for all such groups: obedience to 'authentic' piety (*iltizam*). Being a *multazim* 'naturally' means adhering to the Islamic concept of *jihad*, materialized in the mission of resistance (*muqawama*), which ultimately leads to martyrdom (*shahada*). In Hezbollah's view, resistance is not only military, but is foremost political, social, and cultural; it is conceptually both a spiritual and a military *jihad.* Resistance is thus promoted as a liberation journey, freeing the individual from a life of oppression and humiliation equivalent to death. Failing to resist means submission to injustice and inequality, which not only means disgrace in this life but in the afterlife (Qassem, 2002, pp. 58–59; Saad-Ghorayeb, 2002, p. 125). Consequently, resistance is perceived and experienced as a mission, a duty, and a way of life. It goes beyond combat, and becomes an individual process, carried out through daily practices related to body, sound, signs, and space, transmitting 'religious and community knowledge' (Deeb, 2006).

Interestingly, over the last decade, narratives in the Islamic sphere have increasingly borrowed national references and symbols from Lebanese folklore (the cedar tree, the flag, tourist sites, etc.), proposing a hybrid type of national discourse that mixes secularism with religiosity. This is particularly evident in songs about the Resistance, which have evolved over time to accommodate a youth audience attracted by popular tunes. Such hybrid national narratives are also materialized physically in cultural arenas such as museums, commemorative sites, and places of entertainment that are respectful of religious codes (cafés, restaurants, beaches, fitness clubs, etc.) (Deeb and Harb, 2007).

This dual characteristic explains Hezbollah's institutional success and durability for a quarter of century, against many odds. Its holistic service networks not only provide material resources to their beneficiaries, but they also define a world of social and moral meanings. In what follows, I will discuss how this duality is consolidated at different geopolitical scales, with different sets of actors.

Hezbollah and the Geopolitics of Religiosity

Hezbollah's holistic network of service delivery, as well as the sets of values discussed above, operate within a diverse configuration of stakeholders at the local, national and international level, through which its religiosity is differentially expressed, depending on the nature of stakeholders as well as the types of relationships that link them.

It is undoubtedly through its relationships with the Islamic Republic or Iran that Hezbollah's religiosity is most directly expressed – primarily via its emulation of the Khomeini doctrine of *wilayat al-faqih*. This doctrine is brought into the spaces where Hezbollah operates via the network of institutions described above. These institutions are also partially supported by Iranian funds, though it is difficult to estimate the amount of this support, especially as it has been affected by changes in the governing regime in Tehran.[7] The Iranian government also provides military support to Hezbollah's resistance corps, which has had direct spatial impacts on the natural landscape of South Lebanon and the Bekaa, as well as on the built environment of villages and towns in these areas. However, it is important to note that this relationship is not a one-way street, and that Hezbollah is not a passive recipient of the Iranian model of religiosity and service provision; rather, it modifies and adapts its piety and service-distribution strategies to the local Shi'i Lebanese context and its socioeconomic specificities (Fawaz, 2004). In the case of post-war reconstruction, it is further noteworthy that Hezbollah has elaborated its policies independently from Iran, as demonstrated by the Waa'd reconstruction project in al-Dahiye (Harb and Fawaz, 2010).

However, Hezbollah's governance in Shi'i regions also takes shape in association with another set of actors with whom it has no religious ties. These are the international donors with whom Hezbollah shares common values and norms (modernity, professionalism, and progress). International groups have been very active on the Lebanese development scene, especially in rural and disenfranchised areas. And given that Islamic rhetoric also stresses the importance of these activities, and that its leaders have considerable experience in social mobilization and empowerment, Hezbollah is 'naturally' drawn to these narratives of development aid, which emphasize community and capacity building. In recent years, the association between Hezbollah and international actors has been steadily increasing, especially in the context of municipal governance.

Local councils in Lebanon legally benefit from large margins of manoeuvre with respect to socioeconomic development, but they need donors' aid to access material resources and expertise. I have shown elsewhere how Hezbollah's mayors and municipal councils are highly esteemed by donor agencies, who appreciate their yearning for professionalism, progress, and learning, and who value their efficiency in implementing development policies (Harb, 2009). Thus, Hezbollah's institutions – NGOs and municipalities – have partnered successfully with donors such as UNDP, UNICEF, WHO, ILO, and the EU – as well as with the World Bank and a plethora of international organizations headquartered in Europe. This association strengthens the Party's service institutions, but it also contributes to its image as a professional and efficient development partner – an image that consolidates its 'non-military' identity and helps it further forge its socioeconomic legitimacy, especially at an international level where it is predominantly viewed as a terrorist group (Harb and Leenders, 2005). This 'ideological' rather than 'religious' association also explains the partnerships between Hezbollah and a variety of public, associative, and private institutions at national and local levels. Generally, these partnerships are described in positive terms by Hezbollah's counterparts, who praise its professionalism and efficiency – though some might mention its obsession with exclusive control of projects and with restricting outside interference, especially when it comes to data collection and access to the field (or simply when the stakes are politically too high).[8]

A more direct religious association links Hezbollah with significant sections of the Lebanese Shi'i diaspora, as well as with returning Shi'i emigrants.[9] Indeed, Hezbollah's publics include communities of Shi'a who have emigrated to Europe, the Americas, and Africa in search of better sources of livelihood. Most of these emigrants maintain strong ties with their families in Lebanon, supporting them financially through regular remittances, and often consolidating their social ties by choosing marriage partners from their native towns or from kin networks. These ties are furthered through regular communication which has been facilitated in the last decade by new media technologies. Indeed, Shi'i diasporic communities are extremely well connected to the social, cultural and economic news of their hometowns, and regularly follow up on the political dynamics of Lebanon (Leichtman, 2007, p. 237). With the liberation of South Lebanon from Israeli occupation in 2000, several Shi'i emigrant families returned and established themselves in al-Dahiye. They invested their savings in real estate development and business enterprises which contributed to the economic growth of the area, as demonstrated by the multiplication of high-end residential buildings in the upper-income neighbourhoods of al-Dahiye, as well as by an increase in commercial and leisure services, such as malls, shopping centres, amusement parks, cafes, and restaurants. This middle class is as much characterized by its taste for consumerism as by its commitment to Shi'i piety or to Hezbollah's resistance cause. Nevertheless, its proponents form a significant base for the Party, especially given their contribution to the political economy of the Islamic milieu.

Thus, Hezbollah's religiosity is varyingly supported by a wide range of geopolitical networks which allows it to thrive locally, nationally, and transnationally – albeit in very different ways. While the Party's religiosity is directly associated with the Islamic Republic of Iran and with the Shi'i diaspora, it is also connected to secular and civic stakeholders and their development agendas through religious values and norms that carry universal meanings (i.e., modernity). Accordingly, through direct and indirect ties to geopolitics, Hezbollah's religiosity is able to acquire more political and social legitimacy – hence, more power – which the Party reinvests in its holistic policy networks.

Hezbollah's Strategies of Spatial Production

The spatial production of Dahiye by Hezbollah follows several strategies which have been adjusted over time as local stakes have varied. However, I have identified three strategies in particular which physically materialize the codes, values, and norms of the resistance society and the Islamic milieu Hezbollah is keen on creating: (*i*) *quadrillage* and centralization; (*ii*) the use of spatial planning tools; and (*iii*) religious symbolizations and territorial markings.

Quadrillage and Centralization

The first strategy is basic and relies on spatial control, or *quadrillage*. In the early 1980s, with Lebanon at war, space was organized according to a strict military logic, and control operated along secret lines. Thus, villages and neighbourhoods in the South, the Bekaa, and Beirut were divided into sections and subsections supervised according to a vertical hierarchy that enabled surveillance. This military organization is still valid and forms the basis of Hezbollah's social and spatial action. Within this system, al-Dahiye forms one sector (*qita'*) organized in six quarters (*murabba'at*), subdivided into a number of cells (*khaliyyat*). Each geographic unit is led by a chief who regularly reports to his hierarchical superior. The cell, as the smallest geographic unit, serves as the organization's 'eyes on the street', and is normally led by someone living within a neighbourhood who has intimate knowledge of its social and urban fabric. The cell's chief has a dual responsibility – on the one hand to maintain security and report any dubious actions, and on the other to identify potential beneficiaries of Hezbollah's services.

Al-Dahiye is the 'capital city' of Hezbollah, and all its institutions are geographically concentrated within its limits. During the civil war, a significant number of Hezbollah's institutions came to be located in the neighbourhoods of Haret Hreik and Bir al-Abed, while a few were dispersed in other neighbourhoods. In the mid-1990s, all of Hezbollah's key institutions were regrouped within Haret Hreik, however. The creation of such a centre of gravity was predominantly a strategic decision linked to security concerns. But this 'security quarter' (*al-murabba' al-amni*) was subsequently razed to the ground by Israeli attacks during July 2006.

The necessity of centralization is counterbalanced by the need to decentralize service institutions to gain access to a larger base of potential clients. Hezbollah's institutions have elaborated such modes of operation by assigning volunteers – mostly female – to identify potential clients in varied neighbourhoods. These 'volunteer sisters' (*akhawat muttatawi'at*) also assist the cell's chief by acting as informal liaisons between him, the institutions' directors, and beneficiary families. Because they reside in the neighbourhood, these women are able to identify, through interpersonal contacts, the families that need aid, and match them with the appropriate Hezbollah organization. 'Volunteer sisters' have at least three functions: recruitment, follow-up and evaluation of the service provided, and building durable ties with beneficiaries and ensuring their piety and political commitment (Fawaz, 2004, p. 14). Thus, they act as relays, informing families of available services, maintaining a regular relationship with them (through visits, correspondence, outings, etc.), making sure that services are provided to a high standard, evaluating their impact, and reporting back to the organization concerned. In addition to the role of female volunteers in the daily lives of residents, local clerics also act as mobilizing agents (Charara,

1996, p. 215), recruiting clients in mosques, *husseyniyyat* (public meeting halls dedicated to the memory of Imam Hussein), and *hawzat* (religious institutes).

The Waa'd reconstruction project illustrates well the use of the Party's *quadrillage* strategy and its centralization dynamics. In the hours that followed the cease-fire in August 2006, hundreds of architects and engineers were mobilized to assist Hezbollah in the process of reconstruction. The Party directed them toward Jihad al-Binaa', which organized a meeting in its newly established headquarters (its old offices were destroyed) in the classrooms of one of the Party's schools. I attended that meeting, which aimed to organize volunteers to rapidly collect data about the extent and range of destruction in al-Dahiye and so provide an adequate basis for reconstruction. In the meeting room, a series of maps were pinned up on the walls, showing the footprint of all of Dahiye's buildings, highlighted in different colours: dark red for the ones razed to the ground, light red for those heavily damaged and structurally unsound, and pink for those badly damaged but in need of further assessment. The colours indicated the result of a preliminary assessment conducted by Jihad al-Binaa's engineers during the July war; a more precise assessment was now needed, and a survey questionnaire had been prepared which the volunteers were to administer as rapidly and as accurately as possible.

A Jihad al-Binaa' delegate described to the volunteers the questionnaire's structure and content. It included about forty questions, geared towards a precise and detailed evaluation of damage to every apartment. He then explained that the volunteers would be divided in teams of four to five individuals, who would each be responsible for surveying a section of Dahiye. He referred to one of the maps on the wall, which showed how the area was subdivided into seventy-five subsections – probably a modified version of the Party's spatial grid adapted to the geographic distribution of destruction. The teams were quickly formed and went off to work in the subsections assigned to them by the delegate, who also nominated a team leader with whom he kept in touch.

The survey was completed in ten days. It was then entered into a computer database and served as the basis upon which the reconstruction policy was developed into a two-tier scheme, involving a rehabilitation project (*tarmim*) for the moderately damaged buildings in al-Dahiye, and a full-fledged reconstruction project for Haret Hreik (the Waa'd project), both centrally led by Jihad al-Binaa'.

Using Spatial Planning Tools

A second key strategy that helps the Party consolidate its spatial control over the territory is the use of spatial planning tools – namely, quantitative information about the spaces it controls and their inhabitants. The objective here is to help the Party induce order and logic to its functional distribution. In 1988, Hezbollah set up a small think tank – the Consultative Centre for Studies and Documentation (CCSD) – whose job was to count the Shi'i population in different Lebanese regions, survey available social services, assess their quality, and identify precisely various social needs. On the basis of this knowledge, which forms the scientific grounding of all 'good' social-work strategies, Hezbollah was able to determine which social institutions to establish, and when and where to do so. Since its establishment, the think tank has grown to become a main player in knowledge production on the national scene, organizing conferences, publishing policy reports and studies, and advising Hezbollah's leadership and institutions on key policy sectors.

The production of information by Hezbollah with the aim of territorial control is illustrated by several examples, including the post-war reconstruction of al-Dahiye just mentioned above, where Jihad al-Binaa' used quantitative data to elaborate and justify its policy choices. These data were backed up by a working paper analyzing reconstruction options, authored by the CCSD, in which a vision for reconstruction was proposed and argued.[10] This paper served as a basis for the justification rhetoric supporting the reconstruction policy of Hezbollah.[11]

The Waa'd project for the reconstruction of Haret Hreik also relied on extensive spatial imagery that highlighted how the outcome would be 'more beautiful' than what was being replaced. This approach built on a speech by Hezbollah's secretary general, Sayyid Hassan Nasrallah, that specified how the reconstruction would ameliorate the built environment. The three-dimensional perspectives of the buildings to be rebuilt, produced by Waa'd, were pinned on billboards on the empty building sites, and reproduced on advertising billboards around al-Dahiye during the campaign to launch Waa'd in spring 2007. They show the new buildings rendered in soft tones, emphasizing 'earthy' colours that symbolize 'classy' taste, and that demonstrate the reality of the reconstruction for those who are impatiently waiting to occupy them.[12] This reality is further materialized inside the offices of Waa'd, where a model of the area to be rebuilt shows prospective residents their forthcoming buildings – in addition to providing them with complete sets of drawings (plans, elevations, sections), which they need to approve. Thus, the use of planning tools has been key in the Party's reconstruction policy in al-Dahiye. It has been a key instrument by which Hezbollah has been able to demonstrate its ability to intervene efficiently and professionally in dire times and circumstances. This effort has served to legitimize and assert the Party's power.

Since 1998, the job of rationalizing the use of space has been increasingly taken over by Hezbollah's local governments. In Dahiye, the municipality of Ghobeyri, the largest and best-endowed financially, is currently elaborating a development plan for its neighbourhoods, of which the mayor is very proud. When he showed it to me, he proudly pinpointed the different projects the municipality has undertaken over the past decade. The plan allows him both to demonstrate the town's achievements and to depict them physically to an audience. He also uses it to prove that the projects have been implemented according to an equitable spatial rationality, and that municipal territory is 'modern' – up to par with the rest of Beirut in terms of private investments, infrastructure, and equipment.[13]

Figure 6.4. Waa'd billboard, showing the reconstruction project. The caption reads 'We will build it more beautiful than it was'. (*Photo*: Mona Harb)

The plan also serves to identify real estate values in the municipality and form strategies to develop high-end neighbourhoods, from which the municipality benefits directly in terms of tax returns. The plan thus also contributes to the improvement of the municipality's tax base and to better organization of its fiscal policy. The experience of Ghobeyri, one of the most dynamic among Hezbollah-led local governments, reveals how the production of information by the Party's local councils is associating territorial awareness with the dynamics of economic growth and competitiveness. Hence, spatial knowledge is being consolidated as an operational strategy – even if the effort is still at an embryonic stage of quality and refinement.

Religious Symbolizations and Territorial Demarcations

Dahiye's built environment is comparable physically to other lower-middle-class neighbourhoods in Beirut, such as Tariq al-Jadideh and Basta, in terms of density and urban morphology (e.g., in terms of the width of streets, the size of blocks, and the height and type of buildings). It is also similar in terms of its socioeconomic practices, which rely on the appropriation of public spaces and on symbolic markings. The built environment of Dahiye has thus become physically differentiated from other Beiruti places by the distinctive ways that Hezbollah's institutions (and other institutions belonging to the Islamic milieu) – as well as ordinary individuals – have demarcated space and endowed it with specific meanings, codes, and values. Obviously, religious institutions such as mosques, *husseyniyyat*, and *hawzat* play a major role in marking places. Also, Hezbollah's institutions demarcate buildings and streets and endow them with a specific identity. Just as Haret Hreik is known to house the Party's main institutions and has become identified as its strategic headquarters, streets like Bir al-Abed and Hadi Nasrallah, on which the Party's media centre and its microcredit organization, al-Qard al-Hassan, are situated, have become associated with it. In addition to these physical markers, characterized by their stable, relatively permanent, spatial component, other markers introduced by Hezbollah provide more temporary and contingent features.

Hezbollah has developed its territorial marking strategy over time through its Information Unit, an institution that was reformed after the liberation of South Lebanon in 2000 to become more independent and operate with a larger budget. In addition to maintaining Hezbollah's official image, the Information Unit organizes the public meetings, commemorations, and demonstrations of the Party. It is supported by another institution, established in 2004, the Arts Organization, which is responsible for the graphic design of

Party slogans during these public events, and for coordinating the Party's music orchestra and special projects relating to the memorialization of the Resistance.[14]

Each year, the Information Unit defines its annual plan, at which time it decides on a theme to illustrate through territorial markings, in close coordination with Hezbollah's leadership. The Unit then produces images that are displayed on billboards along Dahiye's main streets, and along the main highways connecting Beirut's neighbourhoods to Dahiye. These images are also disseminated by al-Manar, the Party's television station; and Party slogans are broadcast on its radio, al-Nour. In addition to the annual theme, the Information Unit commemorates a number of events throughout the year – Achoura, Martyrs Day, Liberation Day, Victory Day, Jerusalem Day, etc. – for which it produces specific images and slogans for dispersal through the same media outlets. Often, accompanying songs (*anashid*) are produced in video clips on Manar TV to reinforce the message of the event.[15] These are picked up by Hezbollah's constituency, who listen to them in their cars, on their portable music devices, and on their cellular phones. The Information Unit is also responsible for organizing regular exhibitions in al-Dahiye, in which Resistance accomplishments are displayed, or, more simply, where books related to the Islamic milieu are sold. Since Hezbollah took over the leadership of local governments in al-Dahiye in 1998, territorial markings have expanded to include the renaming of streets and highways, as well as public squares and roundabouts, with names such as 'To the Glory of Hezbollah's Martyr Leaders', 'Imam Khomeini', or 'Palestine and the Resistance'.

This work contributes to transforming al-Dahiye into a territory marked with a socio-political identity – that of the Islamic milieu and the Resistance society. Indeed, all the iconography used by the Arts organization is directly related to the vocabulary and imagery of Shi'ism and of Hezbollah.[16] The Information Unit director put it clearly: 'Our aim is to produce a message [*rissăla*] that transmits *al-hăla* [*al-islamiyya*] that exists in the streets, that says what is this city and that reveals its characteristics and transmits the identity [*hawiyya*] of this *hăla*'.[17] The Information Unit thus contributes to the production of a territory where Hezbollah's norms and values are materialized physically through iconography and imagery. These markings consolidate certain norms and values in the everyday lives of Hezbollah's constituency, building up their imaginary, and they reassert the necessity of social and political mobilization for the cause of the Resistance. They also indicate to other publics – national and international – the power of the Party of God. Indeed, it is noteworthy that in recent years, since the

Figure 6.5. Khomeyni Street, Ghobeyri. (*Photo*: Mona Harb, 2005)

liberation of South Lebanon in 2000, Hezbollah has transformed its visual language, adopting a slick corporate style of graphic design, using the Arabic, French, and English languages.

Authors have shown how the capacity of a group to materialize its ideology is an essential power resource, 'as it renders possible, via the production and the transmission of ideas, of traditions and meanings, the establishment and reinforcement of the legitimacy and rights of the group'(DeMarrais *et al.*, 1996, pp. 15–17). It is important to emphasize as well how these religious symbolizations are rewriting Shi'i history and memory – highlighting, inventing, and subduing elements (Deeb, 2008; Mervin, 2008).

Figure 6.6. Martyrs' images along Main Street, Ghobeyri. (*Photo*: Mona Harb, 2005)

Figure 6.7. Painted wall commemorating Ashura, Haret Hreik. (*Photo*: Mona Harb, 2005)

The three spatial strategies identified here physically communicate codes, norms, and values to the various publics inhabiting the neighbourhoods of al-Dahiye. Thus, they participate in the making of a 'territoriality', by allowing space to be inscribed with meanings. According to the French geographer Roncayolo (1990, p. 190), the concept of 'territoriality' includes two different dimensions: 'On one hand, an attachment to specific places, which is often the result of a long material and mental investment that is expressed through beliefs and grassroots religiosity; on the other hand, principles of organization … that model the territory but can be transferred to other places'. Hezbollah's spatial strategies combine both these dimensions. First, they produce attachment to al-Dahiye through the material production and dissemination of Shi'i beliefs and religiosity, embedded in the social and urban fabric of neighbourhoods through everyday practice. Second, they rely on principles of organization that model space – namely, the Party's holistic policy networks which are responsible for marking, ordering, and controlling space. In other words, Hezbollah succeeds in combining two concepts of territory (Genieys *et al.*, 2000): one perceives the territory as a space of public intervention where resources and collective services are strategically provided *vis-à-vis* other stakeholders; the other interprets territory as a space of identity and a site of belonging.

In addition to spatial appropriation, control, and organization, which are produced by an institutional apparatus, territoriality is consolidated by the inhabitants themselves and by a dynamic private sector. Spatial assertions of self-identity occur on the balconies of houses or apartment blocks (covered by curtains to hide women from eyesight). They occur in cabs, shops, offices, and homes (through the display of memorabilia and various religious commodities). They are inscribed on bodies (hidden by dark or coloured veils and robes or covered with tie-less costumes). And they are translated in sounds and language (via religious greetings and language and through the broadcast of *anashid*).

Al-Dahiye has thus been (re)produced by the combined actions of Hezbollah's networks, private and associative stakeholders, and ordinary residents. These actions have produced a sense of territoriality which allows the Islamic milieu to be 'embedded' in the daily lives of al-Dahiye's residents, while simultaneously providing them with a 'stage' through which their socio-cultural identities can be displayed, gazed at, expressed, and renegotiated. Through this dialectic relation, Dahiye and the Islamic milieu concomitantly operate as a cognitive framework nourishing a collective Shi'i consciousness and generating a feeling of pride and self-worth. This is especially powerful in a context where Shi'i were once systematically marginalized as poor and disenfranchised.

Conclusion: Hezbollah, Religiosity and Spatiality

I close this chapter by proposing that researchers rethink religious fundamentalism in the city in relation to the dialectics linking religiosity to spatiality – extracting lessons from Hezbollah's experience in Beirut. I believe there are at least two elements which need further research and elaboration, and which can inform scholarship and practice.

First, Hezbollah's action in the city reveals how the interrelationship between spatiality and religiosity operates variably with spatial organization and urban governance, as well as with geopolitics. At a neighbourhood level, religiosity generates a physical materialization of pious identities and meanings, producing stronger – albeit, differentiated – territorialities. Although Hezbollah might appear in this chapter as a hegemonic player dominating all venues of everyday life in al-Dahiye, it is essential to note – though this cannot really be measured – that not all residents in al-Dahiye are adepts of Hezbollah's politics, and several organizations operating in al-Dahiye do not belong to Hezbollah's network or to the Islamic milieu. This variability is, hence, mirrored in urban spaces which present different levels of territoriality and religiosity. Thus, some neighbourhoods in al-Dahiye will be less marked than others, or marked differently.

At a national level, politics tend to diminish the embeddedness of practices of religiosity in territory. At a transnational scale, depending on how different strings of religious narratives link up with the geopolitics of regional, international, and diasporic players, pious identities are more or less expressed. In times of conflict – for instance, in Beirut during the period from 2006 to 2008, when Hezbollah and the government were at odds (in a polarized geopolitical context opposing Iran to the United States) – religiosity becomes further galvanized and trickles down to the national and local levels. Here it translates spatially into militarization, sealed-off public spaces, and delineated boundaries, as well as increased territoriality and *entre-soi* spatial practices. However, in times of relative stability – since the Lebanese parliamentary elections in 2009, for example, which led to a consensual government supported by a spectrum of international powers – religiosity takes a subdued role in geopolitics. It then translates nationally and locally into political pragmatism, allowing spatial interactions and territorial appropriations across boundaries and neighbourhoods.

Second, the interventions of Hezbollah in Beirut uncover how spatial strategies of fundamentalist groups do *not* always yield added religiosity and territoriality. In specific sections of al-Dahiye, the spatial strategies analyzed above are not operative. For example, in the newly urbanized high-end neighbourhoods, Hezbollah is not really keen on spatial *quadrillage*, or providing a central core to neighbourhoods, planning for service delivery, or making space with religious symbolizations. A possible hypothesis, which needs further research, is that Hezbollah's territoriality and religiosity vary in relation to social class and urban density. The services provided to higher-income groups seem to require less religiosity and territoriality, and to rely more on a neoliberal market logic, detached from the piety and the overt inscription of socio-cultural norms. Indeed, in these neighbourhoods, labelled 'new Dahiye', public and spatialized forms of religiosity seem to become a subjective matter, lived individually in private spaces rather than expressed publicly as a territorialized collective consciousness. The case is drastically different in middle- to low-income areas, where urban density is high. Here, religiosity and territoriality reinforce each other and create relatively enclosed environments where the identification of self and collective is more whole.

Thus, the analysis of Hezbollah's religiosity, as inscribed spatially in Beirut, not only reveals enduring territorialization effects for the religious, but also the deterritorialization impacts engendered by modernity, mobility, and new forms of communication. As Hervieu-Léger (2002, pp. 103–105) argues, it is within this structural tension between territorialization and deterritorialization – which physically materializes the tension 'between the universal (religion without borders) and the particular (local religion)', and within 'the formation of individual and collective identities in modernity' – that the relationship between religion and space needs to be theorized. Indeed, as shown above, Hezbollah's territorialization strategies are accompanied by deterritorialization mechanisms through which the religious becomes diluted by market logic, or by sectarian politics, or by the dynamics of changing strategic geopolitical alliances.

Notes

1 Other religious fundamentalist groups in Lebanon are active, but have less influence relatively than Hezbollah. Labaki (2006, pp. 9–10) highlights Sunni fundamentalism, which is active via al-Jama'a al-Islamiyya, the Lebanese Branch of the Muslim Brotherhood; and Al-Ahbash, a group established by Sheikh al-Habashi, an Ethiopian cleric living in Beirut. Both groups have parliament representatives. He also mentions the rise of Christian fundamentalism – namely, Baptist and

born-again groups in the Protestant sphere and Catholic groups supported by Spain, Italy, and the countries of Latin America. The main Christian fundamentalist group on today's political scene, however, is the Lebanese Forces, although their religious activities have been reduced greatly since the regime of Bachir Gemayel (Hage, 1992).

2 The Shi'i duodecimal doctrine 'admits a lineage of twelve imams, the last being al-Mahdî, or the Hidden Imam, awaited for since 941' (Mervin, 2007, p. 465). According to Imam Khomeyni, 'in the absence of the imam, the *faqîh* [the theologian-jurist] has the legitimate power of deciding in his name, on religious and spiritual issues as well as on political affairs' (Mervin, 2008, p. 208). On the paradoxes of this doctrine in Lebanon and on the doctrine that follows its application in Iran following the death of Khomeyni, see *Ibid.*, pp. 209–210.

3 The *marja'* is 'the reference to follow with respect to religious precepts and is then a supreme religious authority. The *marja'* has to be *mujtahid* and considered as more knowledgeable by his peers so he can emerge as such and attract adepts that will emulate his advice and pay him religious dues' (Mervin, 2007, pp. 465–467).

4 General Michel Aoun, leader of the Free Patriotic Movement, is a native of Haret Hreik. Following the civil war, Christians did not return to their neighbourhoods; most sold their properties during the war and purchased new homes in the northeastern suburbs, and their children also live elsewhere and have no desire to return. However, some families still attend Sunday mass in the restored churches of Haret Hreik and Mrayjeh and visit their dead in the cemeteries nearby. The municipal councils of both neighbourhoods are also still led by Maronite mayors – a fact Hezbollah often likes to highlight to demonstrate how it values tolerance and coexistence.

5 The term *'assabiyya* comes from the writings of Ibn Khaldûn, and has been translated by 'group feeling' or '*esprit de corps*'. I use it here with reference to its territorial dimension, as best analyzed by Seurat (1985) in the case of Tripoli's neighbourhood of Bâb Tebbané.

6 Shi'i faith allows believers to select their religious reference (*marja'*). In Lebanon, pious Shi'i choose to emulate either Sayyid Fadlallah, the Lebanese cleric based in the southern suburbs; Sayyid Khamenei, the Iranian cleric whose local representative is Sayyed Hassan Nasrallah; or Sayyed al-Sistani, the Iraqi cleric who also has a local representative. No information is available about the size of each group. It is commonly believed that Sayyid Fadlallah has the most followers, as Shi'i appreciate the Lebanese dimension of his piety and relate more to his progressive religious edicts. The Islamic sphere discussed here predominantly groups Fadlallah's and Nasrallah's followers, who despite some noticeable religious differences, are politically compatible. Other believers do not emulate particular clerics, and follow the Quran directly to practice their faith.

7 During Rafsanjani's rule in the early 1990s, restraints on financial support to Hezbollah were reported by several authors. See Chehabi, 2006, pp. 228–230.

8 We have directly encountered this propensity in the case of the partnership we attempted to build at the Reconstruction Unit of the American University of Beirut with Hezbollah in the aftermath of the Israeli war on Lebanon in 2006. In the case of the reconstruction of Haret Hreik, we clearly experienced the strict boundaries put up by the Party's leadership. When we reached a point they considered to be too close for comfort they asked us politely to withdraw. On this experience, see Harb and Fawaz, 2010.

9 This association is mostly based on anecdotal evidence gathered from our continuous fieldwork in al-Dahiye over the past thirteen years.

10 This paper was presented during a discussion entitled 'The Socio-Politics of Reconstruction', organized by the Sociology Café of the Department of Sociology, American University of Beirut, 30 October 2006.
11 Interview, CCSD vice-president, 10 November 2006.
12 This choice was explained to us during several meetings with Waa'd representatives during our visits to their offices in Fall 2007. It was also clearly noted in newspaper articles published by one of the lead architects of the project, Rahif Fayyad.
13 Interview, Ghobeyri mayor, 6 May 2008. For him, urban modernity is about this large-scale educational and recreational complex, which he calls 'a small UNESCO', and about the beautification projects along new highways and streets. But mostly it is about these foreign and local private investments that demonstrate that Ghobeyri has the status of a market economy.
14 Interview, Information Unit director, 10 July 2004.
15 *Anashid* is the name given to songs which are deemed morally acceptable by Hezbollah and more generally by the Islamic milieu. These songs can be emotional, but need to reflect a certain morality and taste.
16 Most slogans are inspired by the Quran or the *hadith*, and/or make direct reference to Shi'i religious history or to more recent political events involving the Resistance's achievements. Colours used reflect those of Hezbollah's flag (yellow and green) and of the Lebanese flag (red, green, white). Imagery refers to military resistance and the protection of land/nation and its people.
17 Interview, Information Unit director, 10 July 2004.

References

Binder, L. (1966) Political change in Lebanon, in Binder, L. (ed.) *Politics in Lebanon*. New York: John Wiley.

Charara, W. (1996) *Al-oumma al-qaliqa. Al-'âmiliyyoun wal 'assabiyya al-'âmiliyya 'ala 'atabat al-dawla al-Loubnâniyya* [The unsettled nation: sectarianism and the Lebanese state]. Beirut: Dar an-Nahar.

Chehabi, H.E. (2006) Iran and Lebanon after Khomeyni, in Chehabi, H.E. (ed.) *Distant Relations: Iran and Lebanon in the Last 500 Years*. London: IB Tauris and Centre for Lebanese Studies.

Deeb, L. (2006) *An Enchanted Modern: Gender and Public Piety in Shi'i Lebanon*. Princeton, NJ: Princeton University Press.

Deeb, L. (2008) Exhibiting the 'just-lived past': Hizbullah's nationalist narratives in transnational political context. *Comparative Culture, Society and History*, **50**(2), pp. 369–399.

Deeb, L. and Harb, M. (2007) Sanctioned pleasures: youth, piety and leisure in Beirut. *Middle East Report*, No. 245, pp. 12–19.

DeMarrais, E., Castillo, L.J. and Earle, T. (1996) Ideology, materialization, and power strategies. *Current Anthropology*, **37**(1), pp. 15–31.

Denoeux, G. (1993) *Urban Unrest in the Middle East: A Comparative Study of Informal Networks in Egypt, Iran and Lebanon*. Albany, NY: State University of New York Press.

DGU (Direction générale de l'urbanisme) (1973) *Livre Blanc – Beyrouth 1985–2000*. Beirut: Lebanese Republic.

Early, E. (1971) The 'Amliya Society of Beirut: A Case Study of an Emerging Urban Za'îm. Unpublished Masters thesis in Middle-Eastern Studies, American University of Beirut.

Emerson, M. and Hartman, D. (2006) The rise of religious fundamentalism. *Annual Review of Sociology*, **32**, pp. 127–144.

Fawaz, M. (2004) Action et idéologie dans les services: ONG Islamiques dans la banlieue sud de Beyrouth [Action and ideology in service provision: Islamic NGOs in the suburbs of southern Beirut], in Ben Nefissa, S. *et al.* (eds.), *ONG et Gouvernance dans le Monde Arabe* [NGOs and Governance in the Arab World]. Paris: Karthala and Cedej.

Genieys, W. *et al.* (2000) Le pouvoir local en débats. Pour une sociologie du rapport entre leadership et territoire [Local power in debates: towards a sociology of the relationship between leadership and territory]. *Pôle Sud*, **13**(3), pp. 103–110.

Gole, N. (2000) Snapshots of Islamic modernities. *Daedelus*, **129**, pp. 91–117.

Hage, G. (1992) Religious fundamentalism as a political strategy: the evolution of the Lebanese forces' religious discourse during the Lebanese Civil War. *Critique of Anthropology*, **12**(1), pp. 27–45.

Hamzeh, N. (1993) Lebanon's Hizbullah: from Islamic revolution to parliamentary accommodation. *Third World Quarterly*, **14**(2), pp. 321–337.

Harb, M. (2003) 'La *Dâhiye* de Beyrouth: parcours d'une stigmatisation urbaine, consolidation d'un territoire politique [Dahiye of Beirut: the process of an urban stigmatization, consolidation of a political territory]. *Genèses*, **51**, pp. 70–91.

Harb, M. (2009) La gestion du local par les municipalités du Hezbollah [Local development by the Hezbollah municipalities]. *Critique Internationale*, **42**, pp. 57–72.

Harb, M. (forthcoming) *Hezbollah: de la banlieue à la ville* [Hezbollah: from the suburbs to the city]. Beirut-Paris: IFPO-Karthala.

Harb, M. and Fawaz, M. (2010) Influencing the politics of reconstruction of Haret Hreik, in Al-Harithy, H. (ed.) *Lessons in Post-War Reconstruction: Case Studies from Lebanon in the Aftermath of the 2006 War*. London: Routledge.

Harb, M. and Leenders, R. (2005) Know thou enemy: Hezbollah and the politics of perception. *Third World Quarterly*, **25**(5), pp. 173–198.

Hervieu-Léger, D. (2002) Space and religion: new approaches to religious spatiality in modernity. *International Journal of Urban and Regional Research*, **26**(1), pp. 99–105.

ICG (International Crisis Group) (2007) *Hezbollah and the Lebanese Crisis*. Middle-East Report No. 69.

Johnson, M. (1986) *Class and Client in Beirut: the Sunnite Muslim Community and the Lebanese State, 1840–1958*. London: Ithaca Press.

Karner, C. and Aldrige, A. (2004) Theorizing religion in a globalizing world. *International Journal of Politics, Culture and Society*, **18**(1), pp. 5–32.

Khalaf, S. (1968) Primordial ties and politics in Lebanon. *Middle East Studies*, **4**, pp. 247–263.

Labaki, B. (2006) Consequences of the decline of Bandung dynamics: extension of globalization and religious fundamentalism in the Middle East, the Arab World and Lebanon. Available at www.bandungspirit.org/labaki2006.pdf. Accessed 19 July 2009.

Leichtman, M. (2007) Shiite Lebanese migrants and Senegalese converts in Dakar, in Mervin, S. (ed.) *Les mondes chiites et l'Iran* [Shi'a Worlds and Iran]. Paris-Beirut: Karthala-Ifpo.

Marin, B. (ed.) (1990) *Generalized Political Exchange. Antagonistic Cooperation and Integrated Policy Circuits*. Frankfurt/Boulder, CO: Campus/Westview.

Marty, M.E. and Appleby, R.S. (1993) *Fundamentalisms and Society*. Chicago, IL: University of Chicago Press.

Mervin, S. (2000) *Un réformisme chiite. Ulémas et lettrés du Gabal 'Amil (actuel Liban-sud) de la fin de l'Empire ottoman à l'indépendance du Liban* [Shi'a Reformism. Ulemas and the erudite from Gabal 'Amil (the present South Lebanon) from the end of the Ottoman Empire to the Independence of Lebanon]. Paris-Beirut-Damascus: Karthala-Cermoc-Ifead.

Mervin, S. (ed.) (2007) *Les mondes chiites et l'Iran* [Shi'a Worlds and Iran]. Paris-Beirut: Karthala-Ifpo.

Mervin, S. (2008) La guidance du théologien-juriste (*wilayât al-faqîh*): de la théorie à la pratique, in Mervin, S. (ed.) *Hezbollah: Etat des Lieux* [The guidance of the theologian-jurist (*wilayât al-faqîh*): from theory to practice]. Paris: Sindbad.

Nasr, S. (1985) La transition des chiites vers Beyrouth: mutations sociales et mobilisation communautaire à la veille de 1975 [Shi'a transition towards Beirut: social change and community mobilization on the eve of 1975], in *Mouvements communautaires et espaces urbains au Machreq* [Community Movements and Urban Spaces in Machreq]. Beirut: Cermoc.

Norton, A.R. (1987) *Amal and the Shi'a; Struggle for the Soul of Lebanon.* Austin, TX: University of Texas Press.

Norton, A.R. (1998) Hezbollah: from radicalism to pragmatism? *Middle-East Policy*, **5**(4), pp. 147–158.

Picard, E. (1997) Le communautarisme politique et la question de la démocratie au Liban [Political communitarianism and the question of democracy in Lebanon]. *Revue Internationale de Politique Comparée*, **4**(3), pp. 639–656.

Picard, E. (1999) *The Demobilization of the Lebanese Militias.* Oxford: Centre for Lebanese Studies.

Qassem, N. (2002) *Hezbollah: al-manhaj, al-tajriba, al-mustaqbal* [Hezbollah: the path, the experiment, and the future]. Beirut: Dar al-Hadi.

Roncayolo, M. (1990) *La ville et ses territoires* [The City and Its Territories]. Paris: Gallimard.

Saad-Ghorayeb, A. (2002) *Hizbu'llah: Politics and Religion.* London: Pluto Press.

Seurat, M. (1985) Le quartier de Bâb Tebbâné à Tripoli (Liban): étude d'une 'asabiyya urbaine [The Bab Tebbane quarter in Tripoli (Lebanon): a study of urban social cohesion], in *Mouvements communautaires et espaces urbains au Machreq* [Community Movements and Urban Spaces in Machreq]. Beirut: Cermoc.

Tonkiss, F. (2005) *Space, the City, and Social Theory.* Cambridge: Polity Press.

Chapter 7

Hamas in Gaza Refugee Camps: The Construction of Trapped Spaces for the Survival of Fundamentalism

Francesca Giovannini

The resounding victory of the Islamic Resistance Movement, Hamas, in the Palestinian political elections of January 2006 was largely foreseeable. Data collected by Palestinian and international think tanks had shown a steady increase in support for its operations and leadership since the early 1990s. This had translated into political power in 2005 during the municipal elections in Gaza, in which 'Hamas candidates won 76 out of 118 seats up for election, nearly two thirds of the vote tally. In another round of voting in May 2005 a poll secured Hamas 33% of the vote' (Lybarger, 2007, p. 74).

Scholars and policy analysts have hinted that the political setbacks and administrative failures of the Palestinian Liberation Organization (PLO) were key to the success of Hamas and its rapid consolidation of power. Certainly, both the willingness of the PLO's leadership to accept a compromise solution with Israel and its inability to respond to the worsening economic and social conditions of the Palestinian population created a public perception of political opportunism and incompetence (Ayoob, 2008; Bowker, 2003; Levitt, 2006). But limiting analysis of Hamas's rise to the missteps of the PLO risks oversimplifying its historic achievement, which is tied to the complexity of the Palestinian struggle, as it is caught between secular-nationalist forces and Islamic-global aspirations.

Under Hamas's leadership, the efforts to Islamize Palestinian society, initiated in the 1960s by the local branch of the Egyptian Muslim Brotherhood, have expanded into a far more explicit political agenda which includes both secular demands for justice and freedom and Islamic goals such as the establishment of Shari'a law and the liberation of Jerusalem from the infidels. The accomplishment of Hamas, therefore, lies not only in its political triumph, but also in its ability to transform a seemingly nationalistic struggle for land and statehood into a conflict around religious fundamentalism and Islamic identity (Levitt, 2006).[1]

How has a relatively new organization like Hamas been able to 'blur the boundaries between a narrow territorial state (*dawla qutriyya*) and a broad Islamic nation (Umma)'? (Mishal and Sela, 2000, p. 32). Gellner argues that 'Nationalism is not the awakening of nations to self-consciousness, it is rather the invention of nations where they do not exist' (cited in Marty and Appleby, 1993, p. 622). According to this rationale, Hamas has been able to win over the minds and hearts of Palestinians by re-imagining the identity and the features of this yet-to-be Palestinian state. It has done so by situating and historizing the Palestinian struggle in the universal search for a renewed Islamic caliphate, and by providing for and empowering vulnerable social groups like the refugees in the Gaza Strip, who had been previously marginalized by the secular power system.

The spatial isolation typified by the refugee camps in Gaza captures simultaneously a symbolic and social fracture within Palestinian society, which has been largely ignored by secular parties like the PLO, but which has been skilfully exploited by Islamic Hamas (Abu-Amr, 1994; Lybarger, 2007). If the provision of humanitarian services to the refugees by Hamas has greatly improved living conditions in the camps, it has also gained the group preferential access to the refugees and their spaces. The camps, traditionally inward-looking and difficult to penetrate, have now been Islamized; as such, they have become the bedrock of Hamas's political and social support and a privileged base for recruitment into its *jihadist* operations. The indoctrination of the refugees has been facilitated by the territorial, spatial, and social features of the camps themselves. Separated from the rest of the urban environment, they provide an ideal place to experiment with, consolidate, actualize, and manifest Hamas's fundamentalist message.

The connubial partnership between Hamas and the camps, however, is more than simply functional; it has also become essential and existential for the political and social assertiveness of the refugees. 'From oppressed wanderers to fighters for freedom and Islam', Hamas's strategy has empowered and emboldened the refugees by providing new meaning and political scope to spaces previously considered depoliticized, hopeless no-man's lands.[2] In contrast to the PLO (and the Israelis), the Muslim Brotherhood first, and even more so now, Hamas, have both acknowledged and embraced the refugees as proper 'citizens' of a will-be Palestinian state on the basis of their political activism and social involvement in an Islamic and national cause.

In this respect, the camps have become both the main conduit for the dispersal of a fundamentalist Islamic message and the centre stage of Hamas's political agenda. As Michel Foucault claimed, 'space is fundamental in any exercise of power, and landscapes are reproduced as sites and outcomes of social, political and economic struggle' (quoted in Keith and Pile, 1993, p. 24). Thus, Hamas has created a new functionality for the camps within its Islamization project by attributing new ideological and physical meaning to them.

This chapter has two main parts. First, it situates Hamas within the fundamentalist discourse as outlined in the work of Marty and Appleby. Second, it illustrates the spatial and symbolic strategies undertaken by Hamas to propagate its fundamental ideology. The chapter will also discuss the meaning of 'being a refugee' in the context of the Palestinian *al-Nakba* or catastrophe, and explore the social and symbolic implications of being confined to a 'camp'. I will further claim that the identities of Hamas and the refugees are now intertwined and have become dependent on one another.

A final note of caution is necessary. While this chapter aims to analyze the surge to power of Hamas, it will not provide a comprehensive analysis of the root causes of the Israeli-Palestinian conflict; nor has it been written to assess the goodness or badness of Hamas's fundamentalism as an agent of the Palestinian quest for freedom and justice. For this reason, I will limit the discussion to Hamas as a fundamentalist organization, without delving into the complex debate around its potential terrorist nature.

Reinventing Palestine: Hamas in the Fundamentalist Discourse over State, Nation, and *Jihad*

> Humanity is standing today at the brink of an abyss, not because of the threat of annihilation hanging over its head but because humanity is bankrupt in the realm of 'values', those values which foster true human progress and development. This is abundantly clear to the western world for the west can no longer provide these values necessary for the flourishing of humanity.
>
> Sayyid Qu'tb (1988, 1964) *Ma'alim fi al-Tariq* [Signposts along the Road]

The current identification of fundamentalism and political Islam, as suggested by the oversimplified hyper-media explanation of 9/11, is faulty.

As a phenomenon, fundamentalism is certainly not new, and the debate around it should seek to reinstate its relevance and restore a sense of its complexity. Yet it should also be noted from the onset that my approach to it is largely based on what Euben calls 'rationalist epistemology', whereby a rational, logical, and positivist framework is employed to shed light on questions in an academic context. Approaching fundamentalism is challenging, therefore, because this rationalist approach is self-limiting. It favours a logical causal explanation for a phenomenon that is instead largely 'guided by and defined in terms of belief in divine truth unknowable by purely human means' (Euben, 1999, p. 4).

The term itself was created at the beginning of the twentieth century to capture the essence of a radical movement within American Protestantism that opposed progressive change (Bruce, 2008). In that context, fundamentalism was conceived as a way of encouraging a communal rediscovery of the original tenets of faith and their concrete application to the material world (Euben, 1999). According to Marty and Appleby (1993, p. 3), the editors of the monumental Fundamentalism Project, religious fundamentalism can be defined as 'a tendency, a habit of mind, found within religious communities which manifests itself as a strategy by which beleaguered believers attempt to preserve their distinctive identity as a people or as a group'. Thus, fundamentalism is an intellectual and emotional framework through which human socialization is filtered and interpreted (Antoun, 2001). Its ultimate goal is to halt the decline of religious values and to challenge and undermine the secular structure by disseminating faith-based principles as an alternative to lay ones (Herriot, 2007).

Although cultural and geographical differences should be considered in the analysis of individual fundamentalist movements, Marty and Appleby contend that they commonly share four major characteristics. Fundamentalist organizations tend to display a Manichean worldview, a zero-sum understanding that favours a dichotomist approach to reality, where the right and wrong sides to any issue are easily discernable. They also refer to 'scriptural texts' and adopt sacred books as superior sources of knowledge, after which behaviour ought to be modelled and shaped. In this way, they advocate that the source of rightness is located in the ancient past, and not yet compromised by the advent of modernity. Thus, the fight that fundamentalists wage against 'modernity' is seen in messianic terms, in the anticipation of God's rematerialization on earth (Herriot, 2007). Fundamentalist movements, however, are also pragmatic. They do not reject modernity as a whole, but rather choose to engage with and retain those parts of it that are useful to the successful accomplishment of their purposes. In the words of Marty and

Appleby (1995, p. 405), 'Fundamentalism is reactive to and defensive towards the processes and consequences of secularization and modernization'. But it is also proactive, because it is 'shaped both by the perceived threats to the tradition and by the nature of opportunities to resist these threats' (Bruce, 2008, p. 13). In this sense, fundamentalism sits ideologically at the juncture of the present and the future, and it is both conservative and revolutionary (Marty and Appleby, 1995).

When applied to the Islamic context, the debate around fundamentalism appears even more complex. Across time and space, scholars have tended to use the term in reference to so-called political Islam. As such, it describes 'the growth of Islam as a religious force and a political ideology seeking to reinstate the Islamic legal code' (Milton-Edwards, 2005, p. 7). But the term has also been used generally to portray a range of emotional anti-Western feelings, and a vaguely defined Arab malaise.

Scholars have also argued that Islamic fundamentalism stands out among fundamentalist movements because of its distinctive features. First among these is that Islamist movements are not exclusively religious, but rather encompass a 'policy of believers' (Weinberg and Pedahzur, 2004, p. 14). Islamic fundamentalism proactively seeks to recentralize the role of religion not only in the abstract but through a pragmatic political strategy that calls for the rejection of 'statehood' and the reintroduction of the caliphate and the community of believers (*Umma*). Al-Azmeh noted, for example, that Islamists and nationalists share the desire for authenticity, defined as an 'idiom by which the historical world is reduced to a particular order of alternating periods of decadence and health. Thus, for authentic Islamic and authentic secular nationalists, the cure for the current illness in the Arab world is a return to the glorious days of the Islamic Empire' (quoted in Shehada, 2004, p. 5). Finally, Islamic fundamentalism asserts the need to undertake this major political restructuring of society to restore the past golden age of Islam by embarking on major revolutions against modernity (through reform or revolution) within individual Islamic countries (Ayoob, 2008; Moussalli, 1999).

Hassan al-Banna, the founder of the Muslim Brotherhood in Egypt in 1928, contributed greatly to the advancement of the agenda of political Islam, which he saw as an alternative to the self-destructive qualities of Westernization and modernization. The return to the Islamic fundamentals has to be sought, in al-Banna's ideal, through an incremental strategy, by instilling gradual changes within Muslim societies via individual personal transformation. His successor, Sayyid Qu'tb, however, embraced a far more revolutionary and radical agenda (Knudsen, 2005). This favoured a cognitive model based on an absolute submission to God's will in place of a system of rationality and positivism (Moussalli, 1999). According to this view, reason is imperfect, unreliable, and misleading. Only by recognizing 'the limits of reason [are we] prepared to acknowledge God's truths as self-evident' (cited in Euben, 1999, p. 71).

For Qu'tb, as for other fundamentalist thinkers, the goal of society is to ensure happiness based on the idea of unity with Allah, in harmony with nature. The 'oneness' of the human life with the divine, captured by the word *tawhid*, rests on the idea that reason (*fitra*) is achievable only through total devotion to God's will and ruling. But such divine governance (*hakimiyya*), Qu'tb argued, could only be achieved through active social involvement, or *jihad*. In the writings of Qu'tb, this comes like a 'liberating force that both sets humans free and enables them to help bring about the kingdom of god on earth' (cited in Euben, 1999, p. 71). *Jihad*, however, requires social and intellectual maturity and a personal and collective transformation of identity; it cannot be executed without appropriate preparation both spiritually and practically. For this reason, Qu'tb suggests *hijra*, or 'separation', from the materiality of the surrounding world. This would allow believers to 'purify the consciousness of foreign values and then reengage society through missionary outreach, and when the moment [is] right join in outright revolution' (Lybarger, 2007, p. 77).

The Islamic Resistance Movement, Hamas, has been undeniably influenced by the intellectual work of Qu'tb and the *modus operandi* of the Muslim Brotherhood. A few scholars still question its fundamentalist nature because of its pragmatism and its skilful use of contemporary media and Western technology (Mishal and Sela, 2000).[3] However, if, in the words of Marty and Appleby (1995, p. 405), 'to qualify as a genuine fundamentalism, a movement must be concerned first with the erosion of religion and its proper role in society', there is no doubt that Hamas should be studied as such.

Hamas was first established in 1987 by Sheikh Yassin, a refugee from Gaza, widely known as a member of the Muslim Brotherhood and the founder of the Islamic Center, the oldest Islamic organization in Gaza. In the rhetoric of its leaders, Hamas's creation was, *per se*, God's will, because 'when all doors are sealed, Allah opens a gate' (Abu-Amr, 1994, p. 66). And through its social activism, the organization has always explicitly pursued the correct teaching of Islam, the transformation of Palestinian society from secular (and corruptible) to Islamic, and the establishment of an Islamic state on Palestinian land based on the supremacy of Shari'a law (Gunning, 2007).

The charter of Hamas itself, released in 1998, clarifies the Islamic-fundamentalist nature of the organization. And while the first article grounds the work, strategy, and programme of the movement in Islam, the second acknowledges the profound ties with the Organization of the Muslim Brotherhood. The reference to the organization of al-Banna is more than symbolic. The original goal of the Muslim Brotherhood in Gaza was, in fact, religious, not political; its work was deemed necessary because of the 'ignorance and lack of commitment to Islam showed by the Palestinians and considered (by the fundamentalists) the greatest challenge to the survival of their community' (Tamimi, 2007, p. 14).

The perception that Islam is under threat by secular forces and modernity is also captured in article 9 of the Hamas charter, which states that 'The Islamic resistance movement found itself at a time when Islam has disappeared from life. Thus when Islam is absent from the arena, everything changes'. The article thus empowers Hamas to fight for the reestablishment of an institutional order based on Islamic precepts. As such, it typifies the Manichean worldview of Hamas, in which the world of peace is solely achievable by the establishment of an Islamic state (Nyroos, 2001). Palestine is portrayed in the charter as an Islamic *waqf* (endowment) 'consecrated for future Muslim generations until judgment day'. In Muslim practice, such endowments are specific places devoted to religious activities, like mosques, Quranic schools, or cultivated land to feed the poorest (*Ibid*.).[4] Therefore, the act of negotiating and compromising away part of the Palestinian land would not only be a political failure but, even worse, a gross, intolerable violation of its Islamic purpose.

By conferring Islamic meaning on Palestinian soil, Hamas has succeeded in strengthening and unifying both elements of secularism (the right to a proper state and the liberation of the usurped land) and Islamism (the spread of Islamic tenets and the establishment of an Islam-based statehood) within a nationalist framework. As noted by Marty and Appleby, fundamentalists share with secular nationalists the need to define the nation, but they do so by referring to an ancient past of religious purity and sanctity. In fundamentalist discourse, 'this land of purity may coincide with, exist within or transcend existing national boundaries' (Marty and Appleby, 1995, p. 622). For Hamas, this entails three different levels of national self-awareness: Palestinian, Arab, and Muslim (article 14). And although the charter acknowledges that all three levels of struggle and resistance are pivotal in the fight against the 'Zionists', there is minimal doubt that Hamas considers the Islamic circle the most important. Given that Palestine contains one of the most sacred Muslim shrines, the fight for its liberation is inevitably 'religious' (article 15) and a 'universal responsibility for all Muslims around the world' (article 7). The reference to the responsibility of every Muslim in every nation to fight for the freedom of Palestine is accompanied by the idea of a total *jihad* (as in the writing of Qu'tb) which could awake the masses, gather support (article 15), and finally defeat the infidels.

The Construction of Fundamentalist Spaces in the Gaza Strip: Hamas's Islamization Strategy in the Refugee Camps

The most evident (and successful) attempts carried out by Hamas to Islamize the Palestinian land have been conducted in the Gaza Strip, particularly within its eight refugee camps.

It should be mentioned that even before the advent of a Hamas-led government, both Palestinians and foreigners referred to Gaza as 'conservative, traditional and religious' (Lybarger, 2007, p. 179). Thus, the massive presence of Islamic organizations is, *per se*, not a new phenomenon.[5] But in recent years, a growing number of analysts have denounced openly the systematic, massive, and explicit efforts at 'Talibanization' led by Hamas in the Strip. Reports speak about forced Islamic conversions of intellectuals and scholars.[6] And they note the growing anxiety among Christians there of possible violent discrimination and religious intolerance. In 2007, Israeli television, in an attempt to draw an explicit comparison with Kabul under the fearful Taliban regime, pointed out the increasing number of bearded men in Gaza, and mentioned that the new moral code enforced by the Islamic Movement in public places was a radical departure from the libertarian management of the PLO. In a series of public events, PLO President Abu-Mazen has also referred to Gaza as 'an Islamic emirate', and pointed to the danger of melding religious zeal with government authority.[7] After all, in the immediate aftermath of its historic victory, Hamas had itself declared that 'secularism is in fact over' in Gaza.

The religious zeal displayed by Hamas in the course of its ascent to power has found fertile soil in the confinement and relative isolation of the refugee camps and in their patriarchal internal organization. As noted by Levitt (2006, p. 107), 'Palestinian refugees live in an environment that by its very nature creates social preconditions that Hamas is able to use to its advantage in the radicalization campaign'. Thus, while the creation of an Islamic state will involve the entirety of the Palestinian society, the connection between Hamas and the refugee camps has become pivotal in the recruitment of *jihad* fighters and the broader Islamization efforts of the organization.

The United Nations Agency for the Relief of the Palestinian Refugees (UNRWA) estimates that 853,000 Palestinians, 78.4 per cent of the 1.2 million people residing in Gaza, are refugees living in camps (Lindholm-Schulz, 2003, p. 69).[8] Originally built as temporary accommodation, these have now become permanent 'villages or slum-shanty growths of uncontrolled urban settlements' that still lack the facilities of proper cities (Rowley, 1977, p. 86). Poverty is structural due to an absent job market and almost nonexistent access to running water.[9] Sporadic electricity provision has also turned these spaces into alienating suburban environments, where violations of human rights are the daily routine.

Scholars have often examined how the social services provided by Hamas grant it a solid constituency within the refugee camps. In fact, the ideological discourse of the Hamas charity network is based on the Islamic principle of *da'wa* ('a call to God'), which stipulates that the correct faith has to be grounded on 'outright proselytizing, charitable giving and social welfare activities' (Levitt, 2006, p. 16). A recent investigation by the U.S. Treasury Department (2001, p. 5) also confirmed that, while 'Hamas has increased its influence in the refugee camps by first providing basic necessities to needy families', it has also carefully balanced and alternated religious indoctrination with humanitarian assistance.

The 'Islamization' of the refugee camps revolves around three major strategies: the creation of Islamic spaces and symbols; a skilful and methodical use of these religious places for humanitarian purposes; and the manipulation of vulnerable groups such as women and children as the main conduit and primary recipients of religious indoctrination.

At the material level, the construction or enlargement of mosques is the most visible sign of this fundamentalist project.[10] Hakim (1986), for example, noticed that Islamic symbolism and the presence of the mosque should be read as the most important elements characterizing the 'Islamic city'. AlSayyad (1991) has pointed out that the symbolism of the mosque has always been strategically used and manipulated by the political elite to highlight or underplay the importance of religion at specific historical junctures.[11] The sanctity of the mosque and its social functionality has been particularly relevant in the case of the Palestinian struggle for statehood. Not only does the mosque, together with its related endowment, receive money from charity collection (*zakat*), and is able to own land and property, but, 'differently from other institutions, it remains open all the time, becoming a sanctuary that can be used as a place for political work and organization away from the eyes and the interference of the Israeli authority' (Abu-Amr, 1994, p. 15). So the construction of mosques within

Palestinian refugee camps is meant both to promote the sacralization of the human habitat and to continue resistance to the presence of the Israeli occupation (Arkoun, 1982).

Moreover, through the skilful use of these religious spaces for humanitarian purposes, Hamas has transformed the mosques into islands of care and protection for the vulnerable. In this sense, Islam presents itself as the solution to human grievances and the defender of human dignity against the indifference of secularism. The organization of recreational activities in the proximity of the mosques (such as tournaments, musical events, and youth educational programmes) and the frequent provision of basic services (like medicine, clothing, and food) in mosque courtyard projects and amplifies the fundamentalist message that only through an Islamic order, will peace (and justice) be attainable.

The third pillar of Hamas's strategy lies in the extensive involvement of vulnerable groups in its Islamization efforts. In Hamas's rhetoric, these groups are protected and empowered within Islam. The delivery of its services, in accordance with Islamic tenets and driven by a religious agenda, is thus directed particularly towards children, youth, and women (Levitt, 2006). Kindergartens, for example, one of the largest operations of Hamas, are a primary vehicle for indoctrination and relief. Their curriculum is certainly controversial; in terms of geography, history, and culture, children study only material provided by the charity organizations, which is charged with political and religious meaning.[12] They also learn martyr's songs, which they have to perform during the final-year ceremonies together with key verses from the Quran. And they are involved in theatrical plays which usually depict the ultimate victory of Islam over evil, represented by the Jewish state. The kindergartens, however, also provide food and health services to the children and their families. The effective intermingling of religious and political manipulation is also present in higher-education curricula, and has already delivered remarkable results. According to Levitt (2004), 'A survey conducted by the Islamic University of Gaza reported that while 49% of young people aged fourteen to sixteen residing in the camps claimed to have participated in the intifada, 73% have claimed they hope to become martyrs'.

The charter of Hamas also assigns a specific role to women and young girls, and redefines gender empowerment and motherhood in terms of Islamic *jihad*. Article 18 states, 'The women in the houses and in the families of the Jihad fighters, whether they are mothers or sisters, carry out the most important duty of caring for the home and raising the children upon the moral concepts and values which derive from Islam. Therefore we must pay attention to the schools and curricula upon which Muslims girls are educated' (cited in Israeli, 2004, p. 82). Great attention has, indeed, been devoted to girls' education, the mandatory wearing of the headscarf, and the banishment of un-Islamic behaviour in public places. But the wearing of the headscarf by women in public places, originally intended to protect women's integrity, is not a mechanism of disempowerment; rather it is a symbol of truthfulness and commitment to both Islamic pride and the Palestinian struggle (Shehada, 2004).

The empowerment of vulnerable groups in refugee camps through an Islamic path goes even deeper. While the PLO carried out its humanitarian operations from the main headquarters in Ramallah, Hamas has decentralized most of its charity programmes by opening local offices in the main refugee camps. It has also recruited people from the refugee community to run and manage the delivery of services within the camps, thus minimizing the sense of disempowerment that the traditional logic of refugee camps generates through dichotomist divisions between providers and beneficiaries (Levitt, 2004, 2006). Furthermore, the involvement of the refugees in Hamas's welfare programmes is not limited to operational aspects; it is also sought in decision-making. Major renovation projects, for example, such as the construction or the substitution of water pipelines or the remodelling of schools and health centres, are done in accordance with popular will, usually expressed through a council of clan leaders representative of the camp constituency (Abu-Amr, 1994).

Thus, the provision of services, combined with the frequent manipulation of religious symbols, has effectively improved the living conditions of the population while also injecting Islamic principles into a previously secular context. Melucci (1995) talks about collective action as a process of identity-building, where a sense of belonging to a specific group (or community) is generated by specific rituals and behaviours, which also help sustain and reinforce the group. In this regard, Hamas has exploited the difficult humanitarian circumstances of the refugee camps to create a specific community of political supporters and religious believers, indoctrinated into Islamic precepts but also bound to Hamas via its provision of services to meet their humanitarian needs.

Moreover, this community has been selected by Hamas on the basis of its specific spatial and social configurations. The separation of the refugees from the rest of Palestinian society constitutes an advantage for Hamas for several reasons. It generates social cohesiveness on the basis of the communal experience of discrimination, structural violence, and alienation from the rest of the Palestinian community. It also reinforces political and economic dependence on the provision of aid from an external agency. But, more importantly, it facilitates the spread of religious zeal based on the fundamentalist ideals of Qu'tb. If the severance from the material world (*hijra*) has to take place for believers to purify themselves, the structural conditions refugees face already condemn them to an unchosen separation, and ideally place them on the right path to undertake a conversion to the precepts of Islam.[13]

Re-imagining Identities and Citizenship: The Reinforcement of Fundamentalism and the Generation of Trapped Spaces

If the presence of Hamas in the camps represents a strategic decision by the organization, it has certainly become essential and self-defining for the refugees. Under the current Constitution of the International Refugee Organization, a refugee is a person who 'owing to well-founded fear … is outside the country of his nationality…'.[14] Such a definition emphasizes the pivotal connection between 'being a refugee' and the spatial idea of 'being (physically) outside the country of nationality'. The word 'outside' here captures more than the simple geographical location of being displaced, because it finely reconceptualizes the image of 'people outside' in both ontological and epistemological terms.

The presence of an 'outside space' means the existence of a demarcating line between a space 'in' and an unknown space 'out'. The geographical location of being outside the once-familiar borders of the once-acknowledged homeland undermines the sense of 'being who one was' and erodes 'trust' both in the current conditions and in the future (Knudsen and Daniel, 1995). Hence, being a refugee symbolizes a 'radical disjunction between this person's familiar way of being in the world and a new reality of the sociopolitical circumstances that not only threatens that way of being but also forces one to see the world differently' (*Ibid*., p. 1).

The physical manifestation of this transition from a traditional past to an unknown present and future is also captured by the existence of 'a camp', a predisposed space demarcated by territoriality and a border and fundamentally stateless and a-national. This is where refugees must search for meaning for their new identity, while struggling to keep their past one alive (Stedman and Tanner, 2003).

Camps also signify victimhood, vulnerability, and dependence on outsiders' altruism and generosity (Knudsen and Daniel, 1995). The organization and the functioning of the camp, together with the basic and more vital provision of food and medicine, clothing, and shelter, are administrated by 'others', whose existence becomes crucial for the survival of the refugees themselves. Thus, living in a camp also becomes associated with the idea of absolute disempowerment for the refugees, who are at the mercy of external aid and generosity (Rushdie, 1992).

The symbolism associated with the depersonified space of the camp and the physical separation of the refugees from the rest of the community is particularly evident in the Palestinian case. At a social level, Palestinians have been marginalized and isolated by the native community, often treated as 'undesired guests' (Lindholm-Schulz, 2003). They have also been accused of being 'the losers' and ultimate cause of the creation of the Israeli state (Stedman and Tanner, 2003).

Differently from the refugee definition applied to any other case in the world, where return to a home country and re-appropriation of familiar space is plausible and foreseeable in a distant future, the refugee identity of Palestinians is intrinsically related to the idea of irremediable and permanent loss of both home and land in the aftermath of a conflict which, for its cataclysmic consequences, has been defined by the Palestinians as 'a catastrophe' (Rushdie 1992).

The forced renunciation of land and houses has been accompanied and inevitably intertwined with grief for the permanently compromised Palestinian identity. 'What happens to landless people? However you exist in the world, what do you preserve of yourself?' asks Edward Said, and as echoed by Siddiq who, in capturing the wandering Palestinian identity deprived of his homeland, has defined the Palestinians overall as a refugee nation (as cited in Rushdie, 1992, p. 171). For all these reasons, the social and ideological impact of al-Nakba, the Palestinian catastrophe, should not be reduced to a mere humanitarian crisis. Rather, it has to be explored as 'a profound and possibly irreparable disintegration of the Palestinian identity' (Knudsen and Daniel, 1995, p. 87).

In the collapse of the Palestinian identity and in the dissolution of Palestine, the creation of 'refugees' and their 'no-man spaces', the camps, have acquired several meanings, and have been accompanied by contradictory political and social positions. While the United Nations has always emphasized the temporary nature of the camps, called for many years just shelters (Marx, 1992, p. 283), the Israelis and their bureaucracy have permanently removed from the now-Israeli state both the presence of 'refugees' and their spaces. In fact, according to Amira Hass (1996, p. 180), 'If the card holder was born in the Gaza Strip, then the space for Place of Birth was filled in with the name of a specific town or village, such as Kgan Yunis or

Jabalia. But if the card holder was born within the borders of what had since become the new Israeli state, then only one word appeared in that space: Israel'. This has been accompanied by the change to Hebrew of the names of all Palestinian cities, together with the (forced) integration of Palestinians into the Jewish state. The maintenance of 'refugees' would, in fact, symbolize a 'right to return' home, where home has in fact been forever dismantled.

For Gazans, too, the presence of refugees and the camps has constituted, and still represents, a challenge. The so-called *muwataneen*, or the natives of Gaza, perceive them as an imposed space, not integrated with the rest of the society, and fundamentally a burden on an already overcrowded Strip.[15]

Over the years, trapped between a political stalemate which is incapable of returning them to their houses and a fundamental human need to preserve their identity at all cost against the corrosion of time and the alteration of space, the refugees in the camps have adopted a variety of survival strategies. In particular, they have developed specific norms, values, and beliefs that foster social unity and cohesiveness within the community, while distancing and protecting them from a distrusting society outside (Roy, 1986). More importantly, if their identity is located in memories of the past, the preservation of this identity has to be based on the revitalization of this past. Hence, the camps have also evolved into 'self-contained segregated communities that continue to reflect in broad terms the social structure of the pre-1948 society' (Bowker, 2003, p. 168). But this inward-looking approach chosen by camp residents to protect their historical identity has further hampered the social relations between the camps and their external environment. It has therefore forced them more and more into a spiral of self-referential narrative and social isolation.

The lack of an assigned social 'functionality' for the camps within the Palestinian struggle for land and freedom has, for many years, relegated them to limbo. In other words, they have become spaces where the residents simply survive in hope of a radical transformation of their outside surroundings, without being able to trigger that transformation themselves. The ascendance of Hamas, however, has radically changed not only the material landscape of the camps by improving living conditions in the areas themselves, but it has also, and more importantly, transformed these 'bare-life environments' into pivotal places from where to conduct the *jihad* for the ultimate liberation of the Palestinian land (Feldman, 2007).

As Fawaz Turki has argued, the transformation of the refugee from an oppressed wanderer to a fighter (*fida'i*) represents Hamas's greatest and most unprecedented success (cited in Feldman, 2007). It is inevitably related to a new understanding of 'Palestinian citizenship' in Gaza. If, for many years, being a citizen in Gaza was, in fact, associated with 'being born' in the Strip, in the new Hamas-led universal struggle for the establishment of an Islamic state in Palestine, a citizen in Gaza is someone politically active and religiously committed to the values and beliefs of Islam.

'Political values like *sumud* (steadfastness) and armed struggle' have therefore supplanted the traditional idea of citizenship based on nativity rights, and in this sense has blurred the boundaries between 'the native/local' and the 'refugee/outsider' (Feldman, 2007, p. 150). By inserting the Palestinian struggle into the global quest for an Islamic *Umma*, and by placing the refugees at the centre of this struggle, Hamas has been able to 'provide a proper political subjectivity that a normal refugee in a different part of the world would not enjoy' (Nyers, 1999, p. 22).

Conclusions

This concise account certainly shows that the Islamic indoctrination of Gaza's refugees has come to fruition and has provided Hamas with an invaluable base of support and a continuous flow of recruits to sustain the fight for liberation and justice in Palestine. While, in fact, Hamas's provision of services is obviously based on the expectation of returned favours and support from the beneficiaries, its presence in the camps should not be understood merely according to a logic of tit-for-tat that solely advantages Hamas.

The relation between the 'Palestinian refugees', who are a unique category of refugees, and Hamas can be described as a multilayered, complex, and dynamic cooperation, benefiting both parties. At the normative level, Hamas's Islamization of space has fundamentally altered the functionality of the camps. From powerless and marginalized spaces, they have emerged as centred places for Islamic propagation, thus granting to the refugee population a new social identity and a renovated political mission.

The skilful use of Islam as a vehicle of social cohesiveness has also allowed Hamas to generate a new concept of *Gemeinschaft*, or community, within the boundaries of the refugee camp – a protected and safe area nonetheless proactive in the reshaping of its environment. Ferdinant Tonnies has argued that *Gemeinschaft* – intended as a 'basic unit of organization characterized by depth, continuity, cohesion and fulfillment' (cited in Knox, 1995, p. 205) – has, however, in general been supplanted in modern times by *Gesellschaft*, a more advanced form of human organization based on positivist knowledge, individualism, and economic transactions.

For Palestinian refugees, however, 'being a refugee in the camp' represents 'a betweenness of place', as Knox puts it. Thus, Palestinians have suffered the loss of their community of origins, together with the traditions, values, and familiarity of space of the *Gemeinschaft*, without having any access to the economic, political, and social over-structures provided by the *Gesellschaft*. Trapped, in other words, between the logic of memories and a process of disempowerment, the refugee camp is a limbo of identity, temporarily constructed and deprived of meaning.

The advent of Hamas, however, and its strategies of Islamization have generated a new dynamic among the refugees; in short, they have filled that vacuum of identity by offering what Larissa Lomnitz calls 'the cultural grammar that codifies the social construction of spaces and places' (*Ibid.*). Islamic tenets become the ordering principles of the camp, upon which daily individual routine is structured, and also a common knowledge and shared set of norms and rules. In a space deprived of social meaning, the Islamization of space offers new cognitive lenses for interpreting reality. Even more importantly, it provides a new scope in which this space can survive, which rests on the propagation of Islam as its overall ultimate goal.

Lefebvre argues that 'space' is materialized by three different practices. Of these, material practice relates to physical interaction within a delineated space. But the second and third practices speak to the representation and symbolism of space through the adoption of common signs and codifications. These are then internalized into individual mental constructions.

It is important to notice that in the refugee camps Hamas has operated at all three levels by promoting and infusing Islamic principles both at the ideological and material levels. This indoctrination has taken place primarily, but not exclusively, at the material level with the building of mosques and religious schools and the recruitment and celebration of martyrs and suicide bombers. But at the symbolic and individual level, the design of educational curricula filled with religious precepts and Islamic doctrine, together with the slogans used during the electoral campaigns and strategic media manipulation, have also facilitated creation of a common knowledge based on the Islamic tenets, through which the reality is then understood and discussed.

The Islamic indoctrination has provided the refugees with a new symbolism and revitalized social identity. At the material level, it has developed a solid welfare state capable of supplying needs and necessities, while at the symbolic and individual levels it has provided a new social scope and sense of further individual empowerment. By connecting the personal struggle of the refugees with the fight to assert ultimate divine truth, it has encouraged the refugee perception of being 'a selected community of believers' for the propagation of Islam. It has thus demanded the contribution and the social involvement of every member of the community in the triumph of its Islamic call.

Notes

1 Article 12 of the Hamas charter, for example, states, 'Nationalism from the point of view of the Islamic resistance movement is part of the religious creed'. The symbolic identification between nationalism and Islamic resistance is a cornerstone of Hamas's strategy and rhetoric.

2 For years, the Israeli and Egyptian militaries considered the presence of refugee camps in Gaza to be a security threat; meanwhile, the PLO leadership mainly manipulated the refugees as a political bargaining chip and has essentially disregarded their humanitarian needs.

3 Doubts are also the result of a debate over whether a fundamentalist organization is also necessarily a terrorist one. This chapter defines Hamas as a fundamentalist organization but takes no position on whether it does or does not engage in acts of terrorism.

4 It should be noted, however, that the idea of Islamic endowments, or *waqf*, is not present in the Quran. It was introduced in the eighth century as a practice in Muslim societies.

5 Already in 2000, a survey revealed that '10 to 40 percent of all social institutions ... and 65 percent of all educational institutions below the secondary level are Islamic' (Roy, 2000).

6 As in the recent case of Professor Sana al-Sayegh, kidnapped in 2007 from the Palestine University of Gaza City.

7 Exemplary in this respect were the decisions of the administration of al-Saraya prison in Gaza to reduce the sentence of all inmates who learned at least five sections of the Quran. Islamic Legal Committees, overseen by Hamas members, have also been created to replace the district attorney's office in the management of the judicial system.

8 There is no universally accepted definition of 'a Palestinian refugee'. UNRWA, however, in order to respond to the rapidly unfolding humanitarian crisis, has generated an operational rather than legal definition. It characterizes a Palestinian refugee as 'any person whose normal place of residence was Palestine during the period 1 June 1946 to 15 May 1948 and who lost both home and means of livelihood as a result of the 1948 conflict' (Lindholm-Schulz, 2003, p. 69).

9 Collecting data on the social and economic conditions of the camps can be challenging. However, journalists and human-rights activists who have sojourned in the camps report that, 'In many places water flows through the pipes only six hours a day or less — brackish water in a weak stream and with a strong odor of chlorine. In 1996, only 27% of the camps' houses were connected to sewage systems, compared with 40% outside the camps' (Hass, 1996, p. 171).

10 Ziad Abu-Amr (1994) noticed that between 1967 and 1987, for example, the rise of Islamic influence has been demarcated also by the impressive increase in the number of mosques built both in the West Bank and Gaza.

11 While at the peak of Arab expansion, for example, the mosque was 'the first building to be laid out or designed in either a new town or an existing town', with the solidification of the power of the caliph, the mosque became a secondary, but nonetheless important, building. Very often it was not central to the sphere of power (and space) of the caliph (AlSayyad, 1991).
12 Levitt (2006) even mentions the use of instruction cards bearing the pictures of suicide bombers.
13 In the common narrative of Palestinian refugees in Gaza, *hijra* is a term used to describe the forced eviction from their houses and their villages. But, in broader Islamic terminology, it also refers to the pilgrimage of the prophet from Mecca to Medina.
14 The constitution of 1939 is the most recent attempt to define refugee status. It includes and enlists the previous international agreements contained in the Arrangements of 12 May 1926, and 30 June 1928, and further specified in the Conventions of 28 October 1933, and 10 February 1938. The quote is from the Convention of 1951, Article 1A (2).
15 Many Gaza families have proposed the relocation of the refugees into the West Bank. Many of the proposers are also owners of the land on which the refugee camps are sited (Hass, 1996).

References

Abu-Amr, Z. (1994) *Islamic Fundamentalism in the West Bank and Gaza.* Indianapolis, IN: Indiana University Press.

AlSayyad, N. (1991) *Cities and Caliphs: On the Genesis of Arab Muslim Urbanism.* Westport, CO: Greenwood Press.

Antoun, R. (2001) *Understanding Fundamentalism: Christian, Islamic and Jewish Movements.* Walnut Creek, CA: Altamira Press.

Arkoun, M. (1982) *Lectures du Coran.* Paris: Maisonneuve et Larose.

Ayoob, M. (2008) *The Many Faces of Political Islam.* Ann Arbor, MI: University of Michigan Press.

Bowker, R. (2003) *Palestinian Refugees: Mythology, Identity and the Search for Peace.* Boulder, CO: Lynne Rienner.

Bruce, S. (2008) *Fundamentalism.* Cambridge: Polity Press.

Euben, R. (1999) *Enemy in the Mirror: Islamic Fundamentalism and the Limits of Modern Rationalism.* Princeton, NJ: Princeton University Press.

Feldman, I. (2006) Home as a refrain: remembering and living displacement in Gaza. *History and Memory*, **18**(2), pp. 10–47.

Feldman, I. (2007) Difficult distinctions: refugee law, humanitarian practice, and political identification in Gaza. *Cultural Anthropology*, **22**(1), pp. 129–169.

Gunning, J. (2007) *Hamas in Politics: Democracy, Religion and Violence.* London: Hurst.

Hakim, B. (1986) *Arabic-Islamic Cities.* London: KPI.

Hass, A. (1996) *Drinking the Sea at Gaza: Days and Nights in a Land under Siege.* New York: Metropolitan Books, Henry Holt.

Herriot, P. (2007) *Religious Fundamentalism and Social Identity.* New York: Routledge.

Israeli, R. (2004) Palestinian women: the quest for a voice in the public sphere. *Terrorism and Political Violence*, **16**(1), pp. 66–96.

Keith, M. and Pile, S. (eds.) (1993) *Place and the Politics of Identity*, London, Routledge.

Khalidi, R. (1997) *Palestinian Identity: The Construction of Modern National Consciousness*. New York: Columbia University Press.

Knox, P. (1995) *Urban Social Geography: An Introduction*, 3rd ed. Harlow: Longman.

Knudsen, A. (2005) Crescent and sword: Hamas enigma. *Third World Quarterly*, **26**(8), pp. 1373–1388.

Knudsen, J. and Daniel V. (1995) *Mistrusting Refugees*. Berkeley, CA: University of California Press.

Levitt, M. (2004) Hamas from cradle to grave. *Middle East Quarterly*, **9**(1), pp. 3–15.

Levitt, M. (2006) *Hamas, Politics, Charity and Terrorism in the Service of the Jihad*. London: Yale University Press.

Lindholm-Schulz, H. (in cooperation with Hammer, J.) (2003) *The Palestinian Diaspora: Formation of Identities and Politics of Homeland*. New York: Routledge.

Lybarger, L. (2007) *Identity and Religion in Palestine: The Struggle between Islamism and Secularism in the Occupied Territories*. Princeton, NJ: Princeton University Press.

Marty, M. and Appleby, S. (1993) *Fundamentalism and the State: Remaking Polities, Economies and Militance*. Chicago, IL: The University of Chicago Press.

Marty, M. and Appleby, S. (1995) *Fundamentalisms Comprehended*. Chicago, IL: The University of Chicago Press.

Marx, E. (1992) Palestinian refugee camps in the West Bank and Gaza Strip. *Middle Eastern Studies*, **28**(2), pp. 281–294.

Melucci (1995) The process of collective identity, in Johnston, H. and Klandermans, B. (eds.) *Social Movements and Culture*. Minneapolis, MN: University of Minnesota Press.

Milton-Edwards, B. (2005) *Islamic Fundamentalism since 1945*. New York: Routledge.

Mishal, S. and Sela, A. (2000) *The Palestinian Hamas*. New York: Columbia University Press.

Moussalli, A. (1999) *Moderate and Radical Islamic Fundamentalism: The Quest for Modernity, Legitimacy and the Islamic State*. Gainesville, FL: University Press of Florida.

Nyers, P. (1999) Emergency or emerging identities? Refugees and the transformations in world order. *Millennium: Journal of International Studies*, **28**(1), pp. 1–26,

Nyroos, L. (2001) Religeopolitics: dissident geopolitics and the 'fundamentalism' of Hamas and Kach. *Geopolitics*, **6**(3), 135–157.

Qu'tb, Sayyid (1988, 1964) *Ma'alim fi al-Tariq* [Signposts along the Road]. Cairo: Dar al-Shuruq.

Rowley, G. (1977) Israel and the Palestinian refugees: background and present realities. *Area*, **9**(2), pp. 81–89.

Roy, S. (1986) *The Gaza Strip Survey: The Jerusalem Post together with the West Bank Data Base Project*. Boulder, CO: Westview Press.

Roy, S. (2000) The transformation of Islamic NGOs in Palestine. *Middle East Report*, No. 214. Available at http://www.merip.org/mer/mer214/214_roy.html. Accessed 23 March 2010.

Rushdie, S. (1992) *Imaginary Homelands: Essays and Criticism*. London: Granta Books in association with Penguin Books.

Shehada, N. (2004) The rise of fundamentalism and the role of the state in the specific political context of Palestine, in Imam, I., Morgan, J., and Yuval-Davis, N. (eds.) *Warning Signs of Fundamentalism*. London: WLUML.

Stedman, J. and Tanner, F. (eds.) (2003) *Refugee Manipulation: War, Politics and the Abuse of Human Suffering*. Washington DC: Brookings Institution Press.

Tamimi, A. (2007) *Hamas: Unwritten Chapters*. London: Hurst.

U.S. Department of the Treasury, Office of Public Affairs (2001) Holy Land Foundation for Relief and Development, International Emergency Economic Power Act. Action Memorandum, 5 November. Washington DC.

Weinberg, L. and Pedahzur, A. (2004) (eds.) *Religious Fundamentalism and Political Extremism*. London: Frank Cass.

Part III: Identity, Tradition, and Fundamentalisms

Chapter 8

Abraham's Urban Footsteps: Political Geography and Religious Radicalism in Israel/Palestine

Oren Yiftachel and Batya Roded

> You are a traitor, absolutely! A traitor to your nation, religion, and country ... a traitor to the very reason your parents came here... Did you forget that this place is written in the Bible as ours not theirs?
>
> Noam Arnon, speaker of Hebron Jews to an Israeli soldier evacuating Jewish settlers from the central Arab market, 8 July 2007

> We have patience and Allah on our side; we have the Islamic Umma, from Pacific to Atlantic behind us; we shall get these invaders out one day; al-Chaleel belongs to the Palestinians, to the Muslims, but not to these criminal infidels.
>
> Abd Al-Hai Arafah, Mufti of Hebron, 14 April 1969

Religious radicalism (often termed 'fundamentalism') has recently resurfaced as a major force in shaping politics, space, and violence. The hub of the current wave of religious mobilization lies in massive urban agglomerations, particularly at the rapidly burgeoning, impoverished, and often informal urban margins of the global South, such as San Paolo, Mexico City, Baghdad, Johannesburg, Cairo, and Istanbul. But these mobilizations are intertwined with older waves of religious politicizations, associated with ethno-national urban struggles, as found in Beirut, Jerusalem, Sarajevo, Belfast, Ahmedabad, Nicosia, or Hebron. This is vividly revealed in the above quotations, where religion, nationalism, and class overlap to shape the political geography of radical religious mobilization.

Our chapter offers a first step in writing about such a political geography or religious radicalism, which, we suggest, is closely linked to the depth of 'urban colonialism'. Religious radicalism, whether state-sanctioned 'from above', or an articulation of resistance 'from below', is intertwined with the process of urban colonialism, in which colonized populations are racialized, humiliated and materially exploited. Based on the experience of Israel/Palestine, and focusing mainly on Jewish spatial politics, we suggest that these urban colonialisms have created fertile ground for the rise of religious radicalism as part of the struggle between collective identities for control of urban space.

To illustrate the argument, the chapter analyzes in brief the dynamics of ethno-religious politics in three key cities in Israel/Palestine, all bearing the symbolic footprints of Abraham, the mythical father of Jews and Muslims: Hebron, Jerusalem, and Beer Sheva. We show that the state's 'ethnocratic' urban geopolitical policies, and the associated nature of urban colonialism, remain the main (albeit not sole) cause of religious conflict and radicalism. Our argument does not claim universality, but is rather aimed at a 'meso' level of generalization, relevant mainly to states forcefully promoting 'ethnocratic' projects. Hence, rather than seeing Israel/Palestine as an exception, we wish to place it as an emblem – a hyper example of processes underway in other cities and contexts.

The resurfacing of religion as a force of mass mobilization runs against the grain of mainstream Western (universalizing) academic analysis. However, when religion does appear in mainstream scholarship and popular discourse, it is portrayed as a 'dark horse', potentially harbouring evil forces such as 'fundamentalism', messianic colonialism, '*Jihadism*', and, of course, global terrorism. We take issue with such approaches which separate religion from the working of modernity and the modern nation-state. We show that religious radicalism often derives from the very identity projects instigated by modern nation-states and the social and economic conditions they have created. Hence, we suggest rethinking the taken-for-granted link between religious 'fundamentalism', globalization, and 'civilizational wars' (Huntington, 1996; Almond *et al*., 2003). To be sure, globalization has had a major impact, not the least in shaping most political frameworks over the past two centuries – including nationalism, capitalism, economic colonialism, and class action. However, we observe that most radical religious mobilizations have been tied to either national territorial struggles or conditions of urban marginality, rather than to globally oriented campaigns.

Prior to expanding on this argument, let us touch on terminology. Despite the closeness of the two terms, we prefer 'religious radicalism' to 'fundamentalism', because 'radicalism' explicitly addresses the *politics* of religion, as religious radicals attempt to impose a new order on society. Because our emphasis in this chapter is on urban politics, we think 'radicalism' better describes, and without bias, the intent of some religious movements to get 'to the root (= radic)' of the social-cosmic order, and impose their own politicized religious vision on pluralistic societies.

Fundamentalism, on the other hand, is a term laden with popular derogatory meaning, fuelled both by the recent neoconservative 'war on terror' and a secularist-leftist disdain of religiosity. 'Fundamentalism' denotes a modern phenomenon driven by a vision of religious purity, often as a response to crisis or threat (Silberstein, 1993). It rests on a reinterpretation of sacred texts (Almond *et al.*, 2003), but not necessarily with a political agenda. As Martin and Appleby (1991) note, fundamentalists seek to '"fight back", fight for "the truth", fight with particularly chosen repository of sources, fight against other and fight under God'. The boundaries of fundamentalist groups are often rigid, and their discourse inflexible and messianic.

By 'ethnocracy' we mean a regime type whereby the state is appropriated by a dominant ethno-national group, and is used to advance its own 'ethnicizing' political and territorial agendas over contested space and power structures. 'Religious movements' are a form of societal organization, aiming to politicize and institutionalize a divine order based on sacred texts and traditions. Religious movements use their 'goods of salvation' (Bourdieu, 1991), and commonly elevate a theocratic 'order of things' in direct competition with other grids of modern societal organization, such as democracy, civil society, and in some cases nationalism.

Israel/Palestine

The political geography of the land began to change dramatically during the British Mandate period with the massive arrival of Jewish immigrants and refugees fleeing persecution in Europe and (later) oppression in the Arab world. In its early decades the Zionist movement was mainly non-Orthodox (often termed 'secular') and nationalist, and was seen by many as a rebellion against traditional Judaism. But at the same time it harboured deep-seated, religiously inspired, and even messianic concepts regarding Jewish salvation through a return to Zion – the promised biblical land (Raz-Krakotzkin, 2002). Accordingly, it laid claim to the entire 'Eretz Yisrael' (Land of Israel/Palestine) between Jordan and the Mediterranean Sea, while Arabs, who were the majority on the land, resisted the Zionist project by establishing a fledgling Palestinian national movement. Tensions between the two movements escalated, and the British decided to leave. In 1947 the Zionist movement accepted a UN partition plan which allocated 55 per cent of the land to a Jewish state, which was rejected outright by the Arab leadership. The ensuing 1948 war saw widespread ethnic cleansing, during which some two-thirds of the Palestinians lost their homes and land, and were driven to regions beyond what later became Israel's internationally recognized border, the Green Line (figure 8.1).

Following the war, Israel was established as an ethnocratic Jewish state (Yiftachel and Ghanem, 2004), and imposed ethnic rule within its sovereign territory, now 78 per cent of Israel/Palestine. Palestinians meanwhile found themselves dispersed and under the jurisdiction of five neighbouring states. Israel then began a concerted project of internal colonialism, known as the Judaization policy, and built more than five hundred settlements and cities in areas previously inhabited by Palestinians, including Jewish Beer Sheva. The state brought 96 per cent of its land under Jewish-Israeli ownership. Palestinian Arabs were awarded Israeli citizenship, but were placed under military government until 1966, and were subsequently marginalized and dispossessed by the nascent state.

The role of religion in state affairs was at that time relatively minor, but still significant; a division of power allowed religious authorities to preside over personal life, public culture, and religious affairs, institutionalizing a permanent theocratic presence in the Israeli regime. Religious parties participated in all government coalitions, and renewed connections with world Jewry, which began to provide financial and political support to the warring state, and strengthened its religious dimensions.

In 1967 Israel conquered the West Bank and Gaza (including Jerusalem and Hebron), and continued its Judaization project by settling nearly half a million Jews over the Green Line – that is, beyond its sovereign area. Here religious groups assumed even greater importance, because the new territories, and especially the West Bank, are dotted with sacred Jewish-biblical sites. These religious elements soon began a massive settlement project, including Arab Jerusalem and Hebron as key targets. As part of this push, throughout the occupied territories (OT) Jewish settlers retained their full citizenship rights, while Palestinians were disenfranchised and subject to military rule. At the same time, Israel unilaterally annexed Old and East Jerusalem to the Israeli municipality, and began a massive effort to Judaize these areas (see figure 8.4). Due to the 'urban annexation', Jerusalemite Palestinians received residency rights and Israeli ID cards (but not citizenship), despite the fact that under international law East Jerusalem remained part of the occupied West Bank.

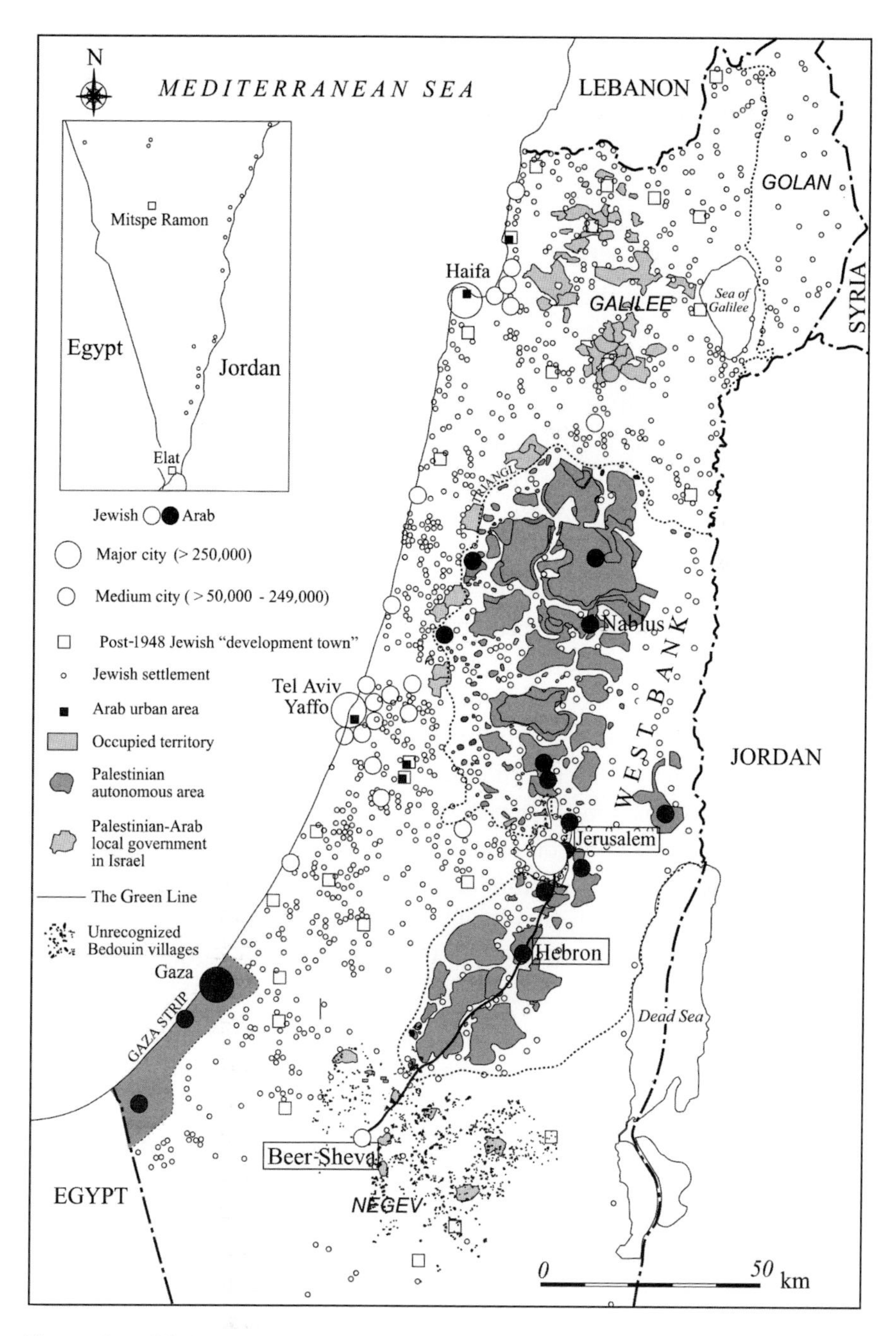

Figure 8.1. Ethnic geography of Israel/Palestine, 2005, and Abraham footsteps. (*Source*: Yiftachel, 2006, p. 74)

Notably, in 2005, after two bloody Palestinian rebellions, Israel evacuated the Gaza Strip, and for the first time willingly dismantled twenty-five Jewish settlements in the area it considers its historical homeland. Yet, the Judaization of Jerusalem and the West Bank has continued. And now, in response to accelerating cycles of mutual violence and terror, Israel is erecting a massive separation barrier ('the Wall'), which effectively transfers 10 per cent of the West Bank to Israel. In 2005 the barrier (and all Israel's colonial settlements in the West Bank) were condemned as 'clear violations' of international law by the international court in The Hague.

Since 1948 the Israeli state has pursued a colonial project of expanding and deepening Jewish control over all parts of the contested land, and against the wishes of local populations. However, we must differentiate between the *various depths* of colonization which have resulted from the gradual, incomplete, and contested imposition of Jewish rule. Whereas in Hebron the Jewish presence is based on a heavily militarized occupation and settlement *vis-à-vis* rebellious, rightless Palestinians, in Jerusalem the edge of the systematic and powerful colonial project is somewhat blunted by the partial civil status of the Palestinians. Meanwhile, in the Beer Sheva region, Judaization has been accompanied by the endowment of the local Bedouin population with citizenship and some legal, political, and urban development rights. While the Judaization logic proceeds in all three cities and results in conspicuous inequalities, we wish to argue that the variation of the political geography of urban colonialism does make a significant difference, *intra-alia*, for the nature of urban religious radicalism.

This brings us to the issue of religious politics, mainly in the form of Orthodox Jewish parties and groups. These groups have steadily increased their power base within the Zionist state, particularly since the 1970s. This has been the upshot of Israel's colonial push, which allowed religious groups to claim a 'frontier' position in the Zionist project both discursively and physically. Later, intensifying Palestinian violence 'confirmed' the religious narrative of the Jewish people in its perpetual struggle against hostile nations, further augmenting religious political power.

Through the middle and later decades of the twentieth-century Zionist ideology – which traditionally treated 'Jewishness' ambiguously as ethnic, national, and religious – became increasingly theocratic. And the influence of religious parties within the Israeli polity reached a peak during the 1990s, as religious parties won some 30 per cent of the Israeli parliament in the 1996 elections. Since then, however, a growing chasm has developed between Orthodox and 'secular' (non-Orthodox) Jews. The joint Zionist framework still holds the two camps together around the goal of containing what they construct as a common enemy; but serious cracks have opened since the mid-1990s. This is reflected in growing cultural and geographic polarization, highlighted by the fact that nearly all West Bank settlers are now Orthodox or ultra-Orthodox, as opposed to the 1980s when the settlements had a far greater presence of non-Orthodox Jews. At the same time, other portions of Israeli society have become increasingly liberal, secular, and globalized.

The tension between 'secular' Jews (70–75 per cent of the Jewish population) and their Orthodox counterparts has now become one of the most volatile issues in Zionism – with the territorial question of controlling Palestinian space at its very heart. In this vein, it is illustrative that the fall of the last five Israeli governments was caused by religious (or radical nationalist) political elements, who vehemently opposed government attempts to advance toward negotiations with the Palestinians, because they would necessitate, in their eyes, relinquishing parts of the sacred homeland. The most notable event was the 1995 assassination of Itzhak Rabin by a religious Jew, which derailed these peace efforts. Yet all four prime ministers since – Shimon Peres, Binyamin Netanyahu, Ehud Barak, and Ariel Sharon – were toppled, or critically weakened, by nationalist-religious elements of Israeli politics in response to what they perceived as tampering with the sanctity of religious-national space.

A process of religious radicalization has also recently emerged among Palestinians in recent times, with even greater intensity and venom. The PLO (Palestine Liberation Organization) was for years the most secular national movement in the Arab world, maintaining a relatively democratic representation among political factions (Hilal, 2006). And since 1994, Fatah has been the main force behind the nascent Palestinian Authority (PA), which received limited autonomy in governing about 40 per cent of the OT, or 10 per cent of historic Palestine. The institutional design of the PA, too, followed a relatively typical secular structure, with a supporting legal and military apparatus (Hilal, 2006; Ghanem, 2000).

During the 1980s, however, the influence of a new wave of Middle East Islamism, in the aftermath of the Iranian revolution, and as a result of a lack of development, widespread poverty, and Israel's brutal measures in putting down Palestinian resistance, created fertile grounds for the emergence of Hamas (the hard-line Islamic resistance movement) and several allied small religious factions. In only two decades Hamas has become the most dominant force in Palestinian politics, using a mixture of Islamist and nationalist rhetoric, and launching a campaign of unprecedented violence and suicide terror against Israel.

Hamas won the 2006 PA elections, and attempted to co-govern with President Abbas of Fatah. But in June 2007 Hamas took total control of Gaza in a bloody act of civil violence. As a result, Abbas fired the Hamas-led unity government, and the Occupied Territories now 'boast' two Palestinian governments – Hamas in Gaza and Fatah in the West Bank.

Towards a Political Geography of Religious Radicalism

We wish to advance several related theoretical arguments, as inspired by a neo-Gramscian perspective highlighting the links between systems of material and political domination and issues of culture, class, and identity (Laclau, 1994; Hall, 1992). This perspective conceptualizes political regimes as seeking to construct a hegemonic status, in which the domination of a particular system of beliefs and values becomes a 'taken-for-granted' truth. The ethnocratic and theocratic mobilizations that are at the centre of our inquiry, are prototype hegemonic projects. At times, these projects conflict (Lustick, 2002), but in other circumstances they may reinforce one another. We are also inspired by (post)colonial scholarship (Samaddar, 2002; Roy, 2009; Shenhav, 2006) to extend the neo-Gramscian framework in two principal ways. First, we note that hegemonic projects may be seriously challenged by the 'stubborn realities' of exclusion and oppression in which the life of the subaltern Other is embedded (Bayat, 2000; Chatterjee, 2004). In other words, and in contrast to mainstream liberal, or critical Foucauldian perspectives, we discern a persistent presence of politicized groups falling 'outside' the nets of control cast by societal powers. In other words, the mechanisms of state co-optation and discourses of governmentality lack the capacity to incorporate these populations, causing long-term instability and presenting challenges to state authorities. Second, we introduce the critical importance of spatial processes to the construction of and challenge to hegemony (Massey, 2005). As shown below, these are not merely backdrops on which the drama of religious radicalism unfolds, but active factors creating the conditions for such drama.

Our argument begins by illuminating the historical moment in which the relationship between ethno-national and religious mobilizations is mutually reinforcing. We claim that in certain 'South Eastern' (non-Western) regions of the world, following the imposition of state nationalism on a pre-existing web of affiliations, religion re-emerged as a supportive, yet subordinant, force within the ethno-national project. The winds of secularism which were carried with the diffusion of nationalism pushed religion to the sideline, and a new conceptual grid was popularized around an 'unbroken connection' between nations and 'their' land. In several regions, such as the Soviet Bloc, Europe, and East Asia, the nationalist order totally replaced religion by a system of centralized anti-religious oppression. But in others, such as the Middle East, South Asia, and Eastern Europe, the national order became dominant, but the shadow of deeply rooted religious traditions remained.

During the period of anti-colonial struggle and the associated nation-building project, some religions reappeared as instruments of the ethno-national projects. We conceptualize these as *'ethnic religions'*, reined in to fortify the process of 'ethnocratic' nation-building both in response to colonial power and, equally importantly, against minorities who staked alternative claims to power and resources. Sri Lankan Buddhism, Zionist Judaism, Indian Hinduism, Palestinian Islamism, and Irish Catholicism are but a few examples. We use the term 'ethnic', in preference to 'national', here to highlight the construction of 'the nation' under ethnocratic regimes. These constructions often work actively against the creation of a civic nation, and are often buttressed, as we shall see below, by religious myths, practices, and institutions.

Here we need to pause and make some qualifications. First, we do not claim that religion is merely an instrument of regime power. We acknowledge its existence as a major societal force with its own grids of meaning, aesthetics, and politics, which can be studied from a variety of angles. Second, we acknowledge that there exists a variety of powerful forces shaping religious radicalism beyond the urban geopolitics on which we focus. And third, the literature on the rise of religious movements accounts for a range of important factors relating to their emergence, such as resources, organization, leadership, ideology, and tactics.

At this stage it may also be useful to advance and make an analytical distinction between religious radicalism 'from above', and 'from below'. The former, on which this chapter mainly focuses, is augmented – explicitly, or, more commonly, implicitly – by the state's identity project, with religious institutions functioning as 'gate keepers' to screen out the 'wrong' groups from full membership and power. The latter ('from below') is generally a form of coping with, and resisting, the oppression applied by the state or other powerful forces affecting people's deprivation and marginality. This often appears in the form of constructing counter-hegemonic religious discourses (Davis, 2006; Finke, 2003; Ram, 1996).

We recognize, of course, that the distinction is malleable, and that circumstances may change the 'above-below' positioning, as in the case of Iran or Afghanistan. Yet, the distinction is helpful in illuminating both the forces generating radicalism and – critically – the active involvement of the state in producing its own radicalism.

Religion and Expansion

The cooperation between ethno-nationalist and theocratic forces is pronounced when states are engaged in (external or internal) 'ethnocratic' colonial projects (Yiftachel and Ghanem, 2004). It may be so in development efforts that direct the flow of capital to a dominant group (often through the exploitation of minority labour); in settlement initiatives that claim ethnic control over contested territories (McGarry, 1998; Newman, 1997); in the articulation of historical, archaeological, and cultural discourses supporting expansionist territorial claims; or in unequal governance systems imposed on certain regions. Examples of state colonial projects abound. Among them are the Sri Lankan Dry Area resettlement; the Russification of the Baltic States; the Malaysian 'new-village' initiative aimed at dispersing Chinese to the south; the Judaization of the West Bank, Galilee, and Negev; the Bantustanization of apartheid South Africa; or the long-term exploitation by Britain of the Celtic fringe (McGarry, 1998).

In such a context, religious frameworks 'ground' sanctity in space by providing a divine (and hence indisputable) narrative of territorial belonging. Isaac (1960) was among the first to write about the inherent spatialities of most organized religions. Smith (2000) and Shilhav (1991) have also shown how religious spatiality is often intertwined with the symbolic and geographical underpinning of ethnic nationalism. As elaborated by Cooper (1992), political power is often behind the delineation and sanctification of space, commonly using a strategic 'selection' of religious narratives and myths. And Jackson and Henrie (1983) have developed a hierarchy of spatial sanctity, at the top of which are sacred sites, followed by the national homeland as a sanctified 'geobody', and specific historical sites reinforcing the collective story. In cases of ethnic conflict, religious narratives tend to become ever more radical, as new interpretations of sacred texts surface, supporting archaeological findings come to light, or new religious zeal emerges to exclude 'less pure' groups from using the 'promised' or divine space (Abu el-Haj, 2001; Silberman, 2001; Mann, 1999; Sibley, 1995).

Akenson (1992) shows convincingly how Protestants in Northern Ireland, Afrikaans in South Africa, and Zionists in Israel/Palestine have relied on ancient texts and narratives of selection, covenant, and territory to justify oppressive forms of racism. In the case of sacred sites, religion provides the state with a particular geography of salvation, which also functions as a popular, strategic, and emotional foundation for expansionism. As perceptively claimed by a recent study:

> The political content of sacrality and the sacred content of power are essential to urban sociology ... and to the analysis of religio-political conflicts. We must understand the sacred as a necessary constituent of power. Sacred centers are not only ideas or symbols, but act as moral sanction for denying the rights of the Other. (Friedland and Hecht, 2007, p. 19)

Further, the apparatus of the modern ethnocratic state conveniently uses religious categories and classifications to create social boundaries and prohibitions, with the aim of maintaining ethnic 'purity' and dominance. The use of Dutch Reform doctrines in the case of the South African apartheid state is well known. Similarly, in ethnic states such as Greece, Armenia, Israel, Serbia, and Iran, the state ranks religious identities, prohibits civil marriages, and allocates unequal resources to members of minority religions. Hence, ethnocratic and religious mobilization have often reinforced one another, to the mutual benefit of both state and church.

Cracks in Expansionist Identity Politics

The argument is, however, more complicated. We wish to raise here the factors of time and historical momentum, which expose an inherent tension – if not long-term contradiction – between the logics of the ethnocratic *state* and religious mobilization. Such tension often emerges precisely as a result of cooperation between the two, as the two 'camps' rely on each other to strengthen their social and political bases and develop rival long-term political projects. This tension has the potential to destabilize political systems, as seen in Sri Lanka, India, Sudan, and Lebanon, to name a few examples. Importantly for the current analysis, such tension often rises in struggles over the production and management of urban space.

There are two central and related elements involved. The first is the metaphysical discourse of destiny. The goal of ethnocrats, who form the mainstay of national and political leadership, is to control a state apparatus. They play according to the contemporary political-geographical 'rules of the game' – namely, that each nation can control 'its own' territory and people, but *only* its territory and people. Given this caveat, ethnocratic elites attempt – with the aid of theocrats – to maximize control over their ethno-national group, whether *vis-à-vis* neighbouring states (as in the case of border disputes or irredentist moves, such as in India, Israel, or Cyprus), or *vis-à-vis* minorities within 'their' states (Mann, 2002; McGarry and O'Leary, 2004).

At the same time, religious movements, now empowered by the state, continue to pursue their own visions of ultimate destiny, redemption, and salvation. These transcend the horizons of the modern state, and challenge its territorial, cultural, and political limits. Theocratic visions abound, but they invariably aspire to lead the population toward a messianic, cosmic order of total and global victory against the infidels. For theocrats, contemporary states are but a necessary and temporary step in the direction of ultimate salvation (Almond *et al.*, 2003).

There is no room here to elaborate on this important point, except to note that it often presents a serious challenge to the modern state, evident in urban politics and the daily discourses of religious communities. Here the cities of Hebron and Old Jerusalem are highly illustrative – both lying beyond the borders of the state of Israel, yet constructed as 'essential' for the fulfilment of Jewish religious salvation. In such locations the embedded tensions surface into open conflict between states and 'their own' religious movements. These tensions do not only revolve around territorial issues, but address a range of matters, affecting all spheres of human life, from the body, dress, neighbourhood, and urban landscapes, to issues of food, festivals, and gender relations.

The second locus of tension between theocrats and states tends to develop around the construction of citizenship. States typically aspire to legitimacy – both internal and international – and hence construct a discourse of equal citizenship, supported by a legal and institutional apparatus. In practice, equal citizenship remains a theoretical and rarely implemented vision. Yet, in contrast, religious movements attempt to replace the discursive and regulative frameworks of equality with a hierarchical system of affiliations based on religious doctrine and customs.

This has adverse consequences on a range of social markers, most notably women (traditionally marginalized and disempowered by religious doctrines) and minorities (either of different religions or of the 'wrong' sects within the dominant religion). In that way, religious movements may attempt to undo a basic construct of the modern state – equal citizenship. When translated to the quotidian practices of government, the fracturing of citizenship ruptures the idea of the 'demos' – a body of equally empowered citizens. It therefore presents a long-term challenge to state legitimacy and stability. To illustrate the ethnocratic-theocratic relations-cum-tension we turn to an old Chinese fable:

> On a cold stormy day, a lizard tries to get into a fox's warren for shelter. Worried, the fox rallies a friend – the stork – to bring water in its beak, drop it at the warren's entrance, and thus prevent the lizard from entering. In return the fox offers the stork warm shelter from the storm. Once successful against the persistent lizard, the stork suddenly realizes that if it continues to pour water, the fox too will be forced out of the warren, and the stork will have it all for itself.

The fable's moral is about carefully calculating who is invited to one's home. For our purposes, we can liken the ethnic state to the fox, the lizard to the neighbouring 'enemy' or a minority of competing nationality, and the stork to religion. The fable illustrates well the dynamic of allies becoming rivals, with a momentum of time and power. It also shows how in such instances space becomes pivotal – with such conflicts over 'home' being grounded within specific geographies of contestation, most notably urban ones. This leads to our next section, which focuses on how the forces of expansion and domination described above are present in the heart of the city.

And the Urban?

The final part of our argument connects the above observations to the urban scene. By virtue of being the growth poles of most societies, urban regions are the point of encounter between diverse groups. It is here that group relations are 'concretized' through the intersection of state, global, and local forces (Lefebvre, 1996). Urban dynamics regularly shape the distribution of material, political, and symbolic resources, turning cities into sites of political contestation, articulation, and mobilization.

But the modern urban scene, by its very dynamism, size, and diversity, harbours a multitude of possibilities. On the one hand, it enables movement and porosity across social and spatial boundaries, unimaginable in rural or traditional societies. The spatial proximity and the quotidian economic and political interactions between groups often create new and dynamic identities and shifting cross-cutting coalitions (Tajbakhsh, 2002; Katznelson, 1996). On the other hand, precisely because of this potential mobility, it is in cities that we find severe forms of social control and surveillance, to combat the 'danger' of social mixing and political dynamism (Wilson, 1995). Hence, a range of control measures is often invented and implemented in urban areas, typically involving housing segregation, uneven land allocation, municipal gerrymandering, uneven investment in infrastructure, jobs, and transport, and efforts to disenfranchise rapidly growing swaths of informal urban settlement (Marcuse, 1995; Robinson, 2006). In polarized cities, deep social (and ethnic) difference, growing economic inequalities, and control mechanisms may thus create *new colonial conditions*.

It is against these urban processes that we find the rise of religious radicalism. Here, immigration, deprivation, and regulative controls interact with national and global powers. We contend that the deeper the 'footprints' of urban colonialism, the more prevalent the rise of religious radicalism. In this formulation, 'urban colonialism' may be understood as the management of urban regions according to a colonial logic, whereby a dominant group controls the political space in order to exploit the region's material advantages and labour force, impose a system of unequal membership based on a ranking of ascriptive identities and economic positions, and manage political opportunity by restricting free movement and participation.

Urban colonialism is a dynamic and inevitably contested process. In cases where colonialism is highly institutionalized, as in Jerusalem, the regime may be defined as 'urban apartheid'. But in most cases, such as Beer Sheva, urban colonialism remains a process in the making – undeclared, and practiced through more subtle discursive, material, and regulative means. But 'the urban' is still a vast field, and it may take some additional distinctions – as 'navigation grids', rather than strict dichotomies – to steer through the messy seas of urban political geography.

First, we should distinguish between cities that are subject to conflicts over territorial collective identity (national, ethnic) and those where the question of sovereignty has been settled (for the time being). Clearly, urban colonialism has a sharper, more volatile, and more violent edge in cities such as Jerusalem, Beirut, Ahmedabad, Sarajevo, or Colombo, where 'the urban' is closely intertwined with 'the national' as a site of struggle over sovereignty or ethnic self-determination. In such cases urban segregation is deeper, and space becomes a zero-sum territorial resource, often subject to bitter struggle. This is different from the more fluid and porous situations in cities composed of massive, often informal, settlements, where conflict arises out of economic and political deprivation. The struggle appears in such cities to be 'within', rather than 'between', national or ethnic collectivities.

Another important distinction should be made between holy urban sites and 'sanctified' urban areas. Holy sites are constructed by elites as key locations of memory and identity, promoted to serve the contemporary identity project. Major holy sites may often form the basis for the growth of an entire urban area, as in the case of the Vatican, Jerusalem, or Mecca. Sanctified urban areas, on the other hand, are those created through the empowerment of religious movements in marginalized urban areas. These are the work of contemporary religio-political entrepreneurs, who often impose a religious order on a pre-existing urban landscape, typically in terms of street life, commerce, dress code, and forms of public morality. These are more dynamic and expansive than holy sites, though less steeped in spatial and historical sanctity. What interests us here particularly is the first type – where religious radicalism 'from above' advances colonialism, which over time generates a counter-movement of religious radicalism 'from below'. Again, the two are not mutually exclusive, but present different core dynamics in the shaping of religious radicalism.

Our argument is now beginning to synthesize. It is in key urban areas imbued with great national, historical, or religious importance that the theocratic agenda brushes most forcefully against the state's civil agenda. This is where attempts may arise to make the ethno-nationalist project more theocratic: by promoting and then colonizing new, old, or invented holy sites; seizing and developing 'enemy' space; constructing walls; and staging provocative events, such as marches, holiday celebrations, and street blockades. Through this spatial and political process, and the associated 'radicalizing religious moves', religious groups attempt to accumulate symbolic and political capital within 'their' ethno-nationalist project, at the expense of the excluded 'others', and in competition with other elements within their nation.

Adding to this basis for conflict is the fact that cities are centres of globalizing economic development, where elites reside in close proximity to marginalized labourers – the unemployed, the informal, and the illegal. In many cities this creates a double movement, whereby conflict is exacerbated by both religious-nationalist radicalism and economic segregation. Religious and radical nationalist groups often agitate against urban minorities through development, planning, or housing initiatives, while the political economy of globalizing urban development typically deepens the deprivation of the same groups. These have spawned phenomena such as ghettos, informalities, and 'floating' populations, for whom organized minority religions offer security and meaning ain the midst of urban turmoil.

It is here, as depicted in figure 8.2, that we find the political geography of religious radicalism, born out of both the spatial struggles *within* the ethno-nationalist project, and the material and political struggles *between* the dominant ethnic nation and ethnic and religious minorities. In cases where these forces persist over time we find a process of 'negative dialectic' operating along several axes, causing, *intra-alia*, the radicalization of religious politics from 'above' and 'below'. The polarization typically occurs in situations of long-term unresolved collective conflict, whereby religious agendas may be gradually introduced to buttress the territorial and spatial

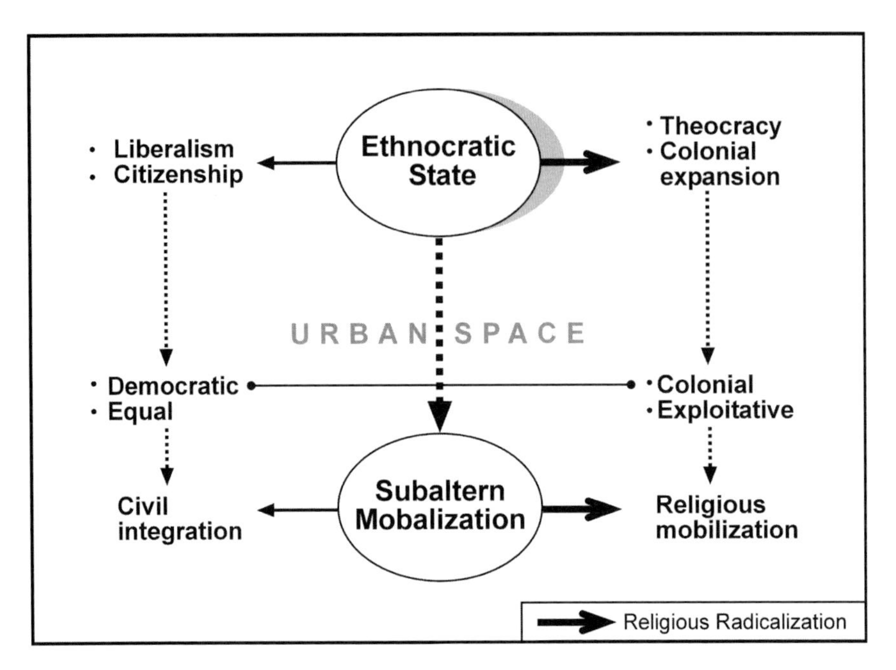

Figure 8.2. Urban colonialism and religious radicalism, Israel/Palestine.

struggles between rival groups. This is the birth of the most radical forms of religious mobilization, as evident in Beirut, Jerusalem, Hebron, Mumbai, Colombo, Baghdad, and Belfast.

It is beyond our scope here to sketch the rich details of the process of 'religiosation', whereby frameworks of identity are imbued with religious significance and redemptive zeal, often as alternatives to state-induced or globalizing trends. It will suffice at this stage to point to good research on the subject which highlights the process both among the state-sanctioned groups occupying 'the frontier' (Ram, 1996; Eldar and Zertal, 2004; Bartholomeusz and de Silva, 1998; Winslow, 1984) and among the resisting weaker groups who search for empowering frameworks to help them cope with urban colonialism and prepare the ground for their own political ascendancy (Budeiri, 1995; Hatina, 2005; Davis, 2006).

We can now turn to an examination of the three Abrahamic cities, highlighting the political geographical forces at work, and focusing on the depth of urban colonialism and religious radicalism.

The Geopolitics of Abraham's Cities

Abraham is the mythical father of both Islam and Judaism. The three cities examined here define Abraham's constitutive biblical journey through the

Promised Land, marking its mythical early 'geopeity'. As such, the three cities possess similar urban religious-national significance.

Abraham, according to the sacred texts, first settled in Beer Sheva. He then travelled to Jerusalem for his son Isaac's sacrifice on Temple Mount – the site where both the Jewish Temples and al-Aqsa Mosque were later built. Abraham, and his wife Sarah, were later buried in Hebron, so the narrative goes, on land purchased in full from local inhabitants.

As shown below, while the three cities are located along a short, 80 km route (figure 8.1), they are also set in different political geographical circumstances.

Hebron: Militarized Religious Utopia

Due to the location of the sacred Abraham's Tomb, Hebron (Hevron in Hebrew, al-Chaleel in Arabic) was considered one of the four 'holy cities' for pre-Zionist Judaism. A small Jewish minority lived within the predominantly Arab city from the sixteenth century onwards, until Jews were evicted following a 1929 riot, in which sixty-seven Jews were killed. Following the 1948 war Hebron was annexed to Jordan.

Shortly after Israel's conquest of the West Bank, during Passover 1968, the first group of religious Jews, led by Rabbi Levinger, invaded an empty city hotel in defiance of government orders. Thus began four decades of an urban colonial project, which has often run against the grain of Israeli policy but has nevertheless received state protection. To date, it has managed to attract some 7,500 Jewish settlers to the city and the abutting colonial town of Kiryat Arba (Swiesa, 2003; OCHA, 2005) (see figure 8.3). Hebron is the only West Bank Arab city (outside Jerusalem) in which Jews settled. During the last four decades it has represented an extreme case of religiously and nationally motivated colonization that has explicitly attempted to push the boundaries of the Zionist regime from ethnocracy to theocracy.

Early settlement rhetoric in Hebron mixed religious promise for a Jewish Abraham's city, nationalistic claims to control the entire Land of Israel, and personal claims to retrieve Jewish property lost in 1929. Religious Jews occupied parts of the city in persistent challenge to the government's attempt to arrive at a more 'strategic' or 'rational' colonial policy that would avoid direct confrontation within Arab cities (Newman, 2001). The settlers remained, however, full Israeli citizens, deep in the occupied West Bank, and received on-going, and highly costly, military protection. By contrast, in 1967 the city's 140,000 Palestinian residents were placed under military rule, with no political standing to affect Israeli policies governing their own city.

With expanding Jewish colonization in parts of the Old City, and the construction of large-scale housing in Kiryat Arba, relations between the two ethnic communities polarized. A violent nadir was reached in 1994 when a Jewish settler massacred twenty-nine Muslim worshippers inside the sacred Abraham's Tomb. Hostilities have since continued, and it is estimated that during the last seven years some 123 Palestinian residents and nine Jews have been killed (B'tselem, 2007).

On 4 May 1999, the Wye Agreement saw the city effectively partitioned: the eastern part was placed under ('temporary') direct Israeli control, while the western parts were vested with the Palestinian Authority (PA) as 'Area A' – with autonomous control. In 2002, following a wave of Palestinian terror aimed at ending the on-going occupation, Israel reinvaded the city. Today, Hebron is governed by the PA but under tight Israeli control, the level of which depends on Israel's self-declared 'military considerations'.

The political geography of Hebron can thus be portrayed as a brutally colonized and occupied city with municipal autonomy for the city's Arab western parts. Economic development has been severely hampered by an occupation regime of road blocks, closures, and curfews and by tight control of Arab building construction. Since 2000, the economy of Arab Hebron, like the rest of the OT, has seriously deteriorated, with parts of the region reporting unemployment higher than 50 per cent and near subsistence levels of physical existence (OCHA, 2005). Meanwhile, the Jewish part has remained well developed, due to near full employment and generous state subsidies which are available to Israeli citizens only.

It is not surprising then that under such colonial settings, religious radicalism has thrived, both 'from above' and later 'from below'. The discourse of Jewish settlers has accordingly evolved over the years. During the 1970s it mixed religious, nationalist, and secularist components, led by the writings of Elyakim Ha'etzni and later by his son Nadav – two of the city's prominent figures. The main argument was that Hebron is the ultimate frontier of Zionism, and that retreat from the city will signal the fall of the entire settlement project. During the 1980s, however, and increasingly during 2000s, radical religious rhetoric has taken centre stage, portraying Hebron as a place where the 'covenant' between God and nation is fulfilled through the sacred deed of settling the Land of Israel (Eretz Yisrael) (now a euphemism for the Palestinian territories). The city has thus also become a site for on-going pilgrimages – organized tours, and festivals attended by tens of thousands of (mainly Orthodox) Jews every year, especially around the Jewish holidays.

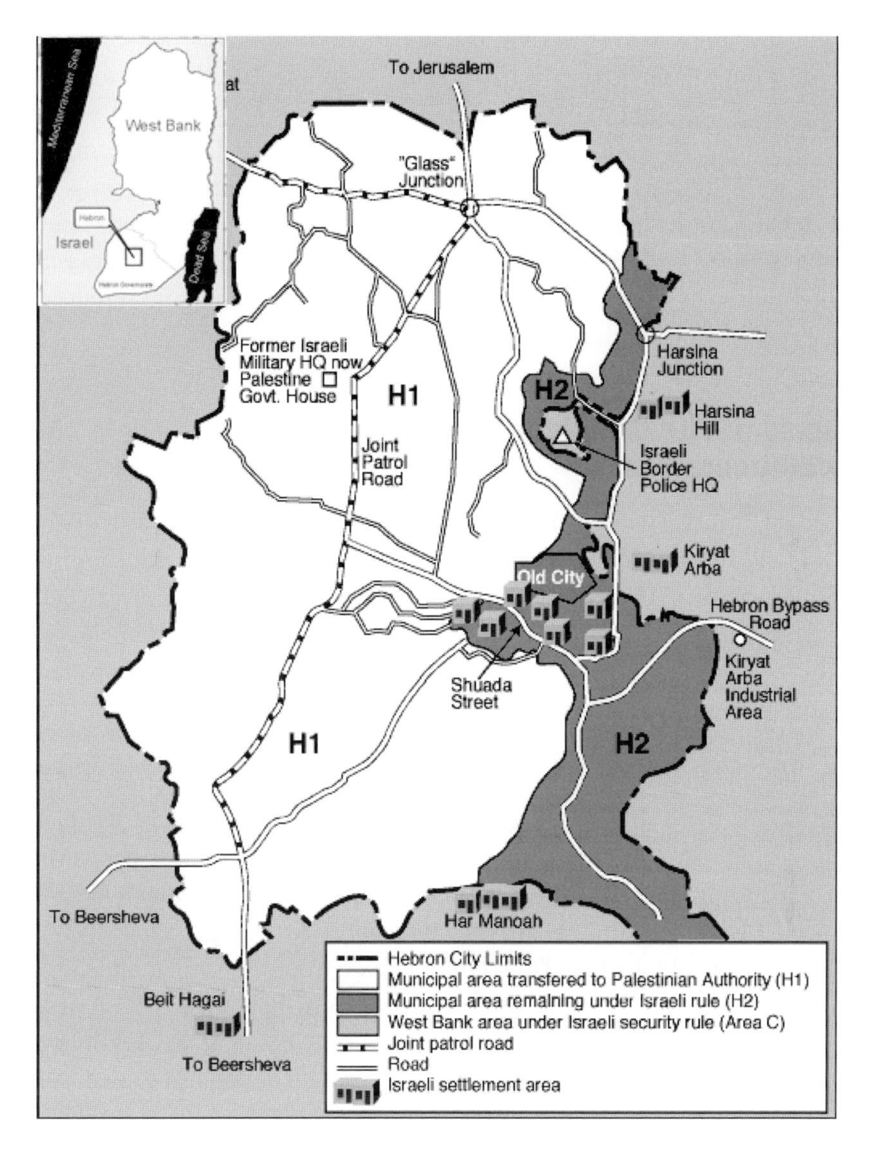

Figure 8.3. Hebron and Kiryat Arba: Jewish settlement in the city of Hebron. (*Source*: B'tselem, 2007)

Local politics have reflected this change. During the 1970s and even the 1980s, nationalist parties polled fairly well in Hebron and Kiryat Arba. However, in the 2006 elections the shift to a theocratic agenda became increasingly clear, with 77 per cent of residents voting for one of Israel's four main religious parties – Mafdal, Gush Leumi, Agudah, and Shas. This was mirrored on the Palestinian side, with Hamas polling 59 per cent, and winning all nine seats in the 2006 Palestinian elections. This was Hamas's strongest political showing, overshadowing even the landslide victory in Gaza.

The multiple dialectics created by the violent Jewish settlement in Hebron well illustrate the political geography of radicalism, bringing to sharp relief various dimensions of conflict: between settlers and state; between settlers and the Palestinian national movement; and between Islamic Palestinian groups, mainly representing the city's poor and refugees, and the more secular and middle-class Palestinian mainstream.

Jerusalem: The Conflicting Embodiment of Two Ethnocracies

Jerusalem ('Yerushalayim' in Hebrew; 'al-Quds' in Arabic), and particularly the Old City and within it Temple Mount, are the epicentres of the Zionist-Palestinian conflict. These small areas, not surprisingly, are also constructed as signalling the deepest attachment of the two nations to their homelands. It is here that Judaism and Islam frame and 'ground' identities around particular places, thereby fuelling the strong senses of geopeity within two national movements.

The city's biblical past tells of two temples built on the Mount as the central places of worship for Israelite and Judaic communities. The temples signal a 'golden past' for the construction of the modern Zionist narrative. 'Zion' concurrently means Jerusalem and the entire homeland (much like al-Sham and Masser, denoting Damascus/Syria and Cairo/Egypt, respectively). Islamic myth sites Mohammad's ascent to heaven from the same temple site, and has long held Jerusalem as one of Islam's most holy places. Its Arabic name, al-Quds, means simply 'the holy'. Since the 1920s it has become a centre of the Palestinian quest for sovereignty and the designated capital of a future Palestinian state. To complicate matters further, Jerusalem is where Jesus is believed to have been crucified, making the city one of the holiest places for Christianity.

Due to its global significance, the 1947 UN partition plan designated Jerusalem as a '*corpus separatus*', to be governed by an international entity. In 1948 Old (East) Jerusalem was captured by Jordan, and its 3,000 Jewish residents were forced out. Following the 1967 war, Israel annexed 70 km^2 from the West Bank, including the 6 km^2 Jordanian city of Jerusalem and a host of nearby territories and villages, creating a large metropolis. The Muslim holy sites remained under the management of the Waqf (the Islamic religious authority).

Over the four decades since, Israel has conducted a massive project of urban Judaization, settling some 180,000 Jews beyond the Green Line (that is, illegally, in occupied territory now titled 'Jerusalem'). This has been accompanied by large-scale land confiscation from local Arabs, thereby producing highly conspicuous gaps between the well-developed and -serviced Jewish development and the largely impoverished and under-serviced Arab neighbourhoods. Segregation remains very high, and movement across ethnic boundaries quite rare (figure 8.4).

Over the same period, the Arab population more than doubled, reaching 240,000 in 2006. Israeli and Palestinian towns and villages, on both sides of the extended city boundary, were functionally incorporated into the metropolis through expanding urban development. This process saw the rise of informal development mostly in the Arab sections. At present, the population of metropolitan Jerusalem is estimated at 1.3 million, nearly twice that of the Old City.

Within the city boundaries, Israeli Jews enjoy full citizenship rights, while Palestinians (Jerusalemites) hold 'Israeli resident' status only, which separates them from West Bank Palestinians and entitles them to a range of welfare and mobility privileges. Despite self-identification as Palestinians, their political status has remained in limbo – being neither Israeli, nor fully Palestinian. These residents of Jerusalem have voted in small numbers (14–16 per cent) in Palestinian National elections, although polling booths were placed outside the city by Israeli decree. They have also been eligible to participate in Jerusalem municipal elections, but have largely shunned them, considering the city government to be an illegitimate, colonial body. Several key Jerusalemites have participated in the Palestinian government, including the late Faisal Hussaini, Hanan Ashrawi, and Ibrahim abu-Tur.

Religion has obviously played a central role in shaping the political geography of Jerusalem, not only because of the high concentration of holy sites, but also because the city's population traditionally had a high proportion of Orthodox Jews. This aspect has been covered by numerous studies (Dumper, 2002; Khemaissi, 2005; Benvenisti, 2002; Yiftachel and Yacobi, 2002), and need not be repeated. The main point here is to discern the link between the depth of urban colonialism and the level of religious radicalism. The latter has increased during the last decade, although not to the levels seen in Hebron. The reason, we argue, is the imposition of somewhat 'softer' urban colonialism, and the associated mobilizations both 'from above' and 'from below'.

Let us elaborate. The first two decades of colonial Judaization in Jerusalem were driven mainly by non-Orthodox ('secular') Jews, most notably Jerusalem's mayor for three decades, Teddy Kollek. His was a prototype ethnocratic project – using the historical and nationalist 'weight' of Jerusalem to expand Zionist development and territorial gains. Since the 1970s, electoral patterns among Jerusalemite Jews have generally reflected a right-of-centre nationalist and religious leaning, as illustrated by the 1984 national elections, when Likud received 34.2 per cent, the group of religious parties 38.5 per cent (about twice the national figure), and Labor a mere 12.6 per cent (less than half the national percentage). In city elections the situation has been even starker, with Orthodox and ultra-Orthodox parties receiving 44 per cent and 47 per cent in 1998 and 2003, respectively.

The Oslo process, which divided Hebron, deliberately avoided dealing with Jerusalem. It was considered one of the issues left for permanent settlement negotiations. As such, a process of 'creeping apartheid' – an increasingly institutionalized yet undeclared political order – has taken root in the city. The two Palestinian *intifadas* staged a record number of terrorist attacks in Jerusalem. But Israel has also used widespread violence and state terror to preserve its dominance (B'tselem, 2007; Cohen, 2007) and violence in Jerusalem has remained, to some extent, a national, rather than an urban, issue. The main urban control methods used by Israel have been severe, but somewhat softer than in the West Bank, revolving mainly around issues of municipal and land policies. These have included widespread land confiscation, denial of planning rights, economic deprivation, and house demolition. These tactics have shaped relations between Palestinians and the city more than state or Islamic terror (Margalit, 2001).

But the conflict over the city has continued to polarize and turn local populations towards religion. In recent years, the job of Judaizing Arab Jerusalem has been carried out almost entirely by religious Jews, either ultra-Orthodox families buying new apartments in rapidly developing Israeli colonies at the edge of the city, or by small radical groups who settle in the heart of Palestinian neighbourhoods. Two such groups, Atteret Kohanim (Priests' Crown) and El'ad-Ir David (City of David), have received most of the recent attention. They are mainly young, ultra-Orthodox, and nationalist ('Hardalim') groups who use highly charged, 'redemptive' religious rhetoric. The religious shift is also noticeable electorally: in the 2006 elections the representation of Orthodox and ultra-Orthodox groups rose to twenty-four seats (24 per cent of the Jewish vote), while that of nonreligious parties declined to 76 per cent.

The separation of Jerusalemite politics from the general national and international hype about the city is, of course, impossible. However, in this chapter we are less interested in the latter, focusing more on religious

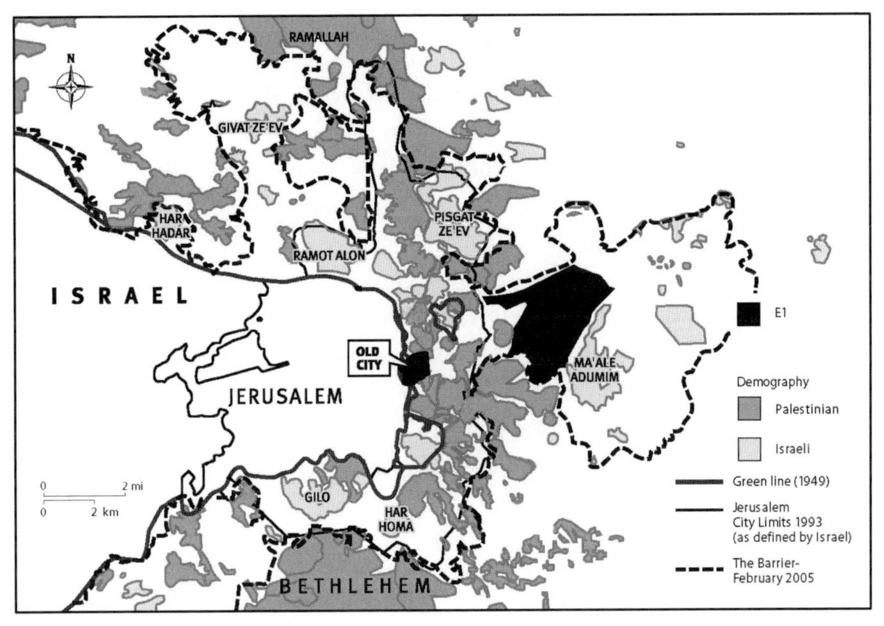

Figure 8.4. The new Jewish neighbourhoods/settlements around Jerusalem: Jewish settlement in the sacred basin. (*Source*: Foundation for Middle East Peace, 2003)

politics in the city and less on the range of global and national religious groups who have mobilized around the sanctification of Jerusalem. But two such 'external' groups should be mentioned, as they feature heavily in the daily workings of the city. The first is the notable Jewish Orthodox messianic group Ne'emanei Har Habayit (Temple Mount Loyalists), which is constantly mobilizing to allow Jews to pray at the Temple Mount, and which also aspires to build a synagogue there. Their proposals are often debated by the city administration, religious courts, and the state government, and during the Camp David summit (2000) Israel even submitted a formal proposal to build such a synagogue.

Among the Palestinians, too, messianism has been increasingly prominent, and is carried out often by external groups. A notable case is the 'northern' branch of the Islamic Movement in Israel Proper, which has planned an effective mobilization and expansion campaign around the slogan '*al-Aqsa fee chatar*' ('al-Aqsa in danger'), and has been coordinating renovations and redevelopment of the Temple Mount site for the last decade. The group, led by controversial Sheikh Raed Sallah of Um al-Fahem, uses its Israeli mobility rights to attend al-Aqsa Friday prayers, which are often closed to West Bank Arabs, *en masse*. The group also actively and openly mobilizes for the Arabization of Old Jerusalem. The two groups exemplify the inseparable connection between cities such as Jerusalem and overlapping fields of power and identity, which propel the forces of religious radicalism in the city.

In parallel, Hamas has risen rapidly to political dominance, winning 47 per cent and four of six Jerusalem seats in the 2006 elections, and using the existence of the holy sites to mobilize large groups of Muslims to special events and regular pilgrimages. This is not entirely new in Palestinian politics, which during the 1920s and 1930s were dominated by Jerusalemite elites who often mobilized support through the use of religious institutions. Hamas often builds on the early connection of Palestinian nationalism to Islam, by scorning the failure of secularist mobilization to secure a state, and by emphasizing a 'return to the roots'. Here, glorification and control of the holy city and its sacred sites provide a route to national and personal salvation, embodied in the popular slogan 'Islamic is the solution'.

Beer Sheva

Beer Sheva (Be'er Sheva' and Bi'r Saba'a in Hebrew and Arabic, respectively) is mentioned in the Bible as Abraham's first place of residence in the Promised Land. It is believed the city was abandoned in the seventh century CE, and was rebuilt only at the beginning of the twentieth century by the Ottomans. The 1947 UN partition plan included Beer Sheva under Palestinian sovereignty, but the city was captured by Israel in 1948 and since has remained within its sovereign territory. During the 1948 war, some 80 per cent of Arabs of the Naqab region (Negev) were driven out, mainly into Gaza, Egypt, the West Bank, and Jordan, leaving only 11,000 concentrated in a special military-controlled zone known as 'the Siyyag' ('the Fence'). This group was eventually awarded Israeli citizenship (figure 8.5).

In the ensuing decades, Israel has invested a great deal of effort Judaizing the previously Arab Naqab (Negev) region through a combination of deeply ethnocratic land, development, housing, and planning policies. Israel appropriated nearly all Bedouin land (with about 5 per cent of the region still under dispute) and built ten new Jewish towns and about one hundred rural Jewish settlements. Here, Jewish immigrants were housed, wrapped in a glorifying national and planning discourse about 'settling the frontier'.

In the 1970s Israel began to implement an urbanizing planning strategy for the region's Bedouin Arabs, attempting to concentrate them in seven modern towns immediately surrounding, but not included in, the Jewish Beer Sheva. This policy relocated about half the Arabs of the south (some 85,000 in 2007 – mainly those with no land claims) though the (significant) lure of modern infrastructure. However, despite some development, the towns have become known for their marginality, unemployment, deprivation, and crime (Abu-Saad and Lithwick, 2004).

The other 80–85,000 Bedouins have remained on their claimed land, in some forty-five shanty towns and villages (figure 8.5). A bitter land conflict has developed, with the state denying their indigenous land rights, and as a result declaring them 'invaders' in their own historic localities. In an effort to force them to relocate, the state has prevented the supply of most services, including roads, electricity, clinics, and planning, and has regularly

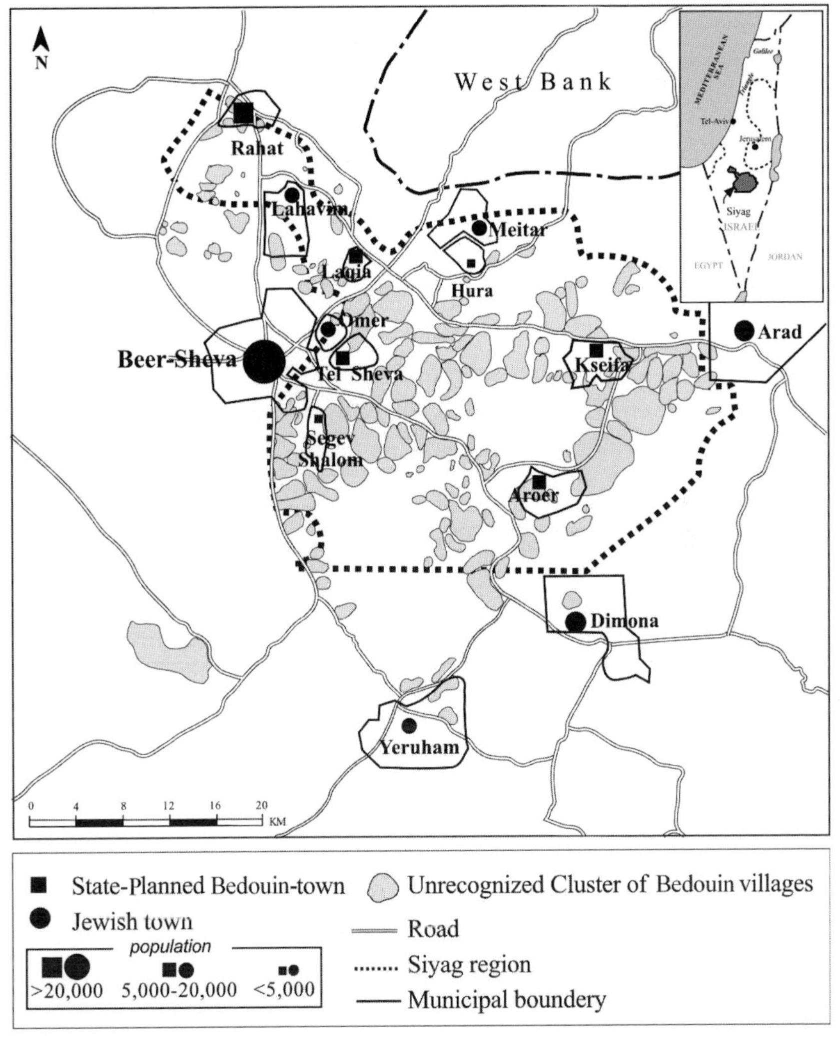

Source: Israel Land Authority (report to metropolitan planning team, 2003).

Figure 8.5. Jewish and Bedouin settlement in the Beer Sheva region.

launched house-demolition campaigns (Yiftachel, 2004). Levels of poverty, mortality, and crime are among the highest in Israel/Palestine, and create a metropolitan geography of stark ethno-class contrast with the neighbouring, well-serviced, Jewish localities.

The Beer Sheva metropolis has therefore come to resemble many Third World cities, with a well-developed modern urban core and a range of peripheral informal localities which suffer from severe poverty and deprivation. This is reflected in the nature of religious politics. Unlike Hebron and Jerusalem, Jewish (internal) colonial policies in the Beer Sheva region have only rarely been couched in religious terms, despite the biblical significance of the place. Instead, they have used discourses of modernization, national (Jewish) territorial control, 'proper planning', and 'law and order'.

The Bedouin Arab challenge to Jewish hegemony is represented through repeated moral panics over demographic and territorial dominance, and over crime and 'primitivism' – but very rarely over religious issues.

Politically, the composition of the Beer Sheva City Council, too, has remained quite stable, with religious party representation hovering around 20–25 per cent for the last two decades (25 per cent in the last municipal elections). The relatively low profile of religiosity also reflects the conspicuous presence of Russian-speaking immigrants in the city, who now outnumber all other Jewish groups and are known for their secularism. City politics have been dominated for years by the centrist and nationalist Labor, Likud, and (recently) Kadima parties. The last two local elections were won by a Labor-affiliated mayor, while Likud remains the largest party in national elections. Religious parties occupy five of the 25 city council seats, but they rarely raise biblical-related issues or pursue radical religious demands.

Recently, however, religion has become more prominent. In a similar fashion to cities of the global South – being a region where sovereignty itself is not contested – these religious politics have mainly emerged 'from below' as a 'weapon of the weak' (Scott, 1985). The Islamic movement, for example, has effectively mobilized Bedouins, whose religious practices were traditionally quite removed from formal Islam, through effective social and educational campaigns, and since the 1980s the landscape has become dotted with mosques. Naqab Islamic organizations are affiliated with the more moderate 'southern' branch, and have rapidly increased their political support, currently holding the mayoral position in five of the seven Arab towns. In the 2006 national elections they won the support of 55 per cent of Naqab Arabs (running jointly with a traditional Arab United List). Among the Jews, too, the main expression of religious politics has been 'from below', though the emergence of the Shas movement, representing the lower-income Oriental Jewish classes, and holding four council seats. Shas has been more concerned with material services and education facilities than with linking religiosity to urban colonialism.

It should also be mentioned that inter-communal violence, so visible in Hebron and Jerusalem, has been rare in Beer Sheva. Only four people are estimated to have died on 'national grounds' during the last two decades, compared to hundreds in the other two cities (B'tselem, 2007). Continuous urban colonialism, however, is apparent in the Beer Sheva region, and while it is less confrontational than in Hebron and Jerusalem, it generates its own politics. Religion has also played an increasing part in recent Arab campaigns in the city, especially around education and places of worship. The latest such issue surrounds the renowned and architecturally significant Beer Sheva mosque, built by the Ottomans to serve the region's population. Despite constant appeals, the city has refused to open it for prayer, and a powerful councillor of the ruling coalition, Eli Bokker (2005), has claimed, 'the region has dozens of mosques in Bedouin localities and towns, and Beer Sheva is now a Jewish city, with the right to protect this urban character'.

As a result, the mosque has been lying idle for decades, and is now in an advanced state of architectural deterioration. In a recent appeal, the Israeli high court ruled in favour of opening the mosque for 'Arab cultural uses' (Adalah, 2005). Yet, despite the ruling, the city is steadfast in its refusal, and has now condemned the building as too dangerous for human use. While the most vocal against opening the mosque have been members of the nationalist Likud and (the mainly Russian) Yisrael Beitenu parties, Jewish religious parties have also joined the choir. Yaacov Margi, a Beer Sheva Shas leader, has claimed,

> ... if implemented, this high court decision could be the last nail in the Beer Sheva coffin ... we have been increasingly surrounded by Bedouins from all sides, and now they attempt to penetrate the heart of our city by opening their mosque... Let us never forget – Beer Sheva was the first Jewish city; this is where Abraham's wells are still in existence after 4,000 years. We should continue and drink the wisdom of our Tora like the water from these wells, and remember that one of these wisdoms is to never, but never, let the Amalek [hostile nations] raise their heads! (*Sheva* [newspaper], 12 April 2005)

Margi's statement is a reminder that in spatial conflicts typical of urban colonialism and contested identity politics, religion is rarely far from the surface. But the process of polarization and radicalism, which has led to massive mobilization and widespread violence in other regions, has so far remained quite dormant in Beer Sheva. Thus, in this city religious politics have begun to make their mark, but have not yet become radicalized.

Instead of Conclusions

Students of religious politics are urged to incorporate the political geography of urban colonialism into their work and understanding. As shown here, the hegemonic systems of control – ethno-nationalism, globalizing capitalism, and increasingly politicized religion – intersect through the 'thick matter' of making and changing cities. It is there that new forms of appropriating and racializing colonialisms are being produced as the foundation for religious radicalism, both 'from above' and 'below'.

But rather than rely on these macro-processes as 'given', scholars are urged to 'breathe life' into the details of urban spaces and configurations of power, rights, and identities which emerge in different types of sacred or sanctified spaces, and which give rise to different forms of domination and counter-mobilization. These, we suggest, provide insightful clues to the rise and nature of religious radicalism, as also depicted in the wise words of the late Hebrew poet Yehuda Amichai:

> *Gods Change, Prayers Remain Forever*
> .... We are all Abraham's children
> But also the grandchildren of Terach, Abraham's father.
> And it's now perhaps time for the grandchildren to do
> To their father what he did to his,
> Break his statutes and idols, his religion and faith,
> But this too will be the beginning of a new religion.

References

Abu el-Haj, N. (2001) *Facts on the Ground: Archaeological Practice and Territorial Self-Fashioning in Israeli Society*. Chicago, IL: University of Chicago Press.

Abu-Saad, I. and Lithwick, H. (2004) *A Way Ahead: A Development Plan for the Bedouin Towns in the Negev*. Beer-Sheva: The Center for Bedouin Studies.

Adalah (Legal Center for Arab Minority Rights in Israel) (2005) 9 Supreme Court Petitions: Religious Rights. Haifa.

Akenson, R.H. (1992) *God's Peoples: Covenant and Land in South Africa, Israel, and Ulster*. Ithaca, NY: Cornell University Press.

Almond, G.A., Appleby, R.S. and Sivan, I. (2003) *Strong Religion: The Rise of Fundamentalisms around the World*. Chicago, IL: University of Chicago Press.

Bartholomeusz, T.J. and de Silva, C.R. (1998) Buddhist fundamentalism and minority identities in Sri-Lanka, in Bartholomeusz, T.J. and de Silva, C.R. (eds.) *Buddhist Fundamentalism and Minority Identities in Sri-Lanka*. Albany, NY: State University of New York Press.

Bayat, A. (2000) From 'dangerous classes' to 'quiet rebels': politics of the urban subaltern. *International Sociology*, **15**, pp. 533–557.

Benvenisti, M. (2002) *Sacred Landscape: Buried History of the Holy Land Since 1948*. Los Angeles, CA: University of California Press.

Bokker, E. (2005) Without Tarbush and Muazins. *Tsatetet*, 26 April.

Bourdieu, P. (1991) *Language and Symbolic Power*. Cambridge: Polity Press.

B'tselem (2007) *Report on Casualties in Israeli-Palestinian Hostilities, September 2000–December 2002*. Jerusalem: B'tselem.

Budeiri, M.K. (1995) The nationalist dimension of Islamic movements in Palestinian politics. *Journal of Palestine Studies*, **24**(3), pp. 89–95.

Chatterjee, P. (2004) *The Politics of the Governed: Reflections on Popular Politics in Most of the World*. New York: Columbia University Press.

Cohen, H. (2007) *The Rise and Fall of Arab Jerusalem 1967–2007*. Jerusalem: The Jerusalem Institute for Israel Studies, Keter (in Hebrew).

Cooper, A. (1992) New directions in the geography of religion. *Area*, **24**, pp. 123–129.

Davis, U. (2006) Whither Palestine-Israel? Political reflections on citizenship, bi-nationalism and the one-state solution. *Holy Land Studies*, **5**(2), pp. 199–210.

Dumper, M. (2002) *The Politics of Sacred Space: The Old City of Jerusalem in The Middle East Conflict*. Boulder, CO: Lynne Rienner.

Eldar, A. and Zertal, I. (2004) *Lords of the Land: The Settlers and the State of Israel 1967–2004*. Or Yehooda: Kinnereth, Zmora-Bitan, Dvir Pub (in Hebrew).

Finke, R. (2003) Spiritual Capital: Definitions, Applications, and New Frontiers. Presentation to the Spiritual Capital Planning Meeting, 10–11 October 2003. Available at http://www.spiritualcapitalresearchprogram.com/pdf/finke.pdf. Accessed 25 March 2010.

Friedland, R. and Hecht, R. (2007) Sacred Urbanism: Jerusalem's Sacrality, Urban Sociology, and the History of Religions. Presentation to the conference on 'Jerusalem Across the Disciplines', Arizona State University, Tempe, 19–21 February 2007.

Ghanem, A. (2000) *The Palestinian-Arab Minority in Israel, 1948–2000*. Albany, NY: State University of New York Press.

Hall, S. (1992) Cultural identity in question, in Hall, S., Held, D. and McGrew, T. (eds.) *Modernity and Its Futures*. Cambridge: Polity.

Hatina, M. (2005) Theology and power in the Middle East: Palestinian martyrdom in a comparative perspective. *Journal of Political Ideologies*, **10**(3), pp. 241–267.

Hilal, J. (ed.) (2006) *Where Now For Palestine: The Demise of the Two-State Solution*. London: Zed Books.

Huntington, S. (1996) *The Clash of Civilization and the Remaking of World Order*. New York: Touchstone Books.

Isaac, E. (1960) Religion, landscape and space. *Landscape*, **9**, pp. 14–18.

Jackson, R. and Henrie, R. (1983) Perception of sacred space. *Journal of Cultural Geography*, **3**(2), pp. 94–107.

Katznelson, I. (1996) Social justice, liberalism and the city, in Morrifield, A. and Swyngedouw, E. (eds.) *The Urbanization of Injustice*. London: Lawrence and Wishart.

Khemaissi, R. (2005) *Towards Expansion of Arab Municipal Areas*. Jerusalem: Floresheimer Institute (in Hebrew).

Laclau, E. (1994) Introduction, in Laclau, E. (ed.) *The Making of Political Identities*. London: Verso.

Lefebvre, H. (1996) Philosophy of the city and planning ideology, in Lefebvre, H. (ed.) *Writings on Cities*. London: Blackwell.

Lustick, I. (2002) In search of hegemony, *Hagar*, **3**(2), pp. 171–202.

Mann, M. (1999) The dark side of democracy: the modern tradition of ethnic and political cleansing. *New Left Review*, No. 235, pp. 18–45.

Mann, R. (2002) The colour of prejudice. *Sydney Morning Herald*, 23–24 February, pp. 4–6.

Marcuse, P. (1995) Not chaos, but walls: postmodernism and the partitioned city, in Watson, S. and Gibson, K. (eds.) *Postmodern Cities and Spaces*. Oxford: Basil Blackwell.

Margalit, M. (2001) A Chronicle of Municipal Discrimination in Jerusalem between West and East. *Palestine Israel Journal of Politics, Economics and Culture*, **8**(1). Available at http://www.pij.org/details.php?id=172. Accessed 25 March 2010.

Martin, E.M. and Appleby, R.S. (1991) *Fundamentalisms Observed: Modern American Religion*, Vol. 1. Chicago, IL: University of Chicago Press.

Massey, D. (2005) *For Space*. London: Sage.

McGarry, J. (1998) 'Demographic engineering': the state-directed movement of ethnic groups as a technique of conflict regulation. *Ethnic and Racial Studies*, **21**(4), pp. 613–638.

McGarry, J. and O'Leary, B. (2004) *The Northern Ireland Conflict: Consociational Engagements*. Oxford: Oxford University Press.

Newman, D. (1997) Metaphysical and concrete landscapes: the geopiety of homeland socialization in the 'Land of Israel', in Brodsky, H. (ed.) *Land and Community: Geography in Jewish Studies*. Bethesda, MD: University of Maryland Press.

Newman, D. (2001) Boundaries, borders and barriers: a geographic perspective on territorial lines, in Albert, M., Jacobson, D. and Lapid, Y. (eds.) *Identities, Borders, Orders: New Directions in International Relations Theory*. Minnesota, MN: University of Minnesota Press.

OCHA (United Nations Office for the Coordination of Humanitarian Affairs) (2005) West Bank Closure and Access – April 2005. Available at http://domino.un.org/unispal.nsf/0/e2294ff978610ff385256ff7005181d2?OpenDocument. Accessed 25 March 2010.

Ram, H. (1996) Mythology of rage: representations of the 'self' and the 'other' in revolutionary Iran. *History and Memory*, **8**, pp. 67–89.

Raz-Krakotzkin, A. (2002) Between Brith Shalom and the Temple: the dialectics of redemption and messianism in the wake of Greshom Scholem. *Theory and Criticism* (*Teorya U'vikkoret*), **20**(1), pp. 88–107 (in Hebrew).

Robinson, J. (2006) *Ordinary Cities: Between Globalization and Modernity*. London: Routledge.

Rouhana, N. (ed.) (2004) *Citizenship without Voice: The Palestinians in Israel*. Haifa: Mada al-Karmel Center (in Arabic).

Roy, A. (2009) The 21st century metropolis: new geographies of theory. *Regional Studies*, **43**(6), pp. 819–830.

Samaddar, N. (2002) *Space, Territory and the State: New Readings in International Politics*. Hyderabad: Orient Longman.

Scott, J.C. (1985) *Weapons of the Weak: Everyday Forms of Peasant Resistance*. New Haven, CT: Yale University Press.

Shenhav, Y. (2006) *The Arab Jews: A Postcolonial Reading of Nationalism, Religion, and Ethnicity*. San Francisco, CA: Stanford University Press.

Shilhav, Y. (1991) Jewish sacred space: ideal space? in Shilhav, Y. (ed.) *Jewish Town in the City*. Jerusalem: Jerusalem Institute (in Hebrew).

Sibley, D. (1995) *Geographies of Exclusion: Society and Difference in the West*. London: Routledge.

Silberman, N.A. (2001) If I forget thee, O Jerusalem: archaeology, religious commemoration and nationalism in a disputed city, 1801–2001. *Nations and Nationalism*, **7**(4), pp. 487–504.

Silberstein, L.J. (1993) Religion, ideology, modernity: theoretical issues in the study of Jewish fundamentalism, in Silberstein, L.J. (ed.) *Jewish Fundamentalism in Comparative Perspective*. New York: New York University Press.

Smith, A.D. (2000) The 'sacred' dimension of nationalism. *Millennium: Journal of International Studies*, **29**(3), pp. 791–814.

Swiesa, S. (2003) Hebron, Area H-2: Settlements Cause Mass Departure of Palestinians. Available at http://www.btselem.org/Download/200308_Hebron_Area_H2_Eng.doc. Accessed 25 March 2010.

Tajbakhsh, K. (2002) *The Promise of the City: Space, Identity and Politics in Contemporary Social Thought*. Berkeley, CA: University of California Press.

Wilson, E. (1995) The invisible flâneur, in Watson, S. and Gibson, K. (eds.) *Postmodern Cities and Spaces*. Oxford: Basil Blackwell.

Winslow, D. (1984) A political geography of sacred deities: space and the pantheon in Sinhalese Buddhism. *Journal of Asian Studies*, **43**(2), pp. 273–291.

Yiftachel, (2004) Contradictions and dialectics: reshaping political space in Israel/Palestine. *Antipode*, **36**, pp. 607–613.

Yiftachel, O. (2006) *Ethnocracy: Land, Politics and Identities in Israel/Palestine*. Philadelphia, PA: Penn Press.

Yiftachel, O. and Ghanem, A. (2004) Ethnocratic regimes: the politics of seizing contested territories. *Political Geography*, **23**, pp. 647–676.

Yiftachel, O. and Yacobi, H. (2002) Planning a bi-national capital: should Jerusalem remain united? *Geoforum*, **33**, pp. 137–145.

Chapter 9

Fundamentalism at the Urban Frontier: the Taliban in Peshawar

Mejgan Massoumi

In recent times religious fundamentalism has re-emerged as a major force shaping global politics, human relations, and violence. Particularly in the post-9/11 era, it has often also been equated stereotypically with Islam and Muslims. On 4 June 2009, however, United States President Barack Obama addressed the Muslim world in a speech at the Great Hall of Cairo University. His remarks were an appeal to Muslims everywhere, citing the Quran on several occasions and praising the scientific and cultural achievements of Islamic civilization and its tradition of tolerance and racial equality. Most notably, President Obama used his platform that day to make an important observation and distinction – that Islam and violent extremism do not equate with one another. He spoke of Islam as a religion of peace, and emphasized his responsibility as president to fight against negative stereotypes of Islam wherever they appear.

Throughout his speech, the president assured his audience that a future of tolerance and mutual understanding is possible. But how much does his optimism reflect reality? The threat posed by fundamentalists is an intercivilizational challenge, and one can only hope that Obama's vision of 'A New Beginning' – the title of his Cairo speech – succeeds. As he, himself, stated, a single speech cannot be expected to change the course of history. Across the world past grievances still have violent repercussions.

Just five days after President Obama's speech, a car-bomb detonated outside the Pearl Intercontinental Hotel in Peshawar, Pakistan, leaving eleven people dead and more than fifty-five wounded. The previous month, two consecutive bomb blasts had devastated a bazaar in the city, killing five and injuring several others. From 2008 to 2009 violence in

Peshawar increased steadily in reaction to a Pakistani military campaign against religious insurgents in the North West Frontier Province (NWFP). The underlying source of this violence was clear. Once a cultural hub for artists, musicians, and intellectuals, Peshawar has been transformed by the presence of one of the most potent fundamentalist groups in the world.

The Taliban arrived in Peshawar in the 1980s when the American CIA and Pakistani ISI armed, trained, and prepared soldiers to fight the occupation of Afghanistan by the Soviet Union. It then used the city as a sanctuary and a base for operations both in its campaign against irreligion/corruption and its fight to displace the warlords who ruled Afghanistan after the Soviet pullout. The Taliban left Peshawar briefly when it prevailed in those efforts and became Afghanistan's recognized government in the late 1990s. But in 2001 it returned to the city, as the United States under President George W. Bush declared 'war on terror', forcing its leaders to once again seek refuge across the border.

The sustained reliance by the Taliban on Peshawar as a transitional and transnational base highlights the crucial relationship between the group and the city. This chapter explores this relationship and examines the practices through which Peshawar's urban landscape has been altered by its presence and by its efforts to institutionalize and embed its fundamentalist ideology in public life.

In general terms, the resurgence of religious fundamentalist groups across the world shows how broad historical forces play out at the local level and affect the lives of individuals in small communities (Antoun, 2008). In his Cairo speech, President Obama alluded to the legacies of colonialism and the Cold War, in which Muslim countries were often treated as proxies, without regard to their own aspirations. This history has had particularly negative repercussions in places like Afghanistan and Pakistan, where small groups of extremists have been able to thrive based on appeals to the lingering sense of injustice that is the legacy of these policies.

It is important to note that Islamic fundamentalism, even the most militant and violent type, is not an isolated phenomenon; it shares many attributes with other contemporary non-Islamic religious movements. The point of this chapter, however, is not to expand on these shared attributes, but to focus on the Islamic case. Moreover, while many scholars have used the concept of fundamentalism to understand the resurgence of religious identities in various contexts around the world, few have discussed its spatial implications or the role of spaces in shaping and strengthening its processes. In this regard, a central premise of this volume is that the connection between fundamentalism and space provides a useful conceptual tool with which to study groups who use religious ideology to mobilize society – and, more broadly, to understand the impacts of these groups on social, political, cultural, and urban transformations.

Towards this end, this chapter is structured in three sections. The first situates the Taliban within the fundamentalist discourse, highlighting the limitations and usefulness of this categorization. The second provides a brief overview of the history of Peshawar and the events which led to the Taliban's presence in the city. It also examines the notion of 'the fundamentalist city' by investigating the specific spatial strategies employed by the Taliban to control, secure, and order public space and livelihoods. Here the idea of the 'fundamentalist city' will not be employed to catalogue Peshawar, but to conceptualize the specific forms of urban exclusion and surveillance that have emerged in the city, and to interrogate the nature of the urban condition that has allowed religious fundamentalism to flourish. The third section of the chapter describes the Taliban's practice of exclusion, exception, and surveillance in Peshawar as a process of enclave urbanism that bears an uncanny resemblance to forms of citizenship in the medieval city (see AlSayyad and Roy, 2006). The chapter concludes with thoughts on the current state of Peshawar and the potential for radical religiosity to recast the urban in very exclusive and distinct ways.

The 'Fundamentals' of Taliban Ideology

The rise of the Taliban can be viewed as part of the broader resurgence of religious fundamentalism across the globe that began in the 1970s (Emerson and Hartman, 2006, p. 132). Scholars have described this resurgence as a reaction to secularism and to the conventions of the modern world – particularly the processes by which religion has been removed from the public sphere.

In an effort to understand how and why this phenomenon has spread, scholars, particularly in the social sciences, began to study the rise of fundamentalisms around the world in the early 1990s. One of the most comprehensive of these efforts was the Fundamentalism Project, sponsored by the American Academy of Arts and Sciences and directed by religious historians Martin Marty and Scott Appleby. In a series of five volumes, with contributions from dozens of scholars from around the world examining various forms and manifestations of religious fundamentalism, the project identified family resemblances and ideological and organizational traits shared by fundamentalist movements. Among the traits identified were the following: a concern for the erosion of religion in society due to increasing modernization and secularization; the selection and reshaping of aspects of religious traditions to distinguish the group from the mainstream; the embrace of dualist views of good and evil; and an emphasis on absolutism and the inerrancy of sources of revelation. In addition, the project discovered these movements espouse forms of millennialism or messianism; are united by a strong belief that they have been selected or 'chosen' to defend their religious tradition; have distinct ideological boundaries that establish a 'with us' or 'against us' mentality; and are based on authoritarian systems of organization (Almond *et al.*, 1995).

While this list of ideological and organizational attributes seems to present a detailed catalogue of traits uniting fundamentalists worldwide, critics of the work of Marty and Appleby point out that while some religious movements may be described as fundamentalist, they seldom fit all the aforementioned characteristics, and vice versa (Emerson and Hartman, 2006, p. 135). As such, several scholars, including some involved in Marty and Appleby's project, have attempted to further define this model into a single, powerful statement that would also prove historically comparative and intercultural. Surveying a few of those who have addressed the Islamic case in particular will help develop such a conceptual understanding here, as well as indicate the complexity and usefulness of applying this model to the Taliban movement.

Many scholars point out that it is not only recently that the concept of fundamentalism has been applied to Islam. But this connection drew increased attention following the Islamic Revolution in Iran in the 1970s (Antoun, 2008; Riesebrodt, 1993). Particularly relevant to the discussion of the Taliban is the observation that fundamentalist movements frequently pit themselves against the ideological orientation of modernism. Some scholars, however, reject the use of the term in relation to Islamic groups because it originated as a description of certain American evangelical Protestant sects of the early twentieth century, and they believe its use is limited in other contexts. Other scholars prefer to use terms such as 'political Islam' or 'Islamism' to describe the resurgence of religious movements in Muslim contexts, and to distinguish Islam as a modern ideology from Islam as a faith. Still others such as Edward Said (1979) have not dismissed the term altogether, but have objected to how it has been employed against Islam.

It is understandable that scholars tend to shy away from the term 'fundamentalism' in such contexts, because it supposes an uncomfortable equation with the experience of colonialism and the specific practices which attempted to contain religious influences in the public sphere. Indeed, in postcolonial contexts radical religious movements have been conceived by the very technologies and instruments of colonial domination, as in the case of the Taliban. Nonetheless, there is widespread recognition that the value of the term lies in the reality to which it refers. Thus, many scholars and observers of Muslim societies find it useful to have a term that can refer to 'the complex cluster of movements, events, and people who are involved in the affirmation of the fundamentals of the Islamic faith…' (Voll, 1991, p. 347).

To borrow again from John Voll, Islamic fundamentalism may be defined as the reaffirmation of foundational principles and the effort to reshape society around those reaffirmed fundamentals. Many Muslims assert the truth in the revelation of the Quran and believe that they have an obligation to implement the fundamentals of that truth in their daily lives and societies. However, the 'fundamentalist' approach to this obligation is marked by an exclusive and literal interpretation of these principles and by the introduction of special reforms to prevent the supposed deterioration of Islamic religious life.

In applying this definition to the Taliban, one must ask an important question: are the characteristics that unite the Taliban really characteristic of Islamic fundamentalism, or are they the expression of some other type of (un)definable political activism? The noted journalist Ahmed Rashid (2000) has pointed out that the Taliban's ideological roots lie in the practice of an extreme form of Deobandism, a branch of Sunni Hanafi Islam which sought to reform and unify the Muslim community living under British colonial rule in India.[1] Within this tradition, however, the Taliban's particular ideology is derived largely from the Jamiat Ulema-e-Islam (JUI), a religious movement formed shortly after the creation of Pakistan in 1947. In the 1980s the JUI had established hundreds of religious seminaries in Pakistan (NWFP) and Baluchistan, where refugees escaping the Soviet occupation in Afghanistan were offered a free education alongside local Pakistanis (Rashid, 2000). This is where the Taliban came into being.

As Rashid and others have pointed out, the Taliban and the Deobandi tradition have a few important similarities and one major difference. Both emphasize Shari'a law as the basis for governing society, and both hold restrictive views of the role of women, rural village customs, holy men, shrines, and Shi'a Muslims. However, they have quite different views on the role of education and the advancement of Muslim society. Where Deobandism prized education and aimed to train a new generation of Muslims to revive Islamic values based on intellectual learning, the Taliban members have little training and knowledge of Islamic history, Shari'a, or the Quran, and do not care to engage with political and theoretical developments in the Muslim world. In describing the Afghan Taliban, Rashid (*Ibid.*, p. 93) points out that there is 'no Taliban Islamic manifesto or scholarly analysis of Islamic … history'. While other contemporary Islamic groups engage with scholarship and global religious debate, the Taliban departs from this intellectual tradition.[2] Their lack of knowledge, not only of Islam but of their own personal history, presents an anomaly within the wider Muslim world. In short, Taliban ideology perversely amplifies certain aspects of Deobandism and begs the question whether other forces have shaped it as a movement.

Although the Taliban, drawn primarily from Pashtun tribes, profess an Islamic ideology, their Pashtun ethnicity has played a major role in their self-definition. This is most apparent in their chauvinism and in the construction of ideological boundaries between them and non-Pashtun groups such as Hazaras, Tajiks, and Uzbeks.[3] Pashtunwali, the unwritten Pashtun tribal code, has also been an important influence on the Taliban's highly traditional and patriarchal interpretation of Islamic ideology. For example, Taliban punishments seem to be drawn more from Pashtunwali than Shari'a law.

In this regard, Rashid (*Ibid.*, p. 112) has argued that the Taliban's enforcement of Pashtunwali-Shari'a law in the late 1990s served only to deepen the ethnic divide in Afghanistan, as non-Pashtuns viewed it as an attempt to impose Pashtun laws on them. However, when the Taliban was ousted in 2001, its leaders were easily able to find refuge across the border in the NWFP and the Federally Administered Tribal Areas (FATA) of Pakistan because the majority Pashtun population there naturally upheld the traditions of Pashtunwali – which includes giving asylum to any person, or people, who requests or seeks protection from an enemy (Johnson and Mason, 2008). Furthermore, once the Pashtun tribes in these areas accepted the Taliban as their guests, Pashtunwali dictated that they be treated with the greatest hospitality, loyalty, and respect (*Ibid.*). The implications of this salient fact – that religious fundamentalism is centred within a single ethno-linguistic group – has not been given much attention by scholars. Its importance lies in the fact that it distils the larger cultural dynamics that have shaped the Taliban as a fundamentalist movement (*Ibid.*).

The ideology of the Taliban is even more fractured than can be gleaned from such an ethno-religious understanding, however. Other groups have also influenced the Taliban's social composition. Most importantly, during the 1980s and 1990s it was heavily influenced by the Wahhabism of its Saudi benefactors, including Osama bin Laden.[4] The creation of a religious police force and a Ministry for the Promotion of Virtue and the Eradication of Vice during the years in which the Taliban governed Afghanistan, mimicked Wahhabi practices. In sum, therefore – whether religious, ethnic, or political – Taliban ideology is a fusion of competing agendas. Crews and Tarzi (2008) have further pointed out that the Taliban have changed their political perspectives and cultural practices as they have embraced different core and support elements, primarily reflecting different sources of financial and material backing.

Despite this fragmented nature, the Taliban do fit many of the ideological and organizational traits of a fundamentalist group as identified by the work of Marty and Appleby and by John Voll's definition of Islamic fundamentalism. Other attributes of the Taliban with fundamentalist overtones include, but are not limited to, the following. In their effort to purify the impure and corrupt modern world, members of the Taliban see themselves as engaged in a militant struggle to overcome and seize that world (Antoun, 2008). Ultimately, they want to reinstate religion – albeit the distorted and imagined version of Islam they practice – in all aspects of public life and purify society of its modern and Western influences. They justify their actions and purpose by seeking credibility through the Quran and the Hadith. But they also employ certain aspects of modernity, such as technology and communication, to expand their reach and maintain their power, even as they reject schooling for women and modern innovations such as television (*Ibid.*). Finally, in their quest for authenticity, members of the Taliban claim the immediate, literal relevance to today's society of society as it existed during the life of the prophet Mohammed. Thus, they seek to distinguish themselves from the modern world by imitating the physical and social life of the hallowed past, whether by their physical appearance (beards, dress, demeanour) or religious practices (diet, prayer, Shari'a law). Their mission is thus contradictory and built on exemplary violence.

In understanding the Taliban it is also useful to question commonly held views of modernity. In particular, such a discussion can benefit from the alternate framework of 'medieval modernity' formulated by AlSayyad and Roy (2006). Building on the work of Agamben (2005), they have explained the rescaling of citizenship and exclusionary power in fragmented, enclaved cities by historicizing new urban practices, and hence 'new forms of discipline and control'. Specifically, AlSayyad and Roy (2006, p. 17) employ the concept of the 'medieval' as a transhistorical analytical category, and argue that modernity in enclaved environments should be understood as 'an inevitably fractured, divided and contradicted project'. The concept of medieval modernity, 'thus reveals the inherent paradoxes of the modern: fiefdoms of democracy, the materialist immediacy of religious fundamentalism, the simultaneity of war and humanitarianism'.

AlSayyad and Roy's argument also stresses the profound and reciprocal relationship between new spatial forms, citizenship, and governmentality as described by Abourahme (2009). Peshawar is at the centre of such a world. It represents the emergence of what Balbo (1993, p. 25) terms 'microstates' and the 'tribalization' of city spaces. Essentially, in such a concept, the Taliban represent a different kind of governmentality. Their unregulated practices translate into a distinct form of regulation, or 'a set of tactics that recreate informality as governmentality' (AlSayyad and Roy, 2006, p. 8). The ability of the Taliban to establish their 'informality' and informal regulation in Peshawar is therefore an expression of the sovereign power to establish the 'state of exception', where questions of citizenship and individual rights can be diminished, superseded, and rejected in the process of claiming an extension to power well beyond that allowed by law in the past (Agamben, 2005). 'In every case, the state of exception marks a threshold at which logic and praxis blur with each other and a pure violence without *logos* claims to realize an enunciation without any real reference' (*Ibid*., p. 40). Peshawar then resembles the fragmented and unregulated informal city – a city in which the *de facto* sovereign power has established both a distinct form of governance based on an 'imagined' interpretation of Islam and particular forms of negotiated citizenship based on its own distinct politics, regime of rule, and institutional dynamics.

Taking the above discussion into account, the Taliban in this case may represent a new style of Islamic fundamentalism – one which rejects any form of religious moderation and refuses to adopt any system of government except that based on a distorted interpretation of Shari'a law where justice depends on punitive violence. Thus, while the Taliban do reflect many of the characteristics that typify other Islamic fundamentalist groups, they are marked by a changing ideological identity and a particularly violent and aggressive approach to protesting and resisting alien influence in the societies in which they take root. Nowhere is this new style of fundamentalism more apparent today than at the urban fringes of the Pakistani frontier. And while Pakistan has moved military forces into the NWFP since 2003 to remove 'terrorists and extremists', its ability and willingness to contain the resurgence of the Taliban has been largely unsuccessful. The area has thus turned into a major battleground for the Taliban, complete with suicide attacks, bombings, and sectarian violence.

As one of the urban strongholds of the Taliban, Peshawar, capital of the NWFP and the gateway to FATA, is often referred to as 'the city under siege', evoking the strict rule the Taliban instituted in Afghanistan from 1996 to 2001. By 2008, according to Jane Perlez (2008), incidents of 'spraying black paint on female images atop signboards; ordering schoolgirls to wear the *burqa* and blowing up stores selling DVDs, among other acts of violence' had become common in the city. A recent study by the Institute of Conflict Management documented seventy-two terrorism-related incidents in Peshawar in 2009 alone, including a mixture of suicide bombings, attacks on government and civic buildings, raids on stores selling Western music and films, and car bombings.[5] These violent interrogations of the urban landscape have created an atmosphere of insecurity, fear, and intimidation which has been successfully exploited by the Taliban, and which has helped legitimate their power in the city.

As the discussion above illustrates, the Taliban must be understood in their specific geographical context, even though it can also be studied as a form of religious fundamentalism. The concept of fundamentalism is useful because it facilitates comparative study of political groups which use religious ideology to define and mobilize society, and because it provides a better understanding of ways that radical interpretations of power and religion alter the character of urban communities. However, before examining the spatial conditions or urban features under which a religious fundamentalist movement like the Taliban may successfully expand its power, I turn to a brief overview of the history of the city of Peshawar before the Taliban resurgence.

Religion, Resistance, and Revolution in the Frontier City

Peshawar is one of the most ancient cities of Pakistan. Two thousand years ago it served as the centre of the Buddhist empire known as Gandhara, which was ruled by the Khushan king, Kanishka (Dani, 1995). During this time, Peshawar was a great centre for Buddhist learning, and King Kanishka built what may have been the tallest building in the world, a giant stupa, to house the Buddha's relics, just outside the Ganj Gate of the old city. Throughout its lifetime, Peshawar also welcomed many invaders and has been part of empires stretching across Asia, including the Afghan, Persian, Greek, Maurya, Scythian, Arab, Turk, Mongol, Mughal, Sikh, and British empires.[6] Located at the base of the Khyber Pass near the Afghan border, the city served as one of the main trading centres on the Silk Road and a major crossroads for the various cultures of South Asia, Central Asia and the Middle East.

Sometime in the first millennium BCE, the Pashtun ethnic group, which now dominates the city, began to arrive in the area from Afghanistan.

Whether or not the Pashtuns lived there before this time is still a subject of debate among scholars, because evidence to prove otherwise has been difficult to obtain (*Ibid.*). For example, Sir Olaf Caroe wrote that a group who may have been the Pashtuns existed in the area and were called the Pactycians by Herodotus and the Greeks – which would place them in the area much earlier, along with other Aryan tribes (*Ibid.*, pp. 76–77). Ancient Hindu scriptures such as the Rig-Veda also speak of an Aryan tribe called the Pakht, who lived in the region (*Ibid.*, p. 77).

Regardless of the specific timing of their arrival, over the centuries the Pashtuns eventually came to dominate the region, and Peshawar emerged as an important centre of Pashtun culture, along with Kandahar and Kabul, and more recently Quetta. In the eleventh century the Pashtuns began to convert to Islam under the Afghan Ghaznavids, a dynasty which later fell to the Mughal Empire. In the late 1600s, however, the Afghans took Peshawar back from the Mughals, and held it until the Sikhs captured it in the early nineteenth century. The British then captured it in 1848, and the city remained under their rule until 1947 (*Ibid.*).

It was during the formative years of British occupation that Peshawar emerged as a centre for Hindko and Pashtun intellectuals.[7] In particular, this period witnessed the creation of the Pashtunistan movement, which sought either to merge Western Pakistan with Afghanistan or to form a greater Pashtun state (Rashid, 2000). While Pashtunistan was never realized, it remained a burden for Pakistan, which has long feared separatist ambitions among its Pashtun community (*Ibid.*).

As a frontier city, Peshawar also became an important outpost of British India, and it was used as a base for British military operations against Pathan tribes in the nineteenth century. However, following the dissolution of the British Raj and the creation of an independent Pakistan, the North West Frontier Province, to which Peshawar belongs, was referred to as the country's Achilles's heel (Dani, 1995). The reasons are historic. While the British had captured most of the Indian subcontinent with relative ease, they faced great challenges in claiming the areas of the NWFP. As the Afghans and Pashtuns resisted British occupation, the British looked to other ways of enforcing their rule in these areas. They settled on a tactic of 'divide-and-conquer', by which they installed puppet Pashtun regimes and divided them into artificially created regions.

Eventually, this tactic allowed full annexation of the NWFP from Afghanistan and the demarcation of the Durand Line. But the Durand Line, marking the edge of the British Raj, split the Pashtun tribal areas, and today, east of the Khyber Pass and up to the outskirts of the city of Peshawar, the

British colonial legacy lives on in the FATA. Many Afghans argue that the Durand Line technically expired when British colonial rule ended, and they claim parts of the Tribal Areas as their own.[8] On the ground, this ambiguity is evident in the fact that Pakistani law is not recognized in FATA, although Peshawar does serve as its administrative centre.

The present character of Peshawar is reflective of this complex history. Until the mid-1950s, its central districts were enclosed within a wall with sixteen gates. This 'old city' is now a warren of popular bazaars. For example, the Qissa Khwani, or storyteller's bazaar, extends from west to east in the heart of the city, and was once used as camping grounds for caravans and the British military. Merchants from Samarkand, Bokhara, Afghanistan, and India converged at the Qissa Khwani bazaar, a hub of Peshawar's trade with Central Asia. Outside the central area, the former British cantonment is today distinguished by its shady boulevards, churches, army quarters, and lavish high-walled homes. The city's post-partition face also includes the affluent University town and the sprawling administrative-residential township of Hayatabad (*Ibid*., pp. 195–208). In terms of architecture, the city's blend of colonial and Mughal architecture, reminiscent of empires past, contrasts with its newer multi-storey residences and often overcrowded and polluted streets.

During the decade-long Soviet occupation of Afghanistan (1979–1989) Peshawar became a centre of relief operations for Afghan refugees, many of whom were Pashtuns from the southern provinces, including Kandahar and Jalalabad. It also provided a command centre for the coalition of guerrilla groups intent on expelling the Soviet forces from Afghanistan. During these years, the city was transformed almost overnight into an exile headquarters for Afghan resistance groups, known as the *mujahideen*, and a base of operations for Western aid and intelligence agencies.

By August 1980, there were 90,000 Afghans reported to be entering Pakistan each month (Perlez, 2008). Not only did this attract great numbers of aid workers and development agencies to the city, but it also had economic benefits for the government of Pakistan. As long as there were Afghan refugees to resettle, the Pakistani government received aid from the West and other donor countries (Hussain, 2007). However, various intelligence agencies also used Peshawar as a portal to obtain sensitive information and train resistance groups fighting the Soviets.

With the active help and support of the CIA and Saudi Arabia, the ISI turned the Afghan resistance into 'Islam's holy war' (*Ibid*., p. 16). The United States and Saudi Arabia poured billions of dollars of covert aid into the '*jihad*' against the Soviets, the majority of which was channelled by the ISI in cooperation with the CIA. By its end, however, the war in Afghanistan provided inspiration to an entire generation of Pakistani Islamic radicals, who considered it their religious duty to fight the oppression of Muslims anywhere in the world. It also gave a new dimension to the idea of *jihad*. The Afghan war saw the privatization of the concept, as militant groups emerged from the ranks of more traditional religious movements and took the path of an armed struggle for the cause of Islam (*Ibid*., p. 21).

During the Afghan war Peshawar, as a political centre for the anti-Soviet struggle, became surrounded by huge camps of Afghan refugees. And many of these refugees remained there through the civil war which broke out after the Soviets were defeated in 1989 and the years of rule by the Taliban that followed. During this tumultuous period Peshawar would replace Kabul and Qandahar as the centre of Pashtun cultural development. Additionally, Peshawar managed to assimilate many of the Pashtun-Afghan refugees with relative ease, while many others remained in camps awaiting a possible return.

Today, Peshawar continues to serve as the urban hub of an outside invader: the resurgent Taliban. Pakistan's war with Islamic militants is now being fought in its alleys and bazaars, forts and armouries, killing innocent civilians. Within the past few years, as the Taliban have tried to impose their strict and orthodox interpretation of Islamic law, the violence has increased dramatically. The Taliban now hold strategic pockets on the city's outskirts, from which they strike at the military and the police. According to the Peshawar Police Department, suicide bombings, bomb explosions, and missile attacks occurred an average of once a week in 2007 (Perlez, 2008).

The proximity of Peshawar to the Tribal Areas makes the city a feasible prize for the militants, a possible future urban base from which to carry out their operations and further their goals. In the summer of 2009, at the height of the Taliban's increased violence, several Pakistani journalists, including Razi Ahmed, tried to capture in words the city's transformation. He wrote:

> This conservative city served as the advance base for the United States-sponsored Afghan mujahideen operations against Soviet forces in the 1980s; once concluded with the Red Army's withdrawal from Afghanistan in 1988–89, it remained a locus of jihadist ideology and fulmination, and has found a new role with the coming of another Afghan war in 2001. But this time round, the deep penetration of militants inside the city and across the Frontier make this latest protracted conflict both indigenous and more bloody. (Ahmed, 2009)

Today, the Taliban use Peshawar as an access point to both Pakistan and Afghanistan. Militant propaganda in the form of cassette recordings, videotapes, and literature is common, calling on young men to join the *jihad*. Fundamentalist Islamic law, espoused by the Taliban in Afghanistan, is also implemented, as movies and music have been banned and women have been forced out of schools. Local madrasas, or religious schools, run by the Taliban, have sprouted across the city, where clerics preach their radical brand of Islam and militant culture in the effort to gain more recruits for the global *jihad*. A recent report from Peshawar describes the city as:

> ... a garrison town. Armoured vehicles belonging to the Pakistan Frontier Corps occupy key positions. Paramilitary forces and anti-terror units patrol the streets. Nevertheless, Taliban warlords freely roam the city in pick-up trucks. Abductions and hit-and-run raids have become routine facts of life. (Sardar, 2008)

Staging Fundamentalism in Peshawar: Strategies of Performance and Practices of Dominance and Oppression

In the opening chapter of this book, Nezar AlSayyad advances the idiom of 'the fundamentalist city' as a conceptual tool to understand the possible connections between religiosity and urbanity, taking into account the specificity of each case. While it is not my objective here to designate Peshawar as a prototype 'fundamentalist city', this categorization is helpful because it permits interrogation of the particular urban condition which allows religious fundamentalism to flourish. The Taliban's appropriation of the cityscape is manifested through spatial strategies that rely on the performance and practices of dominance and oppression. This allows their ideological agenda – the establishment of an imagined form of Islam – to flourish. What then are the ways in which the Taliban regime is instituted and gets institutionalized in cities like Peshawar? What is the role of the state in allowing this informal rule of law? What are the spaces thus produced? Is this a new form of urban warfare, or the effect of increased Talibanization?

In essence, Peshawar today embodies the violent repercussions of Pakistan's dealings with Islamic militants. These dealings have resulted in the systematic transformation of the urban landscape. To help understand this transformation, I will next examine the specific forms of urban exclusion and surveillance that have emerged in Peshawar as a result of Taliban resurgence – what I term strategies of performance and practices of dominance and oppression. Together, the strategies, detailed below, establish the process by which the Taliban have spatially manifested and systematized their fundamentalist ideology in public life.

Deprivatization of Religion in Public Space

Peshawar today increasingly resembles Kabul under the rule of the Taliban in the 1990s. Women are not seen in public, beards for men are mandatory, and an informal Vice and Virtue brigade has ordered women to wear *hijabs* and men to pray five times a day (Perlez, 2008). In a recent incident, the Taliban blew up the shrine of a famous Sufi poet, Rehman Baba. According to a BBC report, it did so because it objected to visits there by women. This spectacle and the reasons given for it are only the most recent iterations of what has been a steady encroachment by the Taliban on the lives of citizens of Peshawar, specifically women. At one local girls' school, a threatening letter from Taliban militants demanded that the students cover themselves in *burqas*, or the school would be blown up (*Ibid.*). When the principal of the school took the initiative to contact the militants and tell them that some of her students could not afford the price of the *burqa*, they responded by saying, 'If the girls can afford makeup, they can afford *burqas*'.

It is crucial to recognize how the deprivatization of religion is not an accidental by-product of the Taliban insurgency, but an integral component of it. In pushing women out of society, out of jobs, and out of educational institutions, and forcing men to grow beards, preventing shopkeepers from selling American and Bollywood DVDs and music, the Taliban are attempting to redefine the public and private spheres in a way that gives tangible vision to their 'counter-modern' world (Cole, 2008). The possibility of educated women and productive citizens in society, the popularity of dual-income families with consumer habits born of globalization, the idea that films and music are forms of interest and leisure which take time away from the practice of religion, and the encroachment of Western ideas are precisely the reference points against which this 'counter-modernity' is constructed.

Street Spectacles

The Taliban are perhaps most well-known for their iconic public spectacles. According to one writer, these involve 'the public punishments of miscreants through mutilation, burial alive, and Toyotas circling through town with corpses hanging from cranes' (Antoun, 2008, p. 153). In Peshawar, this production of spectacle continues with direct attacks on police officers and foreigners, often involving public executions and beheadings (Perlez, 2008). In a recent news article, Pakistani journalist Sajjad Tarakzai (2009) reported that 'People avoid going to the bazaars and public places. Peshawar has the look of [a] war zone. Police and security have established checkpoints everywhere'.

The public display of violence is not a new strategy for the Taliban. In fact, it mimics the fear tactics they employed in the late 1990s in Kabul. Regular public executions at Kabul's Olympic sports stadium became the iconic symbol of the Taliban's brutality. Juan Cole (2008, p. 128) quotes the journalist Jan Goodwin and her account of one of the spectacles in 1998:

> Thirty-thousand men and boys poured into the dilapidated Olympic sports stadium in Kabul, capital of Afghanistan. Street hawkers peddled nuts, biscuits and tea to the waiting crowd. The scheduled entertainment? They were there to see a young woman, Sohaila, receive 100 lashes, and to watch two thieves have their right hands amputated. Sohaila had been arrested walking with a man who was not a relative, a sufficient crime for her to be found guilty of adultery. Since she was single, it was punishable by flogging; had she been married, she would have been publicly stoned to death. As Sohaila, completely covered in the shroud-like burqa veil, was forced to kneel and was then flogged, Taliban 'cheerleaders' had the stadium ringing with the chants of onlookers.

Cole describes this as part of the Taliban's 'implementation of power as public performance' (*Ibid.*, p. 128). In the effort to affirm their power and legitimacy, these violent and gruesome public spectacles purposefully intimidate and instil fear in the minds of local citizens. They serve as reminders of the consequences of failing to obey the regime's informal martial law. Citizens of Peshawar are today literally frightened to enter certain neighbourhoods in their city. Ironically, while the Taliban claim their legitimacy as the purveyors of Shari'a law, Shari'a does not require, and in fact discourages the use of punishment as spectacle (*Ibid.*, p. 129). Cole further argues that this is clearly less about the Taliban displaying their religiosity or piety than it is about displaying their power over people and the environment. In essence, the urban landscape serves as the 'stage' through which the Taliban's socio-cultural identity can be displayed, enforced, and expressed.

The Taliban's informal propaganda machine is another part of this strategy of street spectacle aimed at reinforcing a campaign of intimidation and fear. A reputable Pakistani news agency, *Dawn*, reported that in the Karkhano market in Peshawar, famous for its smuggled goods, shop owners have been forced to remove Western, Indian, and local films and CDs, and replace them with religious music and Taliban propaganda.[9] One of the shop owners, Ahmad Shah, said, 'The Taliban will bomb my shop if I do not keep the Jihadi and religious stocks'. According to the *Dawn* report, the pro-Taliban material ranges from condemnations of the United States to gruesome clips of beheadings and bomb attacks. However, most of the Taliban issued CDs are anti-American *jihadi* hymns geared towards motivating young people to wage *jihad* against 'the aggressor' and to gain more recruits to the Taliban ranks.

Signboards have begun to appear throughout the city's bazaars and shopping districts, reminding citizens that their behaviour is being monitored, that they are under surveillance; some even read 'God is watching' (Filkins, 2008). Until recently, billboards depicting women were censored with black paint, and in some instances, replaced with images of men.[10] This systematic labelling and demarcation of space embed it with specific meanings and values that ultimately identify the city as Taliban territory.

Through these various forms of propaganda, Peshawar is being transformed into a city where the Taliban's norms and values are established as a part of material and physical reality. The tape recordings, CDs, videos, signboards, and warnings are displayed in public space and reassert the Taliban's power over the urban environment.

A Campaign of Urban Violence

A key spatial strategy that the Taliban have employed in Peshawar has been a systematic campaign of violence, which has fostered fear and anger among local people (Antoun, 2008, pp. 147–164). Suicide bombings, air strikes, threats, intimidations, kidnappings, and civilian killings have become typical occurrences, embedding a sense of permanent insecurity in the urban fabric. Residential and commercial property, including shops selling Western films, DVDs, music, American-made products, and beauty products, have been destroyed, literally wiped out, in the city's bazaars and commercial quarters. From 2008 onwards, the Taliban began to target key civic institutions, law-enforcement agencies, mosques, hotels, and individual political figures (Ahmed, 2009).

In the past year, the attack that shook the city most was the bombing of Peshawar's landmark hotel, the Pearl Continental. A recent *New York Times* article (Khan and Masood, 2009) described the bombing, indicating that 'It was by far the largest – using an estimated 1,000 pounds of explosives, the police said – making it the most spectacular against a Western target in Pakistan'. The Pearl Continental was mainly known for hosting Western aid workers, diplomats, and United Nations officials, making it a 'trophy-target' (*Ibid.*). The Taliban are particularly opposed to the presence of foreigners in the country, viewing them as hegemonic forces attempting to further Westernize society.

Ahmed (2009) argues that Taliban violence in Peshawar is part of a punishment strategy similar to the air campaigns employed by powerful states 'to harm enemy civilians in order to lower their morale and motivate them to force their governments to end the war'. It is not ironic, then, that the Pearl Continental shares a location with the Provincial Assembly and High Court – two other 'trophy-targets' in the Taliban's campaign of violence.

Peshawar's urban fabric has suffered the most from the Taliban's brutal insurgency. Residents have had to become accustomed to this emerging pattern of violence in their city. So long as the Taliban justify acts of violence and aggression based on a duty to defend Islam from the secular West, there is no sign that this campaign will end.

The Spatialization of Fundamentalist Knowledge through Institutions

Since the fall of their Afghan regime in 2001, the Taliban have not had any formal institutions to facilitate their efforts. However, several scholars have argued that specific madrasas in Pakistan which espouse Deobandism have undoubtedly helped develop the movement. Perhaps the most popular among these is the Dar-ul-Uloom Haqqania, located on the main Islamabad-Peshawar highway in the NWFP. Rashid (2000, p. 90) describes the school as a sprawling collection of buildings, including a boarding school, a high school, and twelve affiliated smaller madrasas.

The school was founded in 1947 by Maulana Abdul Haq, the father of Maulana Samiul Haq, a religious and political leader intimately tied to the Taliban, and often referred to as the 'father of the Taliban'. In fact, it is well known that this madrasa served as a training ground for the Afghan Taliban leadership. Rashid (*Ibid.*) recounts that in 1999, at least eight Taliban cabinet ministers and several others serving as governors, commanders, judges, and bureaucrats were graduates of the Dar-ul-Uloom Haqqania.

Today, the madrasa is still popular as a centre of religious learning, albeit one that produces more Taliban soldiers than intellectuals. The school has also granted an honorary degree to the Afghan Taliban leader, Mullah Omar – the first of its kind issued from the Dar-ul-Uloom Haqqania. During an interview with Samiul Haq, who runs the school today, Walter Mayr (2008) asked why the school had issued Mullah Omar the degree. His reply was, 'We honoured Mullah Omar for his contribution to peace, just like your universities did with Mother Teresa'. The validity of the comparison between the leader of a radical religious movement and a woman who dedicated her life to serving India's poor is not the purpose of this chapter, however. The larger point is that the madrasa has become increasingly equated with the Taliban, and for the radically religious, this is an alluring attribute. However, the school's popularity may also stem from the fact that it provides young children with free shelter, food, care, and an education. In a city with an increasing population of poor urban youths, this assumption is not too far-fetched.

These spaces of performance and practice of dominance are the physical manifestation of the Taliban's presence in the urban environment – implanting and enforcing their values, norms, belief system, and order on the city and its citizens. Cole (2008, p. 13) would describe this as the Taliban's monopoly of symbolic power in the public sphere. The Taliban believe such inscriptions on the city are justified by their moral duty to protect citizens from the secularizing state. Ironically, this heroic image is almost identical to that of the *mujahideen* who drove the Soviets out of Afghanistan. The only difference is that now the state and its co-conspirators have had to deal with the consequences of what Chalmers Johnston (2000) referred to as 'blowback'. Peshawar is among the first urban sites post-9/11 in which a specific Talibanized identity has recast public life in terms of a religiosity embedded in the social and urban fabric. While other Pakistani cities are threatened by a similar process, some experts have already given this phenomenon a name: 'Talibanization'.

In its popular usage, Talibanization has come to describe areas or groups outside Afghanistan which have come under Taliban influence (see, for example, Kristof and Yefimov, 2009). The term implies a number of conditions: the strict regulation of women; the prohibition of entertainment including music, movies, dance, and television; the enforcement of a specific religious appearance, involving *burqas* for women and beards for men; the aggressive enforcement of regulations of personal conduct, i.e., the Vice and Virtue Brigade; the destruction of non-Muslim artefacts (as was the case most notably with regard to the Buddhas of Bamiyan in Afghanistan); the oppression of Muslim minorities including the Shi'a; the harbouring of Islamic militia operatives – Al-Qaeda among others; and, last but not least, discrimination against non-Muslims, particularly 'outside invaders'.[11]

Considering the breadth of such descriptions, there are clear signs of the Talibanization of Peshawar, and perhaps the spaces and practices I have already described can be added to them. But I would argue that this concept of Talibanization fails to encompass the full extent to which the increasingly fragmented landscape of Peshawar has come to be a space of exception within the larger nation-state of Pakistan. In this sense the notion of

Talibanization limits the scope to which the fundamental and seemingly new urban practices shaping the contemporary city may be explored.

Nasser Abourahme (2009) argues that fragmentation is evident in landscapes of deepening inequality, socio-spatial polarization, and enclave urbanism – 'the segregation of urban populations into self-enclosed "islands" with parallel but distinct realities, physically proximate but institutionally and cognitively estranged'. In the light of this definition, it can be argued that Peshawar is an example of enclave urbanism – of an enclaved city – fragmented and divided by the exclusionary and regulatory spatial practices of the Taliban. This fractured and exclusionary 'microgeography' is rationalized through the diffusion of violence, 'technologies of social control' that enforce the public display of the religious self (*burqas* and beards) alongside the public spectacle of brutal punishments, and the production of fundamentalist knowledge (*Ibid.*). This has allowed the Taliban to transform Peshawar into a 'stage' – a space of performance – in which their rule of law can be expressed, enforced, gazed at, and feared. And in many ways their practices of exclusion, exception, and surveillance bear an uncanny resemblance to forms of citizenship in the medieval city.

The Taliban in Peshawar: Paradoxes within the State of Exception

The concept of medieval modernity discussed earlier (AlSayyad and Roy, 2006) allows an interrogation of the Taliban's presence in Peshawar and the specific spatial strategies they have employed to impose their hegemony. The practices of exclusion, exception, and surveillance employed by the Taliban can be viewed as an expression of their sovereign power to establish a state of exception. Within this state of exception, Talibanization becomes a highly regulated set of processes and practices which subordinates the citizen-subject to a distinct form of religious governance and citizenship.

In its rigidity, Taliban rule in places like Peshawar mimics a medieval political system, in which the relationship between feudal territorial organizations (the Taliban) and monarchic power structures (the state) resembles 'fractal modalities of administration' (*Ibid.*, p. 12). In other words, in Pakistan the state has declared a war to eliminate 'terrorists and extremists', responding to American imperial interests. But in its inability to do so, it has allowed the Taliban to advance their reach and capture Peshawar, informally establishing a 'state within a state'. The Taliban have therefore proved to be adept at co-opting the state at the local level.

Their expansion has often followed a well-known, predictable pattern: well-armed groups of young men enter the city in white pickup trucks carrying Kalashnikovs, calling themselves the Taliban; they attempt to win the favour of the community by taking on local criminal elements and prohibiting un-Islamic behaviour. Having garnered some measure of local support (largely through intimidation and fear), they next set out to solidify their control by marginalizing or killing local notables and government officials, enacting stricter Islamist measures, and establishing an environment that is conducive to their insurgent behaviour. AlSayyad and Roy, quoting Mbembe, would describe this as 'a patchwork of overlapping and incomplete rights to rule … inextricably superimposed and tangled, in which different de facto juridical instances are geographically interwoven and plural allegiances, asymmetrical suzerainties, and enclaves abound' (AlSayyad and Roy, 2006, p. 13; Mbembe, 2003, p. 31).

Such a campaign of urban violence, the deprivatization of religion in public space, brutal and demoralizing street spectacles, and the institutionalization of fundamentalist knowledge and power through madrasa proselytization all signify the Taliban's quest to restore Islam to modern society and fight against the forces of secularization and Westernization. Essentially, these spatial strategies, or Talibanization tactics, are part of the movement's attempt to take society back to medieval times, in which violence and brutality were deployed in the name of peace and order (AlSayyad and Roy, 2006, p. 13). However, their quest for an authentic past combined with their selective modernity (as evident through the spatial strategies described in the previous section) contradict their anti-modern and anti-secular beliefs. And, more importantly, it pushes against the grain of their 'fundamental' opposition to and reaction against modernism. In other words, they maintain modernity as an essential frame of reference, while denying some of its key premises. The irony here, Charles Taylor (2006) explains, 'lies in the fact that precisely these attempts to return to purer forms are the sites of the most startling innovations; what is more, they feed on those innovations that are usually seen as quintessentially modern'.

From the beginning, however, the Taliban (irrespective of their Afghan or Pakistani formulation) have had a contradictory mission: to set up a 'modern' state and project a stable sovereignty in the international community while also attempting to establish that state under an Islamic order, ruled by Islamic law, and built on exemplary violence (Antoun, 2008, p. 153). This dichotomous, paradoxical mission is reflected in many domains, particularly in the public sphere: the suppression of television, but the promotion of radio; the banning of music, but the promotion of radical propaganda; the closing down of cinemas, but the spectacle of violent and brutal 'street theatre'; the deprivatization of religion, requiring women to wear *burqas* in public (and men to grow beards), but also the exclusion of women from the public arena (from schools, offices, etc.). To these might be added the necessity of weapons and modern warfare to advance their hegemony, the use of the iconic white Toyota pickup truck to deploy Taliban troops in the city, and the employment of cell phones and the Internet.

In sum, as part of their effort to return society to the fundamentals of Islam – albeit an imagined or invented Islam – the Taliban make use of the very inventions of modernity. Yet, because the paradigm of medieval modernity puts into context the 'imagined' Islam of the Taliban, this imaginary then becomes an act of construction as well as a reversion to an authentic Islamic past. Cole describes this imaginary in the Islamic context: 'Islamist movements are revolutionary; yet past-oriented ... radical in the criticism of traditional interpretations of Islam and yet deeply reliant on an imagined Islamic past' (quoted in Roy, 2004, p. 81). These paradoxical articulations give a new meaning to the idea of an imagined modernity, and following Roy's analysis, 'the inauthenticity of tradition is revealed as is made evident the multiplicity, even duplicity of the modern' (*Ibid.*). Thus, as evident in the case of Peshawar, this 'fixing and unsettling of tradition' (i.e., the aspiration of an authentic Islam), is a spatial process.

Clearly, the Taliban do not espouse a strict interpretation of Shari'a law. In negating hundreds of years of Islamic jurisprudence, of administrative procedures and methods of reasoning, sources of law and juristic analysis, Rafia Zakariya (2009) argued that the Taliban has redefined Shari'a as a form of spectacle. In responding to a recent video depicting a young woman being flogged by a Taliban solider while a crowd gathered as witnesses, Zakariya vividly described the complex, entangled relationship between modernity and Islam as practiced by the Taliban. It is worth quoting here at length:

> Thus the Qazi [Muslim Judge], arguably the most integral of those involved in justice provision, is nearly always invisible, while the crowd, the victim and those meting out a punishment play a central role. Justice is redefined as a means to subjugate and punish, with the entire collective crowd partaking in its pornographic enactment. There is no mention of the basis of the Islamic law applied, the deliberations which led to the application of the punishment, or any form of legitimacy that would associate the punishment with being Islamic. It is instead the anti-modernity of the whole spectacle, the absence of institutional safeguards, that makes the scene Islamic. The calculation is simple and persuasive: the more

visibly different from the epithets of modernity that the Taliban can be, the more automatically Islamic it becomes.

This is modernity's multiplicity or perhaps duplicity? This is the state of exception.

From the Informal City to the Fundamentalist City

In the current global moment, the resurgence of religion in public life has challenged theories of modernization and development, previously grounded in secularist assumptions that relegated religion to the private sphere. Like many other imaginations of modernity, however, this too has an artificial divide, and as the discussion above points out, it has been challenged by those who propose alternative modernities or multiple modernities. In this chapter I have shown how a religiously fundamentalist group, the Taliban, have managed to embed their fundamentalist ideology in the city of Peshawar, using specific spatial strategies to transform and redefine the social and urban landscape. Essentially, these are spatial re-inscriptions of the city, by which the Taliban have reclaimed religious expression in public life. This is then enabled by the nation-state, which allows for the production/performance of fundamentalism.

However, Peshawar is not simply the space in which the Taliban have established their rule of law. It is also the re-imagination of the city that recasts the urban as an essential locus of the Taliban's hegemonic strategies, and over time, becomes the norm – the established urban order. It is for this reason that the nation-state is crucial for the perpetuation of the law and politics of the Taliban, the entrenchment of their ideology, and the stabilization of their power. This chapter has discussed the ways in which the Taliban have established their own instruments of planning, creating spaces of performance and practices of dominance grounded in the fundamental belief that they are the purveyors of religion, and hence will 'save' the city.

AlSayyad and Roy (2006, p. 8) contend that 'non-state actors have emerged as the de facto state in informal settlements'. In Peshawar, the state's inability and unwillingness to manage, control, or contest the rule of the Taliban permits 'informal' law to trespass the 'formal'. As this chapter has illustrated, the Taliban have established their own distinct politics, regime of law, and institutional dynamics, over which the state has hardly any control. While the citizens of Peshawar are subject to this informality, they also participate in it through their physical appearance (*burqas* and beards) and religious practice (prayer in public and acceptance of the 'imagined' Islam of the Taliban). Spatial assertions of this informality are visible atop signboards that have blacked-out, literally erased, the image of the woman in the public sphere; in the marketplace and shops that sell Taliban propaganda and the technologies of fundamentalism (Kalashnikovs, grenades, cell phones, etc.); and more frequently through the propagation of violence (suicide bombers, targeted attacks on the state's military and governing personnel, street spectacles, and public punishments).

Peshawar has undoubtedly been transformed from a formal city that once boasted of being a centre for Buddhist learning and a tourist destination into an informal, fundamentalist city characterized by an insecurity so intense that the state itself does not know how to manage it. In essence, the city is the 'stage' through which the performance of fundamentalism is brought to life. The actors' expression of religiosity within the theatre of spectacle leaves us with an (un)imaginable ending: a new urban order in the current global moment – 'The Fundamentalist City'.

Notes

1 The Deobandis sought to train a new generation of Muslims who would revive Islamic values based on intellectual learning, spiritual experience, and Shari'a law. They promoted a strict view of the role of women, denied all forms of hierarchy in the Islamic tradition, and rejected minority Shi'a groups (Rashid, 2000).
2 For example, Hezbollah in Lebanon, Hamas in Palestine, or the Muslim Brotherhood in Egypt, which engage with Islamic scholarship and debate and have charters that explain their worldview. See, for example, Tamimi (2007).
3 See Emran Qureshi, 'Taliban', in *The Oxford Encyclopedia of the Islamic World. Oxford Islamic Studies Online*: http://www.oxfordislamicstudies.com/article/opr/t236/e0895. Accessed 18 September 2009.
4 Wahhabism is a sect attributed to Muhammad ibn Abd-al-Wahhab, an eighteenth-century CE scholar from what is today Saudi Arabia. He advocated purging Islam of innovations. For further information, see Esposito (2003).
5 http://satp.org/satporgtp/countries/pakistan/nwfp/datasheet/peshawar_incident.htm. Accessed 25 March 2010.
6 For a comprehensive history of Peshawar, see Dani (1995).
7 Hindko is an Indo-Aryan language spoken in northern Pakistan by about five million people. It is the old language of the historic city of Peshawar (Dani, 1995).
8 Author interviews with Afghan expatriates in California and Washington DC.
9 'Swat's Dancing Girls Fear Returning', July 2009.
10 Dawn Media Group, May 2009.
11 Sources describing Talibanization include, but are not limited to, the following: Rashid (2008), Khan (2005), Crews and Tarzi (2008), and Giustozzi (2007).

References

Abourahme, N. (2009) Contours of the neoliberal city: fragmentation, frontier geographies, and the new circularity. *Occupied London*, No. 4.

Agamben, G. (2005) *State of Exception* (translated by K. Attell). Chicago, IL: University of Chicago Press.

Ahmed, R. (2009) Lahore to Peshawar: the trophy target war. *Open Democracy*, 6 November.

Almond, G.A. and Appleby, S. (2003) *Strong Religion: The Rise of Fundamentalisms around the World*. Chicago, IL: University of Chicago Press.

Almond, G.A., Sivan, E. and Appleby, S. (1995) Fundamentalisms: genus and species, in Marty, M.E. and Appleby, R.S. (eds.) *Fundamentalisms Comprehended*. Chicago, IL: University of Chicago Press.

AlSayyad, N. and Roy, A. (2006) Medieval modernity: on citizenship and urbanism in a global era. *Space and Polity*, **10**(1), pp. 45–67.

Antoun, R. (2008) *Understanding Fundamentalism: Christian, Islamic, and Jewish Movements*. Lanham, MD: Rowman and Littlefield.

Balbo, M. (1993) Urban planning and the fragmented city of the developing world. *Third World Planning Review*, **15**(1), pp. 23–55.

Cole, J. (2008) The Taliban, women, and the Hegelian private sphere, in Crews, R.D. and Tarzi, A. (eds.) *The Taliban and the Crisis of Afghanistan*. Cambridge, MA: Harvard University Press.

Crews, R.D. and Tarzi, A. (eds.) (2008) *The Taliban and the Crisis of Afghanistan*. Cambridge. MA: Harvard University Press.

Dani, A.H. (1995) *Peshawar: Historic City of the Frontier*. Lahore: Sang-e-Meel Publications.

Emerson, M.O. and Hartman, D. (2006) The rise of religious fundamentalism. *Annual Review of Sociology*, **32**, pp. 127–144.

Esposito, J., (2003) *The Oxford Dictionary of Islam*. Oxford: Oxford University Press.

Filkins, D. (2008) Right at the edge. *New York Times*, 5 September.

Giustozzi, A. (2007) *Koran, Kalashnikov, and Laptop: The Neo-Taliban Insurgency in Afghanistan*. New York: Columbia University Press.

Hussain, Z. (2007) *Frontline Pakistan: The Struggle with Militant Islam*. New York: Columbia University Press.

Johnson, T.H. and Mason, M.C. (2008) No sign until the burst of fire. *International Security*, **32**(4), pp. 41–77.

Johnston, C. (2000) *Blowback: The Costs and Consequences of American Empire*. New York: Holt/Owl.

Khan, M.F.R. (2005) *Talibanization of Pakistan: A Case Study of TNSM*. Peshawar: Pakistan Study Centre, University of Peshawar.

Khan, I. and Masood, S. (2009) Militants strike hotel in Pakistan, killing 11. *New York Times*, 9 June.

Kristof, N.D. and Yefimov, N. (2009) Creeping Talibanization in Pakistan. *New York Times*, 23 July.

Mayr, W. (2008) The Taliban at the gates of Peshawar: Pakistan's deal with the devil. *Spiegel International*, 7 July.

Mbembe, A. (2003) Necropolitics. *Public Culture*, **15**(1), pp. 11–40.

Perlez, J., (2008) Frontier insurgency spills into Peshawar. *New York Times*, 18 January.

Rashid, A. (2000) *Taliban: Militant Islam, Oil, and Fundamentalism in Central Asia*. New Haven, CT: Yale University Press.

Rashid, A. (2008) *Descent into Chaos: The United States and the Failure of Nation Building in Pakistan, Afghanistan, and Central Asia*. New York: Viking.

Riesebrodt M. (1993) *Pious Passion: The Emergence of Modern Fundamentalism in the United States and Iran* (translated by D. Reneau). Berkeley, CA: University of California Press.

Roy, A. (2004) Nostalgias of the modern, in AlSayyad, N. (ed.) *The End of Tradition?* London: Routledge.

Said, E. (1979) *Orientalism*. New York: Vintage Books.

Sardar, Z. (2008) Pakistan must cure itself of the Taliban. *New Statesman*, 24 July.

Tamimi, A. (2007) *Hamas: Unwritten Chapters*. London: Hurst.

Tarakzai, S. (2009) Pakistan's Peshawar transformed by Taliban threat. *AFP*, 11 June.

Taylor, C. (2006) Religious mobilizations. *Public Culture*, **18**(2), pp. 281–300.

Voll, J.O. (1991) Fundamentalism in the Sunni Arab world: Egypt and the Sudan, in Marty, M.E. and Appleby, R.S. (eds.) *Fundamentalisms Observed*. Chicago, IL: University of Chicago Press.

Zakariya, R. (2009) The Taliban: A Response to Modernity. 13 April. Available at http://www.altmuslim.com/a/a/print/3018/. Accessed 25 March 2010.

Chapter 10

Taking the (Inner) City for God: Ambiguities of Urban Social Engagement among Conservative White Evangelicals

Omri Elisha

The modern city has always presented something of a moral problem for conservative Protestants in America. Since the early twentieth century, white evangelicals have denounced secular humanist values associated with the institutions and sensibilities of urban elites. Christian fundamentalists at the vanguard of the 'culture wars' have attacked the progressive ideas of evolutionary science and social liberalism on the basis that they are propagated by savvy urban intellectuals dedicated to undermining biblical orthodoxy and the authority of the church. Salvationist preachers have disparaged cities as havens of moral vice, poverty, rampant commercialism, and racial and ethnic conflict. At the same time, however, the ascendance of evangelical revivalism since World War II, and of the Christian Right since the 1970s, would not have occurred without the mobilization strategies forged under conditions of urban engagement. For popular revivalists like Billy Graham, and influential parachurch organizations like the Promise Keepers, the great American city provides an ideal space for massive spectacles of public religiosity and the formation of powerful institutional networks which are used to revitalize the spiritual climate of the entire nation. From this perspective, cities are stages – literally and figuratively – from which the Christian gospel of sin and redemption can be proclaimed to national and global audiences, and the tenets of conservative Protestantism projected upon secular society. As a result of these ambiguities, contemporary white evangelicals tend to regard the city as both the Devil's playground and a key battleground in their struggles for cultural hegemony.

In recent decades, many evangelicals – a large number of whom reside in counties with major cities (Smith, 1998, p. 79) – have come to see urban centres as places of social need, ripe for the missionary interventions of social-outreach ministries and religious charity organizations dedicated to relieving the effects of poverty and urban decline. In light of the popularity of 'faith-based' social services in the wake of federal welfare reform in the 1990s, and the growing prominence of seemingly moderate tones of evangelical engagement to destabilize the reactionary rhetoric of the Christian Right, socially engaged evangelicals are emboldened to revive the spirit of organized benevolence which animated nineteenth-century evangelicalism. While by no means a prevailing trend, the impulse has taken hold even among conservative churches where select pastors and lay churchgoers are committed to the idea of urban social engagement as a legitimate and effective strategy of modern evangelism. Over the years a number of noted white evangelical preachers, theologians, and ministry activists have emerged as vocal proponents of urban social engagement, calling upon middle-class evangelical congregations to overcome their ingrained separatist tendencies and structural indifference to urban social problems, especially with regard to struggling black communities in inner-city neighbourhoods (e.g., Bakke, 1997; Dawson, 2001; Shanke and Reed, 1995; Sider, 1999).

However, the legacy of Christian fundamentalism remains strong among white evangelical churches, impeding the institutionalization of outreach efforts with premillennialist eschatological beliefs, lingering scepticism towards social activism, and racially tinged preconceptions about the nature of urban dysfunction. The culture of separatist withdrawal from 'worldly' affairs that has long characterized churches at the far right of conservative Protestantism retains much of its influence, even among evangelical congregations which assume a more 'world-friendly' and less denunciatory stance towards modern society.[1] In metropolitan regions, socially engaged evangelicals are working hard to counteract the negative implications of fundamentalist practice while upholding the integrity of fundamentalist doctrine. Consequently, the urban social landscape, particularly the inner city, becomes a complex mission field where competing and overlapping religious impulses converge.

In this chapter I discuss specific dimensions of the evolving significance of the city in American evangelicalism by examining the attitudes and activities of a group of white suburban evangelicals in relation to predominantly black inner-city neighbourhoods in Knoxville, Tennessee. My discussion is based on ethnographic fieldwork, conducted over fifteen months from 1999 to 2001, during which I studied the moral ambitions of socially engaged evangelicals affiliated with suburban megachurches and faith-based organizations in the greater Knoxville area.

In a region where conservative Protestant religiosity is prevalent, and where the consequences of suburbanization and 'white flight' have been especially detrimental to inner-city communities, white evangelical practices of urban engagement are fraught with cultural antagonisms and racial politics that highlight conceptual ambiguities, including a pronounced tendency among suburban churchgoers to imagine themselves as spiritual exiles in their own region. One of the striking aspects of the 'exilic frame' (McRoberts, 2003) in the context of evangelical outreach is that it reflects integrationist ideals rooted in the millennialist paradigm of the 'kingdom of God on Earth', while simultaneously reinforcing ideological conflicts, regional power dynamics, and patterns of urban-suburban alienation which outreach efforts are intended to minimize. Expressions of an exilic consciousness among white evangelicals also reveal the extent to which notions of cultural separatism associated with fundamentalist attitudes remain salient, albeit recast in ways which are meant to inspire rather than hinder urban social engagement. Moreover, these ambiguities exist coterminously with a core paradox of neoliberalism as it pertains to American cities – that is, the paradox between the mobilization of civil-society infrastructures for purposes of urban renewal and simultaneous processes of depoliticization as issues of urban renewal become defined almost entirely in terms of private enterprise, personal responsibility, and civic accountability, rather than in larger structural forces (Collins *et al.*, 2008).

In my interactions with socially engaged pastors and churchgoers in the suburban megachurches of Knoxville, I observed a remarkable preoccupation with the prospects of 'reaching out' to poor black neighbourhoods in the inner city. Although social outreach ministries addressed other service needs as well – including support for soup kitchens, domestic violence shelters, free health clinics, rural Appalachian missions, and a host of informal services (e.g., mentoring, hospital visits, etc.) in and around the city – the portrayal of the inner city as a quintessential home-mission field was a recurring theme in the discourse of outreach mobilization and everyday conversation. Much like the many foreign and remote locations of the world that typically inspire missionary fantasies, the inner city was represented as an unfamiliar and intimidating space of otherness, except that it was 'in our own backyard'.

Moreover, the inner city was perceived as a place of risk, not just for its inhabitants, who face daily challenges of poverty, crime, and gang violence, but also for white, middle-class evangelicals who ventured out of their suburban 'comfort zones' to participate in urban ministry (or in very rare cases, who relocated their families to inner-city homes).

Symbolic constructions of otherness, risk, distance, and sacrifice allow socially engaged evangelicals to frame urban social engagement in distinctly evangelistic terms, making it possible for them to promote their participation in urban community development and racial reconciliation as extensions of, rather than exceptions to, the primary evangelical mandate to proclaim the gospel to the world. By the same token, the broader aims of urban community development and racial reconciliation are, for white evangelicals, linked to grand religious aspirations that are concerned less with mending fences and repairing social injustices than with the prospects of Christianizing regional cultures and establishing white evangelical institutions as the chief arbiters of public morality and civic life. In this sense, the efforts of socially engaged evangelicals to break from the patterns of fundamentalist pessimism and social disengagement nonetheless resonate with some of the cultural and political overtones associated with missionary interventions of the fundamentalist variety. I do not mean to argue that socially engaged evangelicals are all closet-fundamentalists (the proverbial 'wolves in sheep's clothing'), but I want to posit that in wrestling with the legacy of fundamentalism in their church communities – a legacy of rigid orthodoxy, apocalypticism, and racism – the evangelicals I studied constructed complex moral ambitions in which the desires of utopian integration and urban renewal come into conflict with the impulse to protect the 'body of Christ' from the sins of the city; to maintain some distance between the righteous exiles in the suburbs and the dark recesses of the inner city; and to distinguish 'the world' as it is from 'the kingdom of God' yet to come.

Setting the Scene

The city of Knoxville is situated along the Tennessee River, in the valley between the Great Smoky Mountains to the east and Cumberland Mountains to the west. The third-largest city in the state, Knoxville has a modest population of roughly 180,000, but a regional influence of much wider scope. The city is the centre of a metropolitan statistical area that encompasses roughly 700,000 people in the surrounding six counties. Known alternately as 'the Gateway to the Smokies', the home of University of Tennessee football, and the 'scruffy little city' which hosted an inauspicious World's Fair in 1982, Knoxville has a reputation (even among proud residents) as a city whose accomplishments have rarely matched its potential. It is also part of a regional culture shaped definitively by evangelical revivalism. Urban and rural county roads are dotted with churches at seemingly every corner, many of them Baptist, Methodist, or Presbyterian. Religious schools and bookstores testify to the powerful institutional presence of Christian institutionalism throughout the region's history. And the Rhea County courthouse, site of the infamous Scopes Trial of 1925, where the so-called fundamentalist-modernist controversies gained public recognition, is only a few hours' drive south.

For much of its history, Knoxville was a major regional centre for commercial trade and manufacturing, capitalizing on industries which relied on river and rail transport and a steady supply of rural Appalachian labourers (MacArthur, 1978; McDonald and Wheeler, 1983). Industrial growth slowed down in the early twentieth century, exacerbating Reconstruction-era racial tensions and class antagonisms between elite industrialists and rural white labourers. In 1933 Knoxville became the headquarters of the newly created Tennessee Valley Authority (TVA), an energy-utility corporation that was a cornerstone of Roosevelt's New Deal in response to the Great Depression. The TVA eventually became, along with an expanded University of Tennessee, one of the driving engines of the city's post-war economy. However, rates of population and commercial growth in the city declined in the 1970s, as a result of the shift towards a post-industrial service economy and aggressive residential and commercial development outside the city, especially in the affluent westward suburbs. As the influx of educated and skilled professionals from around the country allowed the region's high-tech and service industries to remain moderately competitive, its urban centre – including downtown and inner-city Knoxville – suffered tremendous losses in terms of sustained economic and cultural development.

In the 1980s, beset by corporate mismanagement and revenue loss, the TVA was forced to implement massive layoffs, which left Knoxville's central business district virtually crippled and accelerated the process of white middle-class retreat to the suburbs. At this point the city's residents 'increasingly tended to be elderly, black, and poor', while well-to-do suburbanites avoided downtown and 'lur[ed] retail businesses away from the city and into the malls'(McDonald and Wheeler, 1983, pp. 163–164). In the decades since, city officials and developers have worked, with mixed results, to promote strategies for downtown revitalization, including riverfront development, attractions related to science and nature tourism, and a new convention centre. At the turn of the century, controversial annexations of revenue-generating property by the city government were curtailed after an agreement was reached with Knox County officials seeking to limit the city's permissible growth boundaries. According to the agreement, the county provided funds to help finance community development initiatives in the city's federally designated 'Empowerment Zone', part of a multimillion-dollar HUD (Housing and Urban Development) programme to improve housing and employment in inner-city neighbourhoods. These initiatives, which included the demolition of public housing projects to be redeveloped into sustainable mixed-income communities, relied heavily on the participation of banks, property-development corporations, philanthropies, neighbourhood associations, and local churches. A handful of suburban white evangelical churches emerged as important players in facilitating, both directly and indirectly, the progress of public-private partnerships designed to revitalize the economic and cultural (and for evangelicals, spiritual) life of the city.

Among white suburban churches with particularly strong links to these efforts were two evangelical megachurches at the centre of my research: Eternal Vine Church and Marble Valley Presbyterian Church.[2] Located about twenty to twenty-five minutes west of the city centre by Interstate highway (and about ten minutes from each other), both megachurches cater primarily for middle- and upper-middle-class white evangelical churchgoers who live and work outside the Knoxville city limits. Both megachurches attract somewhere between 3,500 and 5,000 worshippers on Sundays and must hold multiple services to accommodate them.[3] In addition, the congregations are located along major suburban arteries and housed in sprawling campuses with facilities for a wide range of programmes held throughout the week, including a variety of social, recreational, educational, and therapeutic ministries. Since the late 1990s, the two megachurches gradually increased their public profile in terms of organized benevolence and outreach to local populations in need. Mostly these have been informal and volunteer-driven efforts, but the megachurches have also established formal partnerships with social-service agencies, charity organizations, and Christian nonprofit organizations throughout the region.

Eternal Vine is the relative newcomer in town, having been founded in the 1980s as a 'new way of doing church'. It became a virtual overnight success, due in large part to its emphasis on 'contemporary' worship (upbeat songs, informal dress, multimedia technology, etc.) over traditional liturgy, and its commitment to a doctrinal minimalism suited to the 'post-denominational' sensibilities of mainstream evangelical churchgoers. The megachurch is especially popular among college students and young and middle-aged professionals with families. Marble Valley Presbyterian predates Eternal Vine by nearly two centuries, and has for the most part retained its traditional liturgy and worship aesthetics. The congregation is older in composition, and its membership includes many wealthy Knoxville families who have belonged to it for generations. But its growth to megachurch-status in recent years has been spurred by a steady influx of younger members and the willingness of church elders to expand their ministry programmes to reflect changing preferences and lifestyle concerns.

In general, Eternal Vine and Marble Valley Presbyterian serve as vivid illustrations of two salient characteristics of the greater Knoxville area today: 1. the ongoing physical and cultural suburbanization of West Knoxville, accompanied by racial and socioeconomic separations mapped conspicuously along the urban-suburban divide; and 2. the increasing prominence of white conservative evangelicalism through new institutional formations, including suburban megachurches (and other faith-based organizations) with the social, financial, and political resources to exert influence over a variety of social and philanthropic enterprises. The evangelical establishment that has taken shape in Knoxville in recent decades is led by baby boomers who have inherited the deep theological and political conservatism of Bible Belt Christianity, yet who at the same time seek to promote ambitious and innovative strategies of 'engaged orthodoxy' (Smith, 1998).

The field of urban social engagement counts for a relatively small amount of the energy and resources committed to ministries sponsored or supported by the two megachurches, but the individuals involved in such efforts are motivated by complex and at times contradictory moral ambitions which reflect larger conceptual tensions within their home congregations. They are at once adherents of a religious worldview chiefly oriented towards traditionalist standards of personal piety and missionary zeal, and proponents of new institutional and cultural approaches to personal piety and missionary zeal afforded by modern church-growth strategies and tailored to resurgent interest in urban revitalization, racial reconciliation, and civic voluntarism in the post-welfare era.

Like most Southern cities, Knoxville still wrestles with the ghosts of its segregationist past. Although its history of racial oppression was not as severe as elsewhere in the South (which some attribute to the fact that the region did not rely as much on the slave economy prior to the Civil War), the legacies of racial violence, segregation, and protest remain evident in the form of lingering interracial tensions and profound social and geographic alienation. In 2000, African Americans made up nearly 17 per cent of the urban population (and less than 9 per cent of the county population), and lived mostly in impoverished sections of East Knoxville or in public housing projects near downtown.

In general terms, the relatively low frequency of sustained contact between urban blacks and suburban whites has led to mutual resentment, suspicion, and racial stereotyping. From the removed position of the suburbs, the inner city is a site of racial otherness and a place where indiscriminate violence, dysfunctional families, and moral corruption perpetuate a self-defeating 'culture of poverty'. Conversely, from the perspective of the urban black community (including a number of community organizers whom I interviewed), suburbia is a cultural space of privilege and self-aggrandizement, where white, middle-class people conceal, or at best ignore, their racial prejudices behind a veneer of civility, or channel their prejudices into seemingly well-intentioned efforts to 'uplift' the inner city without acknowledging the persistence of systemic social injustices. Given that conservative Protestants in America have had an especially difficult time overcoming racism in their own churches and institutions (Emerson and Smith, 2000), church pastors and dedicated laypeople in places like Knoxville experience the challenges of racial reconciliation most acutely. Although many regard new ministry initiatives and intrasectoral partnerships in Knoxville as unprecedented and promising opportunities for interracial cooperation and urban revitalization, they are also seen to reinforce regional power dynamics in terms of how resources are allocated, managed, and negotiated, thereby intensifying existing social strains (Elisha, 2010). In order to better understand the effects of new forms of 'faith-based activism', one must take into account the political economy of a region as well as the nature of the motivating paradigms that religious activists and volunteers bring to the table as they seek to influence matters of urban community development and governance (Bartkowski and Regis, 2003; Lichterman, 2005; Wuthnow, 2004).

Urbanism and the Millennial Kingdom

White evangelicals committed to urban social engagement often describe the imperative in primarily evangelistic and spiritual terms. In a volume entitled *Loving Your City Into the Kingdom*, featuring short essays by prominent conservative Protestant ministry leaders, Jack Hayford writes, 'to "take a city for God", by the biblical definition, is not to gain control of city hall, but to break the dominion of oppressive spirits that obstruct the advance of God's Kingdom', and 'to win the *hearts* of the people of a city to know the love of

Christ' (cited in Haggard and Hayford, 1997, p. 18). This statement invokes the idea of 'spiritual warfare', a dominant paradigm through which many conservative evangelicals and charismatic Christians interpret the religious, moral, and geopolitical conditions of the world. However, it is also indicative of the remarkable degree to which the discourse of outreach mobilization invokes theological notions of 'kingdom', which are noteworthy for the distinctly millennialist overtones that they introduce into the field of urban social engagement. Influential white evangelicals writing on the subject of urban ministry emphasize visions of social and systemic integration, in which cities and urban neighbourhoods are re-imagined as microcosms of the kingdom of God, or, as one author put it, 'kingdom playgrounds' where adults 'become children and learn to play again … invent ingenious methods to feed and clothe the poor … create new economies in destitute neighbourhoods, and build homes and businesses and hope where despair has reigned' (Lupton, 1989, p. 88).

Utopian idealizations of harmonious, Christ-centred communities hold tremendous appeal for evangelical churchgoers, culturally predisposed as they are to anticipate a millennial future when, among other things, the virtues of Christian faith and compassion will reign supreme. Such idealizations coexist awkwardly, however, with the apocalyptic framework of premillennialism held by conservative white evangelicals like many of those attending Eternal Vine and Marble Valley Presbyterian. One of the foundational principles of Christian fundamentalism, and a powerful influence on the religious right, premillennialism (and the elaborated doctrine of premillennial dispensationalism) posits that the thousand-year reign of God's kingdom on Earth will occur only *after* the 'Second Coming' of Christ (at which point the faithful will be lifted into heaven) and the seven years of 'Tribulation' under the rule of the Antichrist (Weber, 1987). The implication of this eschatology for its adherents is a tendency to view the world through a narrative of moral and spiritual degeneration, and a belief that the current historical age or 'dispensation' is characterized by inevitable decline necessitating messianic intervention and divine judgment. 'Premillennial dispensationalism gave fundamentalists the theological rationale for withdrawing from political involvement, shunning efforts at social reform, and abandoning the surrounding culture to its inevitable descent into perdition' (Smith, 1998. p. 145). Premillennialist thinking further reinforced the impetus for 'white flight' on the part of conservative white churches which relocated to the suburbs in response to social and economic changes in American cities in the 1950s and 1960s (Dochuk, 2003).

Compounding the awkward coexistence of utopian and apocalyptic millennialism is the sense of conflicting temporalities associated with the concept of the kingdom of God. Although contemporary evangelical theologians maintain that the kingdom of God, in all its 'fullness', is a future reality that is only at best *approximated* by improving human conditions in the present, they also recognize intrinsic ambiguities that lead Christians to see the kingdom of God as simultaneously 'now' and 'not yet' (e.g., Yancey, 1995, p. 250).[4] On the one hand, evangelical churchgoers internalize an understanding of the kingdom as 'other-worldly', as a transcendent, celestial realm that may be imminent but remains fundamentally unattainable through human effort or design. On the other hand, the kingdom of God functions in the religious lives of evangelicals as a motivational paradigm for various forms of evangelistic and humanitarian action, predicated to some extent on the possibility that the kingdom is fundamentally immanent, even tangible, in the here and now. When, during the Lord's Prayer, churchgoers recite the lines 'thy kingdom come, thy will be done, *on Earth as it is in heaven*', the words do not fall on deaf ears. They are absorbed and internalized along with other religious imperatives, including the missionary mandates of the New Testament. Pastor Jerry, senior pastor of Marble Valley Presbyterian, stressed the relevance of this passage to the everyday lives of churchgoers during a Sunday morning sermon:

> When we pray, 'Your kingdom come, your will be done on Earth as in heaven', we are praying for God's love to seek and find us, and through us to seek and find others, beginning with those closest to us, and reaching out to those whom we would never even notice or care about, had God not first sought us and found us, and poured his love into our hearts.

This kind of explication reinforces a practical theology in which evangelism is conceived no longer as simply the need to preach the gospel in order to usher as many willing recipients as possible into the coming kingdom, but also as a project of essentially ushering the kingdom into the world. For socially engaged evangelicals, this corresponds with the desire to do more than merely recruit new converts to the faith. Their moral ambitions are infused with the spiritual and cultural aspirations of transforming urban social and institutional networks so that the values that become dominant are in accordance with the principles of God's kingdom.

How do socially engaged evangelicals express these lofty aims? During a workshop for outreach volunteers, the staff pastor in charge of outreach at Eternal Vine explained what he thought the city of Knoxville would look like if local churches worked together in the spirit of unity that he and many others felt was woefully lacking in the region. He described a community with no racial or economic segregation, with low levels of crime, drug use, and unemployment. He imagined a 'citywide church' where people would celebrate Christian holidays together regardless of social status or ethnic background; a godly society where 'people who dress nicely will go to church with people who don't dress nicely'. Like other socially engaged evangelicals I spoke to or observed, his vision had a utopian edge to it and reflected ostensibly egalitarian sensibilities. Yet, at the same time, such descriptions rarely ever went so far as to suggest the possibility of a world where social inequalities would altogether cease to exist. A conservative antipathy towards reformist or progressive agendas continually reasserted itself in the kingdom visions of socially engaged evangelicals, especially since the values of personal piety and moral individualism assume greater weight in the white evangelical worldview than the social ideals that such visions would seem to imply.

Tom Frick was a prominent figure in Knoxville's non-profit community during my research. He had formerly been a member of *both* Eternal Vine and Marble Valley Presbyterian, and ran an urban ministry organization that facilitated mentoring programmes, leadership and job-training services, and community-development projects in the inner city. During an interview, Tom discussed his hopes to see the city of Knoxville become a model of the kingdom of God, adding that such an ideal could only be realized if churches, community organizations, civic associations, businesses, and politicians collectively committed themselves to what urban ministry specialist Ray Bakke (1997, p. 60) has described as 'a conscious theology of place'. The outcome of such an approach, according to Tom, would be a city defined by multiple levels of social, physical, and spiritual integration:

> If I had my ideal dream – you know that would never happen – but it would be to have four or five blocks of folks here that are all different sizes, shapes, poverty levels, economic levels, cultures, all having a common central theme of Christianity, living in the middle of the city and infiltrating the rest of the city, through the political arenas, the business arenas, the school arenas, all those different levels, mixing it up based on who they are, fleshing their faith out in that way. For me, that would be the ultimate utopia.

Here we see a combined articulation of four key themes of evangelical urbanism: diversity, proximity, civic pride, and Christianization. Yet we also see Tom's curious admission that this 'ultimate utopia' most likely 'would never happen', a remark motivated less by passive resignation than by an underlying sense of the inherent moral shortcomings of the human condition prior to Christ's return. Caught squarely in the conceptual gap between the 'now' and 'not yet' of the kingdom of God, Tom conveyed a religious agenda that is practical, immediate, and emplaced, but at the same time subordinated to theological inclinations rooted in longstanding conservative evangelical traditions and, significantly, that are represented in local churches – including Eternal Vine and Marble Valley Presbyterian – which Tom Frick relies upon for volunteers and donor support for his organization. This does not, of course, mean that he has adopted a fatalistic approach to the work of urban ministry. On the contrary, he was known in the city as a tireless promoter of urban social engagement and racial reconciliation among evangelicals in Knoxville. However, it does suggest that his efforts were shaped by, and not merely exceptions to, the peculiar dialectics of engagement confronted by activists working within the premillennialist milieu of white evangelicalism.

Engaging the Exilic Frame

Routinely invoking verses from the Bible is among the most recurrent motivational practices that socially engaged evangelicals use to justify their efforts and mobilize fellow churchgoers. One of the passages that I came across frequently among evangelicals in Knoxville was Jeremiah 29, in which the prophet addresses the Jewish people in exile in Babylon. In verse 7, Jeremiah writes the words of God instructing the Jews to 'seek the peace and prosperity of the city to which I have carried you into exile. Pray to the Lord for it, because if it prospers, you too will prosper'. The verse was cited in sermons, Bible studies, urban-ministry workshops, and public prayer services held in connection with a 'citywide prayer movement' coordinated by an interdenominational collective of local pastors. In each of these contexts, Jeremiah 29:7 provided a key scriptural reference point for the belief that Christians – as 'people of God' like the Jews of the Old Testament – are called upon to offer intercessory prayers for the welfare of cities in which they find themselves, and by extension, to be positively engaged in local communities with whom they exist in social and spiritual symbiosis.

I rarely heard evangelicals reflect directly on the particular relevance of the notion of 'exile' as it pertains to their religious lives, which is intriguing given the kinds of questions that such an association invites. Do white suburban evangelicals see themselves as 'exiles' like the Jews of the sixth century BCE? What kind of exiles are they? Where are they exiled from, and where or when do they expect to return? Given that contemporary references to Jeremiah 29 (and other similar passages) are often interpreted as instructional and authoritative but not necessarily literal – that is, they need not reflect immediate temporal circumstances to remain essentially true – such questions may not have straightforward answers. Yet the analogy drawn from the verse is an especially poignant one with regard to how white suburban evangelicals in Knoxville imagine their relationship to the metropolitan region as a whole. The symbolic link, however implicit, between the city of Knoxville and the ancient city of Babylon summons images of displacement and oppression. It conveys a scenario in which righteous people of faith are estranged from the sinful city. What we see here then is another example of deep conceptual ambiguity. Renewed engagement in the field of urban ministry on the part of suburban evangelicals entails both the realization that the fate of the city is irrevocably tied to their own, and the recognition that the city remains a problematic place where they do not intuitively belong.

A sense of identification with an exilic consciousness is not uncommon in American Christianity, although its articulations vary depending on a congregation's theology and demographic composition. Sociologist Omar McRoberts (2003, p. 65) has argued that as a conceptual framing device the sectarian thematic of 'in but not of the world' assumes significance because it 'moves beyond the common belief that religious adherence ensures one's place in the hereafter and addresses the earthly isolation one must endure in the meantime'. The extension of this frame into matters of local and practical concern assigns religious meaning to everyday experiences of cultural alienation, especially among minority and immigrant churches. The exilic frame further reinforces a collective sense of mission among church congregants, whether they believe their mission is to change 'the world' from which they stand apart, or to guard themselves, along with all the converts they can muster, from its inherent evils. Aspects of both imperatives influence the missionary identities of affluent white evangelical congregations like Eternal Vine and Marble Valley Presbyterian. Even as the evangelicals I observed in Knoxville constructed large existential distances between themselves and nearby spaces of urban decline (which serve as convenient objectifications of 'worldliness' in all its depravity), the exilic frame also reinforced the interventionist ambitions of socially engaged pastors, churchgoers, and ministry advocates.

The case of Damien Richards offers some insight into the personal and social implications of this religious construct. Damien is a wealthy middle-aged businessman and a prominent lay member of Eternal Vine.

When I met him, he lived with his family in a large house in one of the richest sections of West Knoxville. Before that, he and his family lived briefly in a restored home located near a public housing project in an inner-city neighbourhood with chronic problems linked to drugs and gang violence. Explaining to me his original motivation for moving to the inner city, Damien said, 'I had kind of done a lot of what I wanted to [in business], and realized boy, I could sure transfer a lot of those skills and relationships – my network in town became pretty strong – into an area of need and use that leverage to help make a difference there'. Damien further explained what he believed was the necessity of relocation for suburban evangelicals who hope to 'make a difference' in struggling urban communities:

> If you really want to have a significant impact, you have to live there, especially as someone who would be characterized as an affluent person. You know, [people are always] going in and helping, then leaving. That's great, but you don't live here. The school isn't your school. The crime isn't your crime. The drug trade and how it affects everyone isn't affecting you, because you leave. So you need to live there, and [then] all of a sudden the community's issues become your issues and you become enmeshed in the community through that.

The exilic consciousness conveyed in Damien's comments is complex and multilayered. The suggestion that he, as both a faithful Christian and a resourceful entrepreneur, could help stimulate urban revitalization by becoming 'enmeshed' in a community from which he was otherwise completely detached, implicitly reproduces the moral category of righteous exiles elevated above the murk and moral corruption of the city. Yet it also indicates that the exilic separation is essentially undesirable, something to be overcome through active immersion in the lives (and problems) of urban dwellers. Indeed, Damien often referred to Jeremiah 29 when he talked about urban ministry and tried to get others involved. He considered it to be a 'hugely important scripture for the church' because it is 'counter-cultural to how mainline evangelicalism has operated' in recent history:

> If you really look at the evangelical church … the mentality is: 'The world is going to hell, you need to prepare yourself, remove yourself from the danger of the world, and be a separatist'. Jeremiah 29:7 is completely the opposite. It says the world is what it is – it's got evil, it's got good. You may not even be in a place you want to be, but God's word to the people was 'seek the peace and prosperity of the community to which I've called you'. Immerse yourself, marry these people, learn

> their language, get in the midst of them. That's huge, because very few churches are encouraging their members to engage culture... Most churches have an Us vs. Them mentality. You can't transform culture [that way] because everybody's on the defence – you create distance. But if you are into engaging people in relationship and say, 'I'm allowing Christ to live through me in whatever arena he's called me', you can transform culture.

Damien's criticism of evangelical separatism resonates with the convictions of a growing segment of the white evangelical population, those who support the doctrines of Christian fundamentalism – including end-times prophecies – but seek to advance those doctrines through social and cultural engagement rather than withdrawal. His comments are also consistent with the activist reading of Jeremiah 29 presented by Ray Bakke, white evangelical author and urban-ministry specialist admired and emulated by many socially engaged evangelicals. For Bakke, the purpose of Jeremiah's letter was to urge the Jews to accept their exile from Jerusalem as divine providence rather than victimization, and raise their children in Babylon accordingly. 'God is not asking the exiles to lead a passive, patient existence in the enemy city. He's asking that they actively work for Babylon's *shalom*, that is, peace with justice' (1997, p. 85). Bakke relates this to his own experience of being called by God to minister to the 'dirty, ugly, flat, corrupt city' of Chicago in the 1960s, overcoming his reluctance to send his children to public schools in poor neighbourhoods, and eventually embracing and learning from the 'chaotic and even violent environment' of the inner city (*Ibid*., p. 86).

In the mid-1990s, Damien restored a three-storey Victorian house in the inner city and moved there with his wife and three young children. A nearby housing project was soon to be razed to make way for the development of mixed-income housing with the help of federal grants (through a programme known as HOPE VI) and the support of intrasectoral partnerships between public agencies and private interests such as local businesses, philanthropies, and faith-based organizations. A strong proponent of urban renewal through privatization, Damien was optimistic about these initiatives. He believed that encouraging integrated development, private enterprise, home-ownership, and civic accountability was the best way to improve the inner city and limit the role of the welfare state. His affinity for the politics of neoliberalism was bolstered by his belief that religious institutions, and not secular or government agencies, should be at the forefront of systemic urban renewal efforts, promoting the virtues of evangelical piety, personal responsibility, and moral community. He sought out the assistance of black religious leaders and community organizers, hoping to shape the direction of community initiatives at the grassroots, and serve as a bridge figure between the needs of the inner city and the resources of white evangelical churches, businesses, and philanthropies with which he remained closely associated.

One of his closest associates in the inner city at the time was Pastor Terrence, a widely respected minister of an independent black church. A preacher known for having given up a life of drugs and crime, Pastor Terrence had been struggling for years to establish social programmes providing job training, leadership opportunities, and other vital services for inner-city youth and families. Joining forces with Damien became an opportunity for Pastor Terrence to expand his support base among faith-based organizations, foundations, and affluent churches in the region. In particular, Damien's status as a church elder at Eternal Vine facilitated a mutually beneficial relationship between the megachurch and Pastor Terrence's church. While the black church received financial support towards the building of a new sanctuary and community centre (located prominently near the HOPE VI redevelopment site), the megachurch increased its public profile as a congregation dedicated to racial reconciliation, relieving urban poverty, and serving the general well-being of the city. Gaining this kind of cultural legitimacy is an essential strategy for white socially engaged evangelicals working in urban communities, one that highlights the immersive and missionary aspects of urban ministry and, by its successes as well as its failures, also highlights the oppositional dynamic presupposed by the exilic frame.

The partnership between Damien Richards and Pastor Terrence was productive for a number of years, as they spearheaded social and charitable initiatives in the surrounding neighbourhoods. They were often featured in local news media, and their partnership was hailed by civic organizations as a model of collaboration across lines of race and class, the likes of which many believed were previously inconceivable in a city like Knoxville. However, the working relationship between the two men eventually strained because of differences in organizational management and divergent approaches to issues of social justice and grassroots empowerment. These strains were further exacerbated by concerns among Pastor Terrence's congregants that the autonomy of their congregation was becoming compromised by the priorities of affluent white evangelicals. As tensions rose, Damien gradually diminished his role alongside Pastor Terrence, and shortly thereafter moved his family back to a home in the suburbs. He carried on in his commitments, however, starting his own non-profit faith-based organization to develop a wide range of professionalized ministries and distribute financial resources to smaller organizations in ways that were intended, in Damien's words, to offer 'the biggest bang for the buck'.

Reflecting back on the partnership, Damien acknowledged that one of his toughest obstacles in the inner city was overcoming the suspicions of inner-city blacks, who worried about having their lives controlled by well-intentioned but overzealous white people from the suburbs, who maintained their conservative, middle-class agendas. He also admitted that such concerns were not so unreasonable, and that he could have approached certain aspects of ministry planning and implementation with more patience and sensitivity and less of a zeal to 'fly in' and impose his own cultural standards. At the same time, Damien was frustrated by the fact that potentially significant plans to improve the quality of urban community life were stymied by stubborn distrust and political resistance at the grassroots.

A failed initiative in which he was closely involved was an organized effort to turn over the management of a large inner-city public school to a private corporation (Edison Schools Inc., founded by Knoxville native Chris Whittle). This particular setback had more to do with local politics than with racial tensions or urban-suburban discord, but it was nonetheless an especially disappointing development for Damien. Like a handful of other white evangelicals in Knoxville who relocated from the suburbs to the inner city, Damien wanted to send his children to local public schools but could not commit to the idea because of the poor state of the public school system, and opted instead to keep his children in high-performing private schools. The Edison School plan was geared towards neoliberal principles of privatization and corporate efficiency which resonate well with contemporary middle-class evangelical sensibilities, but it also represented for Damien the last, best prospect for removing the obstacles that hindered his ability to integrate fully into the everyday life-world of the Babylonian city.

Notwithstanding idiosyncratic elements of Damien's story, his experiences in and of the inner city and his personal reflections about those experiences are illustrative of broader cultural and political dynamics at play in the ideologically charged field of urban social engagement as undertaken by conservative white evangelicals. Elsewhere, I have explored these dynamics further through a case study involving Marble Valley Presbyterian's financial patronage of a different inner-city black church and the resulting conflicts over issues of moral and financial accountability and the politics of resource distribution (Elisha, 2010). In that study and in the present discussion, I have sought to analyze the operative paradigms that (*a*) motivate socially engaged evangelicals to get involved in urban ministry; (*b*) inspire (and complicate) multiple forms of practical action that they undertake; and (*c*) help them to make sense of their experiences of immersion, risk, perseverance, success, and/or failure. The exilic frame is more than a powerful rhetorical device, and it should not be casually dismissed as merely a rationalization for ethical dispositions of an entirely different nature. Among white evangelicals who, like Damien Richards, are drawn to black inner-city neighbourhoods from which they feel existentially alienated, the continuous reconstruction of the exilic frame is itself a strategy of social engagement in that it helps to define the very conditions (and limits) of mission-oriented action.

Concluding Remarks

The white evangelicals I encountered in Knoxville often expressed their support for urban ministry, and their nascent fascinations with the inner city, with pithy statements like 'We need the poor'. These statements were not meant to highlight conditions of structural interdependence, but rather to give voice to sentiments of spiritual longing and aspiration projected against the landscape of urban poverty and racial otherness that the inner city represents to them. Their 'need' for the urban poor is a direct extension of their fear of the city, a fear which is cultivated for the very purpose of being conquered, in the name of advancing God's kingdom.

Overcoming fear is the ultimate test of faith, and few local ministries offer such a poignant combination of personal piety and uncertain risk for suburban churchgoers as those that call upon them to enact deep and sustained engagement in the social worlds of black inner-city neighbourhoods. And yet, in a manner reminiscent of the racial stereotypes of colonial missionaries, and persistent to this day in the cultural politics of the Christian Right, the divide between normalized 'whiteness' and marginalized 'blackness' is an ontological distance which white evangelicals traverse with caution. By portraying the inner city as a place of 'dark' despair and dysfunction, desperately awaiting faith-based interventions, white evangelicals reproduce racial anxieties and moral hierarchies. Furthermore, the moral ambitions that propel socially engaged evangelicals into the life of the city reflect an amalgam of ambivalent desires and ambiguous preconceptions that frustrate as well as facilitate, the already formidable tasks of urban social engagement and the millennialist models of integration and Christianization on which they are predicated.

Much has been made in recent years of the supposed decline of the Christian Right as a force in American politics and public culture. This may be attributed in part to an apparent centrist shift among white evangelicals, signalled by their increased openness to bipartisan politics, internationalism, and cultural pluralism (Gushee, 2008). It can also be taken as evidence of the waning influence of Christian fundamentalism as it has been conceived for much of the last century, especially among younger evangelicals. The

GOP's[5] political rhetoric of 'compassionate conservatism' at the turn of the century, and ongoing bipartisan affirmations of the importance of 'faith-based organizations' as welfare providers in local communities, emerged in a cultural moment when conservative Protestants were gradually distancing themselves from the combative, reactionary politics of renunciation and (dis)engagement once embodied by the fundamentalist vanguard. Among other consequences of this distancing has been the broadening of the white evangelical ministry agenda to encompass forms of urban social engagement once considered too controversial, too liberal, and too 'worldly' to be undertaken seriously.[6]

In spite of these developments, we should be careful not to downplay the enduring influence of fundamentalism on the ideological sensibilities of most white evangelical churchgoers. The seeming moderation or 'softening' of evangelical politics is significant, but is not necessarily indicative of substantive changes in the content of the evangelical moral and religious imagination, especially with regard to biblical orthodoxy, eschatology, and hot-button 'moral' issues like abortion and homosexuality. Moreover, while increased attention to issues of social welfare, poverty relief, and urban renewal among white evangelicals is historically meaningful, it does not necessarily mean that socially engaged evangelicals in otherwise conservative churches are adopting progressive social or political values like those typically associated with welfare activism. Evangelicals by and large remain strongly committed to conservative moral and economic ideals, and they live their lives according to social ethics of relationalism, in which spiritual transformations achieved through interpersonal relationships and individual voluntarism are conceived as the primary engines of social change, rather than broader notions of social reform (Smith, 1998). While such commitments have obviously existed separately from and preceding the ascendance of Christian fundamentalism, and therefore cannot solely be characterized as echoes of that movement, it is more than likely that certain fundamentalist tendencies retain their influence in evangelical churches with conservative pedigrees, delimiting the range of flexibility when it comes to strategies of social engagement.

Thus, by way of conclusion, I propose that rather than seek to isolate decisive shifts to the right, left, or centre in American evangelicalism, we would do well to recognize the coexistence of coterminous, yet competing impulses active within evangelical communities. Although congregations clearly vary in terms of the regnant value systems that may define them as ideologically conservative, moderate, or progressive, the ethnographic perspective presented here suggests that in matters of social ministry (among other aspects of congregational life), evangelical pastors and churchgoers work with a plurality of active conceptions that must be negotiated even as they serve the purposes of mobilization. What compounds the challenge of such negotiation are the myriad conceptual ambiguities that stem from the intrinsic polyvalence of religious constructs such as 'kingdom of God' and the exilic frame. It follows that we may best understand the relationship between what evangelicals say and what they do by taking into account the indeterminacy – and not just the absolutism – that is inherent in particular modes of evangelical discourse and action. As socially engaged evangelicals throughout the U.S. become involved in ongoing processes of urban revitalization, gentrification, and neoliberalization, the religious and moral significance they assign to the city as a cultural category will continue to evolve, but the remnants of a more antagonistic past will continue to loom large as well.

Notes

1 Following Woodberry and Smith (1998, p. 36), I use 'conservative Protestant' as an umbrella category for Christians who 'emphasize a personal relationship with Jesus Christ, believe in the importance of converting others to the faith, have a strong view of biblical authority, and believe that salvation is through Christ alone'. The term 'evangelical' refers here to the 'moderate wing' of conservative Protestantism which emerged amidst the revivalism of the post-war period, while 'fundamentalist' refers to the ultra-conservative and typically more sectarian segment of the conservative Protestant population. The term 'socially engaged evangelicals' is my own and is meant to describe evangelical pastors and churchgoers who are actively involved in social ministries of one kind or another (such as organized benevolence or welfare activism), and make conscious efforts to mobilize their churches.

2 The names of all Knoxville-based churches, non-profit organizations, and individuals in this chapter have been changed. In some cases, specific identifying details have been altered as well.

3 A megachurch is defined as a Protestant congregation where the number of weekend worship attendees regularly exceeds two thousand (Thumma and Travis, 2007). There are well over a thousand megachurches in America today, mostly located in Sunbelt cities and suburbs. They are predominantly evangelical in theology and worship style, though they also emphasize ministry innovation and organizational flexibility. Influenced by for-profit corporate management principles, their growth strategies emphasize cultural relevance and responding to the practical needs of their target demographic, usually middle-class, white-collar professionals with families. While such methods are controversial even within evangelicalism, megachurch leaders and congregants view them as valid means of restoring 'authentic spirituality' and 'true biblical values' to modern Christianity.

4 A sense of the duality of the kingdom of God as being *immanent* as well as imminent was evident in the evangelistic efforts of early American revivalists and public preachers. Describing the 'Great Awakening' of the eighteenth century, H. Richard Niebuhr wrote in 1937: 'The good news was not only that there was light

beyond darkness, but even more that in Christ men had been given the opportunity to anticipate both threat and promise, *to bring the future into the present and receive a foretaste of the coming kingdom, a validation of the promise*' (Niebuhr, 1988, p. 140, emphasis added).

5 GOP or Grand Old Party is the nickname of the Republican Party.

6 Evangelical parachurch organizations dedicated to issues of social reform and poverty relief include Ron Sider's Evangelicals for Social Reform, Jim Wallis's Call to Renewal, and Ray Bakke's Internal Urban Associates, to name just a few that have gained recognition since the 1990s. Organizations such as the Evangelical Environment Network reflect the further broadening of the evangelical agenda to include climate change and environmental justice. Rick Warren of Saddleback Church has gained widespread praise for his ministries dedicated to relieving global poverty and the effects of HIV/AIDS in Africa. This is a small sample of social ministries and institutional networks that have been made possible by the larger cultural shift taking place in American evangelicalism. Although they are not new to the national scene, the relatively small constituency of white socially engaged evangelicals whom such organizations represent has been emboldened by changes in the cultural climate of conservative Protestantism.

References

Bakke, R. (1997) *A Theology as Big as the City.* Downers Grove, IL: InterVarsity Press.

Bartkowski, J.P. and Regis, H.A. (2003) *Charitable Choices: Religion, Race, and Poverty in the Post-Welfare Era.* New York: New York University Press.

Collins, J.L. *et al.* (2008) *New Landscapes of Inequality: Neoliberalism and the Erosion of Democracy in America.* Santa Fe, NM: School for Advanced Research Press.

Dawson, J. (2001) *Taking Our Cities for God: How to Break Spiritual Strongholds.* Lake Mary, FL: Charisma House.

Dochuk, D. (2003) 'Praying for the wicked city': congregation, community, and the suburbanization of fundamentalism. *Religion and American Culture*, 13, pp. 167–203.

Elisha, O. (2010) Evangelical megachurches, racial reconciliation, and the Christianization of civil society', in Clemens, E. and Guthrie, D. (eds.) *Politics and Partnerships: The Role of Voluntary Associations in America's Political Past and Present*. Chicago, IL: University of Chicago Press.

Emerson, M.O. and Smith, C. (2000) *Divided By Faith: Evangelical Religion and the Problem of Race in America.* New York: Oxford University Press.

Gushee, D.P. (2008) *The Future of Faith in American Politics: The Public Witness of the Evangelical Center.* Waco, TX: Baylor University Press.

Haggard, T. and Hayford, J. (eds.) (1997) *Loving Your City into the Kingdom.* Ventura, CA: Regal Books.

Lichterman, P. (2005) *Elusive Togetherness: Church Groups Trying to Bridge America's Divisions.* Princeton, NJ: Princeton University Press.

Lupton, R.D. (1989) *Theirs is the Kingdom: Celebrating the Gospel in Urban America.* San Francisco, CA: Harper Collins.

MacArthur, W.J. Jr (1978) *Knoxville's History: An Interpretation.* Knoxville, TN: East Tennessee Historical Society.

McDonald, M.J. and Wheeler, W.B. (1983) *Knoxville, Tennessee: Continuity and Change in an Appalachian City.* Knoxville, TN: University of Tennessee Press.

McRoberts, O.M. (2003) *Streets of Glory: Church and Community in a Black Urban Neighborhood.* Chicago, IL: University of Chicago Press.

Niebuhr, H.R. (1988) *The Kingdom of God in America.* Middletown, CT: Wesleyan University Press.
Shank, H. and Reed, W. (1995) A challenge to suburban evangelical churches: theological perspectives on poverty in America. *Journal of Interdisciplinary Studies,* **7**, pp. 119–134.
Sider, R.J. (1999) *Just Generosity: A New Vision for Overcoming Poverty in America.* Grand Rapids, MI: Baker Books.
Smith, C. (1998) *American Evangelicalism: Embattled and Thriving.* Chicago, IL: University of Chicago Press.
Thumma, S. and Travis, D. (2007) *Beyond Megachurch Myths: What We Can Learn from America's Largest Churches.* San Francisco, CA: Jossey-Bass.
Weber, T. (1987) *Living in the Shadow of the Second Coming: American Premillennialism, 1875–1982.* Chicago,IL: University of Chicago Press.
Woodberry, R.D. and Smith, C.S. (1998) Fundamentalism et al: conservative protestants in America. *Annual Review of Sociology*, **24**, pp. 25–56.
Wuthnow, R. (2004) *Saving America? Faith-Based Services and the Future of Civil Society,* Princeton, NJ: Princeton University Press.
Yancey, P. (1995) *The Jesus I Never Knew.* Grand Rapids, MI: Zondervan Publishing House.

Chapter 11

Postsecular Urbanisms: Situating Delhi within the Rhetorical Landscape of Hindutva

Mrinalini Rajagopalan

> On the next day we arrived at the city of Dihli (Delhi), the metropolis of India, a vast and magnificent city, uniting beauty with strength. It is surrounded by a wall that has no equal in the world, and is the largest city in the entire Muslim Orient.
>
> Ibn Batuta, fourteenth century[1]

> The Muslims and Christians did not come from outside India. Their ancestors were Hindus. By changing religion one does not change one's nationality or culture. [The Muslims and the Christians have deliberately cut themselves off from a common national identity and from the] traditions, mores and manners, which bind the people here to this land itself. So much so, that they refused to accept Rama and Krishna [Hindu deities] as their ancestors and kept themselves aloof from the traditional festivals.
>
> Atal Behari Vajpayee, 1961[2]

This chapter focuses on the appropriation of urban space and urban history in the assertion of religious claims to a larger nation-state. More specifically, it focuses on the redefinition of the urban landscape of Delhi via the ideology of Hindutva – the political movement to reclaim the nation-state of India as a Hindu homeland with diminished rights for citizens who belong to minority religious groups, such as Muslims and Christians. The tension between the two modalities of national imaginaries and urban geographies, which is borne out by the two epigraphs above, reveals the complex dynamics at the heart of Hindutva politics and rhetoric. Ibn Batuta, the Arab scholar who briefly lived in Delhi in the fourteenth century, described the city as one of splendid beauty, well-regarded in the Muslim world. And after his time, Delhi's splendour would only grow more magnificent under the reign of successive Mughal emperors. Indeed, to this day, Delhi is described as a city of wondrous mosques, forts, palaces, and tombs – all of which are a legacy of rich Islamic empires, many of which chose the city as their capital.

This physical heritage, however, stands in sharp contrast to the Hindutva vision of India's nation-space, which is imagined as purely Hindu. This sentiment of national purity is articulated by Atal Behari Vajpayee in the second epigraph. Ironically, the comments are extracted from a speech he gave at a conference for National Integration in 1961. At the time, Vajpayee was a young Turk of the VHP (Vishwa Hindu Parishad), but later he would go on to become the premier of India.[3] An unabashed supporter of the Hindutva ideology, he would also be in power during the 2002 massacre of Muslims in the western state of Gujarat. In this speech, Vajpayee argued for a pan-Indian identity based on a common Hindu past, emphasizing that the terms 'India' and 'Hindu' are in fact synonyms, and that differences between Hindus, Muslims, and Christians are contrived and historically baseless. India is thus imagined as a product of a Hindu culture that has remained unchanged from the ancient past to the modern present.

It is the friction between this particular Hindutva imagination of India and the Islamic cultural heritage of Delhi that I seek to explore in this chapter. Delhi occupies a precarious position within the master narrative of Hindutva due to its long history as the centre of Islamic power, the rich visual culture left behind by this history, and its subsequent absorption into the framing of India as a 'secular' nation. But I also argue that the attempts by agents of Hindutva to reclaim Delhi are not simply aimed at a renegotiation and re-presentation of decades of Islamic culture, but that at the scale and site of Delhi the project of Hindutva is also fundamentally a recalibration of secularism in India.

Writing in the early 1990s and in the wake of the dissolution of the Soviet Union, Martin E. Marty and Scott Appleby wondered what kinds of hegemonies might emerge to fill the vacuum left behind by the demise of Marxism. At the time, the sectarian conflicts in Europe (in Ireland and the Balkans) and in other parts of the world (e.g., Sri Lanka, Israel, and the Sudan) were ample indication of the resurgence of ethnic and religious identities as a means to legitimate claims to sovereignty in the postmodern world (Marty and Appleby, 1993). It was in response to this changing social, political, and cultural climate that Marty and Appleby proposed the rubric of fundamentalism as a means to investigate the many different manifestations of radical religiosity in various contexts around the world. Fifteen years after the Fundamentalism Project that they directed, the participants in the Cities and Fundamentalisms group have sought to re-evaluate the possibilities as well as the perils of reviving the term fundamentalism in the contemporary world. In addition, the contributors to this volume seek to understand how the rapidly urbanizing geography of the contemporary world has contributed to, intersected with, and impeded the project of radical religious movements.

This chapter is divided in two sections. The first, 'Hindutva: Epistemes and Memes', offers a broad introduction to the ideological tenets of Hindutva and to some of its central cultural motifs. It begins by discussing two key elements of Hindutva rhetoric: national geography and militant masculinities. It then provides an overview of the various theoretical frameworks that have been employed to study Hindutva, such as nationalism, communalism, and fundamentalism, and investigates the role of the urban within the political landscape of Hindutva. The second section, 'Delhi, Hindutva, and the Politics of the Past', locates these theoretical arguments within the urban milieu of Delhi. It looks at the various interventions made by Hindutva ideologues to reclaim Delhi's Hindu past and offers an example of the ways new memorial sites are validating Hindu heroes of the past, thereby re-imagining the scope of patriotic duty and citizenship. In effect, this chapter traverses the scale of nationalist imaginations and urban intervention in order to offer a reading of the many strategies by which Hindutva constructs, transmits, and manages its ideological position in contemporary India.

Hindutva: Epistemes and Memes

Hindu National Space and Militant Masculinities

As a political movement, Hindutva lays claim to India as a Hindu nation and challenges the founding secular vision of it as a modern nation-state which has granted equal rights to religious minorities. Hindutva's origins can be traced to a book of the same name written by V.D. Savarkar in 1923 (Savarkar, 1969). As an ideology, Savarkar's articulation of Hindutva clearly differentiated it from Hinduism. Hindutva was neither a religion nor a philosophy, but was defined more nebulously as 'being Hindu', or Hindu-ness. Membership in the nation projected by this ideology was premised on three essential elements: *pitrabhoo*, the requirement to have been born within the country of India;[4] *jati*, the inheritance of Hindu blood by birth;

and *punyabhoo*, an allegiance to the sacred geography of India with all its holy Hindu sites (Basu *et al.*, 1993).

As thus defined, Hindutva included certain – but only certain – other religions, such as Buddhism, Sikhism, and Jainism. Since most holy sites for Christianity, Islam, and Judaism fell outside the national space of India, these religions were seen as outside the inclusive parentheses of Hindutva. Moreover, Islam, over a period of time, has been carefully fashioned as the nemesis of both Hindutva and India in general. As Thomas Blom Hansen (1999) has rightly noted, the early conceptualization of Hindutva was based on the congruency between a geographical territory (ironically defined by the colonial borders of British India) and an imagined culturally homogenous people (i.e., Hindus). The management of India's geography as essentially Hindu is one of two central themes that frame this chapter.

As a tool of political and cultural organization, Savarkar's conceptualization of Hindutva was first applied through the foundation of the RSS (Rashtriya Swayam Sevak Sangh, or the National Volunteers Organization) in 1925. Since then, the RSS has been able to transmit, consolidate, and proliferate Hindutva ideology by using a distinctly spatial strategy. Its organization has been based on the logic of locally based centres, or *shakhas*, where young men are indoctrinated into a disciplined and unified brotherhood of codes, duties, and services. The *shakhas* and the activities that they sponsor, including education, physical exercise, and mentorship, have been an ingenious method for the RSS to make inroads into India's urban slums and impoverished rural areas (Brosius, 2005). From its founding, a key mission of the RSS has been the creation of a new Hindu man, whose physical strength and militant masculinity would be matched by a fervent loyalty to the Motherland (Hansen, 1999; Basu *et al.*, 1993). While this notion of masculinity was part of Savarkar's early articulation of Hindutva, it has only been fully operationalized as a socio-spatial strategy by the RSS. Indeed, M.S. Golwalkar, one of the most prominent leaders of the RSS, recently framed the duties of this new Hindu man thus:

> Let us shake off the present-day emasculating notions and become real living men, bubbling with national pride, living and breathing the grand ideas of service, self-reliance and dedication in the cause of our dear and sacred motherland... Today more than anything else, mother needs such men – young, intelligent, dedicated and more than all virile and masculine. (cited in Hansen, 1999, p. 83)

In her analysis of the centrality of the trope of masculinity to Hindutva rhetoric, Sikata Banerjee (2005) identified two dominant models for the ideal

Hindu male: the Hindu soldier and the warrior-monk. Banerjee further argued that rather than being completely erased from the narrative of nationalism, Hindutva recasts the Hindu woman as 'heroic mother, chaste wife, and celibate masculinized warrior' (*Ibid.*, p. 3). The representation of the ideal Hindu male subject as defined by militant prowess and physical valour is the second theoretical armature of this chapter.

Let me illustrate Hindutva's persistent representation of India as a Hindu territory in need of protection by valiant Hindu males with an image from the *Organiser*, a newspaper that served as the main organ for various Hindu cultural and political organizations between the 1920s and early 1960s (figure 11.1). The image shows a Rajput warrior dressed in armour and presumably about to embark on a battle. As per Hindu tradition, a woman (who, by this representation, could be either his wife or sister) ties a sacred thread on his wrist to ensure his well-being in battle. In between them is a platter with a lamp, incense, and fruit offerings – the requisite paraphernalia of an *aarti*, a Hindu custom performed before important events, intended to ensure an auspicious outcome for its recipient. More important, however, is the script that accompanies the image, which translates thus:

> Rama waged a battle with Ravana for Sita; the Mahabharata was fought to protect Draupadi's honour and Tanaji sacrificed his life at the battle of Sinhaghad in order to fulfil Jeejabhai's wishes. Even today – on the other side – our sisters and mothers call to us! From Kashmir and Hyderabad, the motherland cries out for help. Will we hear their cries?[5]

Published in 1948 (one year after India gained independence), this full-page representation is rife with images of Hindu masculinity as the guardian of Hindu female honour and the Hindu nation (the two being conflated with one another). Historical figures such as Tanaji and Jeejabhai (seventeenth-century Maratha warriors) are conflated with characters from Hindu mythology such as Rama, Ravana, and Sita. And Hyderbad and Kashmir (two majority Muslim states at the time) are called out as being on 'the other side', where Hindu female honour and the Hindu nation are in peril and in need of protection by the modern-day Rajput warrior.

Hindutva: Nationalism, Fundamentalism, or Communalism?

In the preceding paragraphs, I have provided a brief sketch of the ideological modalities that define Hindutva. However, it remains to be asked, how – or

ORGANISER

HINDUSTHAN'S LEADING ENGLISH WEEKLY

Vol. I : No. 36 | Delhi : Thursday, Shrawan Shukla 15, 2005: August 19, 1948 | Four Annas Air Mail -/4/6

KRISHNA

Figure 11.1. Front page of the *Organiser*, 19 August 1948.

even if – Hindutva can be understood within the theoretical rubric of fundamentalism. The term fundamentalism has rarely been used by scholars in the discussion of Hindutva. Instead, the dramatic emergence, and indeed success, of the Hindutva project since the late 1980s and early

1990s has been spoken of either as a manifestation of chauvinistic nationalism (Hansen, 1999; Gould, 2004; Jaffrelot, 1996; van der Veer, 1994), communalism (Thapar *et al.*, 1993; Sarkar, 2002), or increasingly as a crisis of secularism (Ganguly, 2003; Needham and Rajan, 2007). Indeed, Thomas Blom Hansen (1999, p. 12) has categorically stated that 'Hindu nationalism is not an anti-Western religious "fundamentalism". What Hindu nationalists desire is recognition of themselves and India by the Western powers, but a recognition through assertion of cultural difference and assertion of India's sovereignty and self-determination'. And he argues that in the context of Hindutva, issues of cultural purity and unity trump strict adherence to religious codes or scriptures, thus disqualifying it for consideration as a 'fundamentalism'.[6]

Countering the claims of Hansen and others, who view Hindutva as a form of nationalism fuelled by radical religious sentiments and agents, however, are other scholars, who make the compelling argument that in its ideological conceptualization, Hindutva is markedly different from other twentieth-century nationalisms of the subcontinent, which rooted their claims for sovereignty in the anti-colonial struggle (Sarkar, 2002). Thus, while the methods employed by Savarkar's and Golwalkar's 'secular' counterparts like Gandhi, Nehru, and Ambedkar were radically different from one another, they were united in their stand to end colonial rule as the first step towards national sovereignty. In contrast, Hindutva stalwarts, such as Savarkar and Golwalkar, remained ambivalent towards, and sometimes even supported, colonial domination. Second, unlike other (more mainstream) nationalisms, which were unified in their struggle to modernize Indian society via the emancipation of women, the abandonment of the caste system (in Hindu society), and the alleviation of mass poverty, advocates of Hindutva argued that social equality would emerge as an organic outcome of a nation-state based on Hindu moral norms. Thus, instead of anchoring their vision of a sovereign India in emancipation from colonial rule, the founders of Hindutva propagated the view that Hindus were a vulnerable group who had been oppressed and humiliated by an imagined, precolonial Other (Sarkar, 2002). For example, in his 1923 book *Hindutva/Who is a Hindu*, Savarkar claimed that it was the pressure of a common enemy – i.e., the Muslims – which had created the need for a collective national Hindu identity.

The application of the term fundamentalism to contexts such as India also presents discursive problems in its own right. First, there is the etymology of the term, which arises from a very specific twentieth-century American Protestant movement – one self-identified as anti-modernist and which resisted the separation of church and state.[7] In addition, there has been apprehension about applying the term fundamentalism to non-Western contexts and phenomena, for fear of reproducing pejorative Orientalist notions of cultural essentialism. The hegemonic appropriation of the term by popular media and culture, and its conflation with popular images of terrorism and anti-feminist, anti-modern, or anti-Western rhetoric (to produce easily recognizable categories such as Christian fundamentalism or Islamic fundamentalism), only furthers such problematic divisions.

A third reason that fundamentalism has rarely been applied to the discussion of Hindutva is the ambivalence of its definition with regard to the experience and legacy of colonialism in many modern nation-states. In postcolonial contexts, such a discussion would be incomplete without reference to the manner in which radical religious identities have been conceived and instrumentalized via the very tools and technologies of colonial domination (Pandey, 1990). For example, historians and feminist scholars agree that the conceptualization of Hindu masculinity was forged partly in reaction to the colonial representation of Indian men as effeminate and lacking in physical prowess (Banerjee, 2005; Sinha, 1995). Arguably, a sound understanding of Hindutva must therefore take into account India's colonial past as well as the persistence of colonial frameworks in the present.

Bearing in mind the particularities and complexities of the Indian context, it is interesting to revisit Marty and Appleby's definition of fundamentalism, which was framed in opposition to three basic tenets of the modern: preference for *secular rationality*; adoption of *religious tolerance and relativism* in terms of ethics and morals; and *individualism*. However, fundamentalism need not be defined in complete opposition to the modern world or its various manifestations. Indeed, many of those who might be defined as fundamentalists actively employ products and processes of the modern, such as media technologies and scientific tools, for their own purposes. An alternative reading of the relation between modernity and fundamentalism might thus be less *antagonistic* and more *transactional*. Yet, rather than trying to fit Hindutva into *a priori* definitions of fundamentalism, nationalism, or communalism, it may be most beneficial to understand Hindutva as an insidious, albeit effective, critique of secularism itself. Which is to say that in order to understand the recent success of Hindutva, it is imperative to pay attention to the way the movement has actively appropriated secularist rhetoric as well as the apparatus of the secular nation-state to further its own agendas.

Marty and Appleby's definition of fundamentalism relies on the rejection of secular rationality, wherein secularism is defined as the clear separation between church and state (which also implies a strict distinction between the private and the public spheres). However, this abstract concept of secularism, which is deeply Eurocentric in its provenance, needs revision in the case of postcolonial India, where the state has been entrenched within various religious matters since independence. For example, the Indian state has continually played a role in the reform of religious law (as in the outlawing of Hindu customs such as the immolation of widows or *sati* and the establishment of anti-dowry laws); it has mandated national holidays for various religious festivals; and it has allowed secular courts to arbitrate cases that deal with religious law (such as property and alimony settlements for Muslim divorces that technically fall under Shari'a law) (Needham and Rajan, 2007).

A cornerstone of the conceptualization of Indian secularism has in fact been the role of the state as the protector of minorities (religious, ethnic, and caste-based) and their rights. Several programmes of affirmative action were written into the Indian Constitution to uplift the economic and social status of minority groups and ensure them a share in the fruits of India's progress. In recent decades, however, disillusionment with, and abandonment of, many of the socialist ideals that guided India in the decades after independence have been matched with cynicism towards secularism as a viable path of progress. Hindutva ideologues in particular have made the case that secularism has been seized by political parties like the Indian National Congress as a populist strategy to assuage minorities (namely, Muslims) and afford them privileges unavailable to the majority (namely, Hindus) (Nandy, 2007). In this respect, Hindutva has framed itself not in opposition to secular rationality, but as an *alternative secularism* that represents the pragmatic needs of the *majority population* (the Hindus) over the rights of the individual citizen. Indeed, the European concept of individualism (that underlies secularism) deserves revision in the Indian context, where colonial schemata of classification once privileged communal identities (Hindu, Muslim, Christian, etc.) over the individual (Tejani, 2007).[8]

Hindutva's self-fashioning as the politics of the majority essentially aims to render the issue of religious tolerance moot – which is to say the state should no longer be burdened with the responsibility of affording equal rights to all its citizens, but rather, religious minorities should accept their role as second-tier citizens of a Hindu nation-state. When viewed in this light, Hindutva does not abide by the definition of fundamentalism as an anti-secular or anti-modern phenomenon, but rather as a rational model of republicanism that simply represents the rights of the majority demographic over all others. It is precisely Hindutva's appropriation and recalibration of secularism that has made it such an insidious and powerful force to reckon with in contemporary India.

Postsecular Urbanisms

In recent years, considerable scholarly attention has been paid to the intersection of urban space and resources and Hindutva projects. Much of this work has focused, however, on the role of the urban during moments of violence – as in the analysis of the 1992 Bombay riots and the violence towards Muslims who are seen as undeserving of scarce urban resources (Appadurai, 2002); psychological anxieties between Hindu and Muslim communities that sometimes erupt into urban violence (Kakar, 1996); the conflation of consumer goods and middle-class urban desires with a chauvinistic Hindu pride (Rajagopal, 1994, 1996); and the transnational networks that support localized urban violence (Rajagopal, 2009). While these debates have no doubt enriched and expanded the discussion of Hindutva in productive ways, they have focused mainly on dramatic and episodic moments of violence.

Poststructuralist scholars of the urban such as Henri Lefevbre and Michel de Certeau have urged an examination of the urban that includes not simply the spectacular and extraordinary but also the everyday. Attention to the everyday practices of negotiation and seemingly prosaic tactics through which urban space is appropriated is indeed necessary to reveal the pervasive nature of Hindutva in urban India. Let me illustrate this with an example. In 2005 I noticed that the city of Agra had built several traffic roundabouts, presumably to ease congestion in its narrow streets. Agra is most famously known as the site of the Taj Mahal, and no doubt its largest industry is tourism. Historically (and not unlike Delhi), Agra was also the urban invention of Mughal kings, particularly Akbar and later Shahjahan, who built it as an imperial capital in the fifteenth and sixteenth centuries. Conversely, Agra has little if anything to boast of in terms of a pre-Islamic or Hindu history. However, on my visit, I noticed that all the traffic roundabouts were decorated with large statues of either Rajput or Maratha warriors atop horses, brandishing weapons of military might.

This iconography (another example of which will be discussed in detail later) is easily recognizable to most Indians as the symbol of those 'Hindu' warriors (for example, Prithviraj Chauhan, Rani of Jhansi, or Shivaji) who allegedly fought the 'invasion' of the Mughals and other Islamic groups. These urban interventions and their attendant narratives of martyrdom, national duty, and valour create a visual geography of 'Hindu-ness' in a city that formerly lacked any evidence of this. In this spectacle it matters little if Prithviraj Chauhan or Rani of Jhansi ever visited Agra, or if this was the site of a heroic battle. What is transmitted is the notion of Mughals as non-indigenous rulers whose aggression was resisted by Hindus who felt a sense of nationalist loyalty even in the medieval era.

The insertion of Hindu iconography in a city like Agra can be best understood as a stealthy process by which the urban is re-inscribed through incremental acts of appropriation. As far as I know, there have not been any riots or protests staged around the building of these roundabouts, nor the installation of statues depicting medieval Hindu warriors there. These interventions have occurred without the so-called clash or tension predicted by the notion of ethnic violence or religious fundamentalism. However, this does not take away from the fact that the statues have significantly remade the visual landscape of Agra, and that these changes are part of a larger ideological project to reclaim the past in very precise ways. It is in consideration of these types of intervention that I propose the framework of postsecular urbanism. In moving away from the conventional understanding of the city as a product of economic or political forces, I argue that contemporary religious movements actively appropriate and use urban space in order to further their ideologies. Rather than suggesting a collapse of the public sphere in the city, I thus suggest that postsecular urbanism seeks to re-inscribe the public in very particular and exclusive ways.

Delhi, Hindutva, and the Politics of the Past

Hindutva Interventions in an Otherwise Islamic City

I now turn to the case of India's capital, Delhi, and the ways Hindutva politics have appropriated its cityscape and history to suit the ideological agenda of an imagined Hindu India. This project is precarious for several reasons. Among these are Delhi's long history as the centre of Islamic power, the rich urban and architectural culture left in the wake of centuries of Islamic rule, the later appropriation of the city as the capital of British India and the building of New Delhi, and, finally, its postcolonial transformation into the administrative centre of a secular and independent nation. Conspicuous by their absence also are any urban or architectural sites that could be defined as solely 'Hindu', and that could serve the Hindutva narrative of Delhi, or India, as an essentially Hindu territory.[9]

Delhi's urban origins can be traced to the late eleventh or twelfth centuries – i.e., they are coterminous with the earliest Turkish conquests of the northwestern regions of the subcontinent. The first Islamic dynasty in the area, the Sultanate, was followed by the Tughlaks, and later by the Mughals. With each of these empires, Delhi grew in size as well as grandeur.

A central conceit in the narration of Delhi's history is the succession of eight imperial cities built in what today encompasses its metropolitan region. The first these cities was Qila Rai Pithora. Dating to 1052 CE, it was the fortress compound (rather than an urban centre) of the Rajput ruler, Prithvi Raj Chauhan. The area of Qila Rai Pithora was then taken over during the first Turko-Islamic conquest of the northern Indian plains by Muhammud of Ghor in 1191–1192. The fortress was later expanded by the Sultanate kings during the thirteenth century, and also became the site for Delhi's first mosque – the Quwwat-ul-Islam.

In the early 1300s Delhi's second city, Siri, was built by the Sultanate ruler Alauddin Khilji about two miles north of the expanded Qila Rai Pithora. Between 1321 and 1354, three more urban centres – Tughlakabad, Jahanpanah, and Ferozabad – were then built by the Islamic emperors of the Tughlak dynasty. Dinapanah, the sixth city, was built at the beginning of Mughal rule by the emperor Humayun. It also served as the imperial centre for the Suri kings from 1540 to 1556.

A lull in activity (due to the building of other Mughal capitals such as Fatehpur Sikhri and Lahore) followed between 1556 and 1638. But it was broken by the Mughal emperor Shahjahan's decision to build the grandest of all cities. Shahjahanabad survives today as one of the finest examples of Mughal urbanism, and it encompasses many of Delhi's magnificent monuments, like the Red Fort and the Jama Masjid (Friday Mosque). The eighth city, of course, is New Delhi, begun by the British colonial government to mark the transfer of the imperial capital from Calcutta to Delhi in 1911.

The portrait of Delhi's urban evolution as encompassing a series of imperial cities should be understood less as a recounting of historical fact than as an attempt at narrative effect. It came into circulation during the colonial period and echoed the historical motifs of other cities (such as Rome being the city of seven hills, etc.). Indeed, many of these 'cities' – such as Qila Rai Pithora or Siri – would better fit the definition of garrison towns or fortified military enclaves. Despite these inaccuracies, however, the continued imagination of Delhi's past as a series of mostly Islamic imperial cities, coupled with its lack of 'Hindu' built resources, has become a locus of Hindutva anxiety.

To elaborate: little archaeological evidence remains of what might have preceded the 'Islamic' foundations of Delhi in the twelfth century. A stepwell (Surajkund), the remains of a medium-sized fort (Lal Kot), scattered edicts from the Mauryan period (many of which were in fact relocated to Delhi by Tughlak and Sultanate rulers), and other minor sites seem to suggest that Delhi was little more than a garrison town, used infrequently as an outpost by mercenary armies. This absence of a Hindu past is made even more conspicuous by Delhi's surplus of Islamic architecture, spanning a period of almost seven hundred years (which is particularly troublesome to Hindutva's claims that the geo-body of India is essentially marked by Hindu culture and heritage). Nevertheless, Delhi's pre-Islamic past – heretofore too insignificant to be of any note in the imagining of the nation's protohistory – has in recent decades been brought into high relief by Hindutva loyalists who claim that Delhi was once a thriving centre of a nation which was essentially wiped out by Islamic imperialism and violence.

Hindutva attempts to construct an important Hindu past for Delhi have ranged from the archaeological search for a Hindu city below the many Islamic urban strata, appropriations of mosques as Hindu temples, and the building of new memorial sites for the erstwhile Hindu heroes of the city. Yet, it should also be pointed out that in the case of Delhi (at least at the time of writing) Hindutva strategies for re-inscribing the city as Hindu have mostly focused on symbolic appropriation and historical syncopation. Unlike the case of Ayodhya, where Hindutva ideology espoused a violent destruction of Islamic monuments, in Delhi the emphasis has fallen either on reclaiming Islamic sites as Hindu or excavating Hindu sites and cities that supposedly lie buried beneath years of Islamic built fabric.

Territorial Reclamations

One of the most poignant attempts to establish a Hindu history for Delhi has been archaeological excavations to ascertain if modern Delhi is in fact coterminous with Indraprastha, the capital of the Pandava Kingdom mentioned in the Hindu epic the *Mahabharatha*. While this apocryphal notion has been in circulation at least since the colonial period, it received renewed purchase following Indian independence when the Archaeological Survey of India sought to excavate sites which were mentioned in the *Mahabharatha* and its companion epic the *Ramayana*. Archaeologists who led the excavations in Delhi during the 1950s and 1960s were intent on discovering a Hindu city that would serve as the urban origin for all of the Islamic cities which followed, and they showed few reservations in using mythological and religious texts as the basis for their explorations.[10]

It should be emphasized here that these excavations were neither sponsored by Hindutva ideologues, nor conducted under the aegis of satisfying Hindutva agendas. In fact, the archaeologists were civil servants of a secular state, who saw their project as simply recovering the ancient past of a city. And yet, despite the fact that all these efforts failed to prove that Delhi's ancient and pre-Islamic substrata belong to a Hindu empire, the myth that the city's origins can be traced to Indraprastha still survives, and is reproduced not only in popular narratives but also in official maps and archaeological accounts.[11]

The search for Delhi's Hindu origins brings to light the fragility of many Hindutva claims while also revealing the problematic appropriation of seemingly secular apparatuses to maintain them. For instance, the archaeological search for Indraprastha has not been based simply on finding a few objects or signs that would establish that a Hindu period did predate several centuries of Islamic culture and imperialism. Rather, these excavations have been motivated by the expectations of finding evidence of an imperial city to match (and possibly surpass) the grandeur of all the Islamic cities built between the twelfth and the eighteenth centuries. In other words, Hindutva's desire is to extend the serialization of Delhi's history as a compendium of distinct cities backwards to include a Hindu *ur*-city that would serve as the progenitor for all the imperial capitals that followed. It is also important to note the conjecture that the possible existence of Indraprastha – which started in the colonial and possibly precolonial period – has only been given scientific weight, and therefore credibility, in the postcolonial period. Indeed, it may not be a stretch to assert that only the use of secular state-operations (such as archaeology), has allowed the long-standing myth of Indraprastha to be transformed to supposed representational reality.

If the ongoing search for Indraprastha can be explained as the work of agents with Hindu proclivities appropriating the instruments of secularism, more recent efforts to mark Delhi as Hindu territory have taken a more aggressive face. For example, on 15 November 2000, a crowd of between 150 and 200 people gathered outside the Quwwat-ul Islam mosque in Delhi. The group largely comprised members of various cultural and political organizations (such as the Vishwa Hindu Parishad) allied in support of Hindutva.[12] As they protested outside, they demanded access to the mosque to perform a Hindu purification ceremony (*dev-mukti yagna*), which they claimed would 'liberate' the Hindu icons that had been 'trapped' in it. They cited as fact the popular myth that the mosque was built following the destruction of twenty-seven Hindu and Jain temples more than seven hundred years ago. The act of ritual purification would thus allow the protestors to reclaim the Quwwat-ul-Islam as a Hindu space in the contemporary moment.

The monument around which this protest unfolded is an important one for several reasons. First, the Quwwat-ul-Islam is one of the earliest extant mosques in India, and marks the origins of several centuries of Islamic rule over the subcontinent. Built by the first Turko-Islamic rulers of Delhi in the late thirteenth century, the campus of the mosque continued to grow well into the fifteenth century due to contributions by successors to the original Sultanate dynasty of Delhi. The larger complex also contains within it the magnificent Qutb Minar – the original minaret of the mosque. The significance of the mosque and its larger campus – which includes the Qutb Minar and such additional structures as graves, colleges, etc. – also extends into the national imaginary of India's glorious past. As one of only two World Heritage Monuments in Delhi, it garners handsome revenues for the state and is one of the primary tourist attractions of Delhi. The Qutb Minar, in particular, has been absorbed into the canon of Indian architecture, as codified by its appearance in school textbooks, on national stamps, and ephemera, which validate it as an icon of national as well as local urban antiquity. For example, in 1948, when independent India issued its first series of stamps featuring a range of national archaeological sites, the Qutb Minar was prominent in the iconography of the newly decolonized nation-state.

And yet, even as structures such as these have been absorbed into the 'secular' rhetoric of India, a great deal of ambivalence remains around their origins. This is a monument from which defaced sculptures of Hindu gods and goddesses supposedly tumble, from which plaster falls away during restoration work to reveal figurative representations of animals and birds, and that in all probability once belonged to a faith other that of its Muslim builders. And while it may seem that the secularism of the nation-state has no recourse but to create a silence around the religious tensions of the past, it is precisely these silences which have allowed the religious right to manipulate the past to serve their own agendas.

It is, of course, old news at this point that the spectre of Islamic iconoclasm, particularly the destruction of Hindu sites at the hands of medieval Islamic rulers, has found particular utility in the Hindutva agenda of inscribing India as Hindu homeland. Indeed, contemporary redemption for acts of medieval Islamic iconoclasm (it matters little if there is historic evidence for them or not) was the primary motive for the destruction of the Babri Masjid in Ayodhya in 1992. It is a trope that has been employed with immediacy by the followers of Hindutva. But particularities surrounding the recovery of the Qutb Minar and the Quwwat-ul-Islam mosque are different from the destruction of the Babri Masjid and other sites, because this monument operates at the scale of the nation. In other words, the protests by Hindutva supporters on 15 November 2000, outside of the Qutb complex were not simply about reclaiming an important Islamic building complex as a Hindu space, but also recovering a well-known object of the secular Indian nation as a Hindu object.

The territorial recovery of India as a Hindu nation has been central to the mandate of Hindutva, because its ideological framing rests on belief that the Indian subcontinent is also the geography of Hindu religious and historic sites. Scholars such as Romila Thapar have attributed this to Hindutva's founders, who drew inspiration from German Romantic nationalism and its conflation of race with the national landscape. For example, among its regular articles and editorials, the *Organiser* (the main newspaper of the Hindu Mahasabha mentioned earlier) featured a weekly column known as 'Holy Bharat', dedicated to the cataloguing and description of the historical landscape of the new nation-state of independent India.[13] Muslims (and less frequently Christians) appear in this historic sequence, as well as geographical *mise-en-scĕne*, only as iconoclasts attempting to destroy and erase the magnificent monuments that Hindu hands have built, and no reference is made to Islamic monuments as being part of Indian architectural heritage. The cataloguing of Hindu historical sites across the length and breadth of India allowed the *Organiser* to align the boundaries of the modern nation-state (a product of colonial whimsy rather than territorial allegiances) with the *terra infirma* of a historic Hindu nation. In contrast, Islamic iconoclasm was given a new terrifying relevance, in that the violence perpetrated on Hindu monuments was also now cast as violence against a Hindu nation. The mutilation of one corpus (that of the monument) thus began to represent the mutilation of the body of the nation at large.

I will return here to my previous point about secularism as the discursive antonym of religion, and particularly radical religious practices. While it is tempting to position the protestors outside the Quwwat ul-Islam mosque as agents of fundamentalism rallying against a nonconfessional secular state, it may be more useful to see how secularism is itself based on and perpetuates 'moralistic' codes of social conduct that reproduce marginality according to agendas very similar to religious structures.

Ann Pellegrini and Janet Jakobsen (2000) have provided a useful critique of the artificial divide between secularism and religious movements by drawing attention to the high moralism of the Enlightenment that is the very foundation for secularism. They argue that secular states have emphasized the disciplined body (with its attendant sexual desires and social codes of conduct) on the grounds of moralism rather than religion, and in so doing have perpetuated the very same epistemic divisions that were the hallmark of religious power in the past. Thus, religion and secularism appear to be less antitheses of one another than hopelessly interconnected strands of modernity that uphold certain (disciplined) bodies as morally better than others. Further, analysis of them reveals why modernity has not allowed certain bodies, such as women or gays, full equality no matter which side of the religious/secular divide they fall.

We may draw the same theoretical thread towards the Muslim body in postcolonial India. Their representation as iconoclasts or lascivious threats to female honour (as implied by the text accompanying figure 11.1 from the *Organiser*), not only writes them out of the Hindu nation, but also denies them a place within a modern secular nation.[14]

Masculinity and Martyrdom in the City

Hindutva's representation of Muslim bodies as aberrant to the historic as well as contemporary imaginary of India has required the corollary construction of an idealized Hindu body. This next example, illustrates the emergent rhetoric of masculinity and valour as a key element for citizenship in a robust Hindu nation.

In December of 2000, the Minister of Urban Development and Poverty Alleviation, Jagmohan, presented a motion in the Lok Sabha (Lower House of Parliament) for permission to build a commemorative monument for Prithviraj Chauhan in Delhi. Prithviraj Chauhan, also known as Rai Pithora, has become a prominent figure in Hindutva histories. Credited with building Qila Rai Pithora, which was later taken over by the Sultanate kings as a site for one of the first of Delhi's many cities, Chauhan's defeat at the hands of Turko-Islamic conquerors marked the beginning of centuries of Islamic rule over the subcontinent. Prithviraj Chauhan, however, occupies an ambiguous position in this historical narrative because of his decision to free Mahmud of Ghori (one of the first Islamic 'invaders') after Ghori captured North India. Chauhan's decision to free the Islamic conqueror has been perceived as a mark of valour and generosity. But it has also been resented as a fateful lapse of judgment, which resulted in the return of Mahmud of Ghori with a stronger army, the defeat and death of Chauhan, and the conquest of the region around Delhi. It is partly the ambiguous nature of this 'Hindu' king of Delhi which was at stake in the commissioning of the monument in 2000.

As Jagmohan claimed, this new monument would function as part museum, part information centre, and part commemorative symbol for the medieval Rajput warrior. Jagmohan – a member of the BJP, the political party which was then in power and is part of the larger Hindutva machine – envisioned the monument as having a didactic role in that it would broadcast the importance of Prithviraj Chauhan to the citizens of Delhi as well as the rest of India. It would also allow the Delhi Municipal authorities to take over and protect the ramparts of the fort, which were allegedly begun by Prithviraj Chauhan in the late twelfth century. It must be noted that the ramparts which Jagmohan was claiming as the exclusive structural remnants of the Lal Kot or Qila Rai Pithora were in fact rebuilt by Sultanate rulers (like Alauddin Khilji), who added to, or built on top of the fort they inherited from Prithviraj Chauhan.[15] And when Jagmohan proposed this project, the extant ramparts of Quila Rai Pithora and Lal Kot had been under the protection of the Archaeological Survey of India (ASI) for well over seventy years, and Jagmohan's project to rebuild the walls and suggest a new use for the fort were technically antithetical to the mandate of the ASI. As such, the project (which Jagmohan had hoped to complete with the assistance of the Delhi Development Authority, DDA) was initially opposed by the ASI.[16] Nevertheless, the project went ahead, and the following extract is taken from the motion passed by Jagmohan on the floor of the Lok Sabha, reasserting its importance.

> In this connection, I wish to make it clear that the Government attach great importance to the project and development of Qila Rai Pithora complex, which was conceived by it in November, 1999. It was inaugurated by Union Home Minister, Shri L.K. Advani.
>
> Located at the most strategic part on the Aravali range at the entrance of Delhi, the Qila was witness to many ups and downs of history of India. It was built with thick rubble stone and was also known as 'Lal Kot', which, in fact, is the first city of Delhi.
>
> Qila Rai Pithora/Lal Kot constitutes an inspiring saga of our past. Here, history speaks through bricks and battered stones; heritage stares in our eyes through heavy but broken and sunken walls; and herein can also be heard the voices emanating from our past omissions, when we did not fully remember that if we were not cohesive and concerned, then things could go wrong, despite the great and glorious strands of our culture and civilisation.
>
> The project has two basic objectives in view. First, to preserve, protect and strengthen our architectural and cultural legacy. Secondly, to weave history and heritage in the new urban fabric that is being presently spun in Delhi and to

> develop a large park around the Qila Rai Pithora/Lal Kot complex and create a glorious backdrop of Rajput style of garden that it deserves. It would attract lakhs [hundreds of thousands] of people from all over the country, particularly the youth and acquaint them with the great acts of valour of our leaders.
>
> An equestrian statue of Samrat Prithviraj Chauhan, 18 feet high, mounted on a four feet high pedestal, is being set up at the terrace of the Conservation Centre which is a part of the overall development of the complex. This complex would also serve as vast area of community green for a large population of low and middle income group people living in the neighbourhood.[17]

In 2005, the Qila Rai Pithora Complex was almost complete. The gardens and landscaped area around the central museum-like structure had been built and were being regularly used by small groups of picnickers, joggers, or strollers. The central structure which was to serve as the information centre, with its circular gallery chronicling the urban achievements of Indian civilization, was still receiving informational plaques, etc. However, for the purposes of the present discussion I want to examine a few of Jagmohan's motivations for building the centre, as expressed to the Lok Sabha in December 2000. First, the minister argued for the complex based on the need to revive in the memory of Indians their forgotten histories, particularly the danger that befalls a community when it does 'not fully remember that if [they] were not cohesive and concerned, then things could go wrong, despite the great and glorious strands of our culture and civilisation'. The warning inherent in this message renders Prithviraj Chauhan as a hero, the last Hindu king of Delhi who died resisting the Islamic 'invaders' to the region. However, he is now also painted as a failed hero, one whose personal valour and courage were of little utility given the lack of unity among the Hindus, which allowed foreign rulers to conquer the subcontinent with little effort.

Jagmohan is expressing the commonly held notion that the disunity and feudalism which prevailed in India was the real reason why medieval 'Hindu' civilization fell to the Islamic Sultanate.[18] But the message (and by extension the monument) also serves as a prescriptive call to arms for Hindus to amalgamate as a homogenous, cohesive, and even chauvinist group against non-Hindus. This definition of Hindu identity is very much in line with Hindutva ideals of unified and aggressive community. Jagmohan's concern, in particular, that the 'youth' of the country must be educated about the valour of past leaders such as Prithviraj Chauhan is telling about the contemporary mission of the project. While it can be argued that the 'monument' has not gained the kind of reputation that

Jagmohan had intended for it (in that it rarely receives more than a few dozen visitors per day), the 18 foot equestrian statue of Prithviraj Chauhan is clearly visible from both the major roads that border the site. In this sense, the statue of the Hindu militant king has become a clear visual icon in a city which has very little in the way of Hindu monuments.

In analyzing the broader success and appeal of the project of Hindutva, scholars have argued that it is only through various strategies of public intervention and mass participation that the basic tenets of Hindutva are accepted, and indeed internalized, by Hindus in the nation-state. These strategies include the proliferation of 'Hindu' images and icons in the public domain, the deliberate collusion between religious Hindu ideology and middle-class aspirations, and socially visible performances of Hindu identity.[19] For example, in 1991, when L.K. Advani (the same politician who inaugurated the Prithviraj Chauhan Complex) organized his *rath-yatra* (a chariot journey mimicking the mythology surrounding the Hindu deity Rama) around the country, the message of replacing the Babri Masjid with a Hindu temple was disseminated through posters, stickers, and audio-tapes which his political party had produced and distributed to the public at large. The historian K.N. Pannikar has argued that the image of the Hindu deity Rama that was depicted in these posters and other visual paraphernalia was startlingly different from the traditional depiction of the deity as benevolent and tranquil. Instead, they showed Rama as an angry warrior, poised with his bow and arrow as if to launch an attack (Pannikar, 1993). Further, the incendiary remarks made by Advani against the Muslim community of India were meant to 'infuse a sense of shame and humiliation among the people for Hindu society's alleged failure to protect its shrines from desecration by Muslim conquerors' (*Ibid.*, p. 70). This representation of Hindu historical and mythological figures – whether Rama or Prithviraj Chauhan – converts them into religious warriors whose use of violence and aggression is not just sanctioned but unequivocally justified as long as it is in the defence of 'homeland'. In this representation of heroism, violence is not only justified, it is presented as imperative to the very survival of the nation.

The canonization of Prithviraj Chauhan as an idealized masculine citizen of Delhi replicates similar patterns in other cities and towns, such as the emergence of Shivaji as the hero of Mumbai, or Lord Rama in Ayodhya by Hindutva supporters. Indeed, a comparison between these representations reveals a startling, if somewhat predictable, symmetry. For example, Prithviraj Chauhan and Shivaji bear similarities for their anti-Islamic (as opposed to anti-colonial) feats and their skills as aggressive warriors (as opposed to the posture of nonviolence made famous by Gandhi or the shrewd diplomacy of Nehru in later years). Another similarity between Prithviraj Chauhan and Shivaji arises from the overt emphasis on their non-Brahmin identities – a populist strategy which deliberately engages lower-caste groups as actors in national politics. Martha Nussbaum (2007) has convincingly argued that the Shiv Sena's glorification of Shivaji in Mumbai (as well as the larger state of Maharasthra) via legends, while constantly reiterating his lower-caste affiliations, has allowed a broad demographic of Hindus to embrace him as their hero.[20]

In recent years, scholars have drawn attention to the way traditional definitions of citizenship based exclusively on the framework of the nation-state are being recalibrated by emergent forms of urban citizenship. (Sassen, 2003). Appadurai and Holston (1999) warn that these may not have anything to do with a replication of citizenship as articulated by nation-states, and in fact may even challenge the national paradigms of belonging. Similarly, AlSayyad and Roy (2005) have argued that new patterns of urban citizenship have been marked by emergent enclave cultures that rest upon protectionism, vigilantism, and increased segregation (both perceived as well as physically manifest) of atomized communities in the urban realm. Within the specific context of India, scholars have perceived the transformation of the urban milieu resulting from Hindutva politics as either a decosmopolitanization (Appadurai, 2002), or a provincialization (Varma, 2004). Given these debates around the formation of alternative urban citizenships, is it possible to understand the Qila Rai Pithora memorial as a site for the representation of an idealized Hindu citizen? Like Shivaji for Bombay, the representation of Prithviraj Chauhan can be seen as an atavistic strategy whereby Delhi's origins are reclaimed as Hindu, while also reinforcing the ideals of Hindu patriotism as key to good citizenship in the contemporary city.

Conclusion: The Religious Reinscription of Urban Space

Conventional definitions of modernity have relegated religious belief to the realm of the private and have attempted to abrogate its public appeal by privileging scientific temper and rational thought. By confining religion to the inner sanctum of the private, the realm of the public may be left untainted by the forces or desires of religious affiliation and expression. Undoubtedly, and like many other imaginations of modernity, this is an artificial divide that has been questioned by those who offer valuable insights into alternative or multiple modernities. I have argued here that the central tenet of postsecular urban space is the recovery of the public as the religious. Contemporary expressions of radical religiosities are not simply about the enduring success of a set of religious beliefs or a movement against modernity; rather they are efforts to reclaim the very sacred foundations of modernity, such as the incontrovertible divide between sacred and secular, public and private, and religious and scientific. The urban realm is central to this project of religious expression. Where better to infiltrate the public with the religious than in the city, the ultimate signifier of the modern, heretofore understood through other fictive devices such as cosmopolitanism or a secular public domain? In other words, the postsecular city as a project does not aim to annihilate the public, but rather to continue its robust survival along the lines of a religious public.

The two examples of urban reinscription offered in this chapter also point to one other important strategy of Hindutva. In his analysis of Hindutva as an ideology, Sarkar has claimed that history has been exceptionally important to the project of Hindutva. Like other authoritarian regimes (such as those of Hitler, Stalin, and McCarthy) which sought to control history in one way or another, Hindutva has also consistently given high priority in its larger ideology. It is in the manipulation and transmission of this distorted history that the urban realm takes on a new significance. Whether through the staging of protests outside of the Quwwat-ul-Islam mosque as a means to claim it as a Hindu monument, or the installation of visual icons that symbolize the historical martyrdom of Hindu warriors at the hands of Muslims, history is mobilized as a spectacle via urban space and urban forms.

In a recent essay, Charles Taylor (2006) argued that radical religions of the contemporary world can only be fully understood by looking closely at the modern means through which they mobilize themselves. In other words, while Hindutva may legitimate itself on the basis of historical narratives of belonging, it does so today via the products and technologies of modernity, such as print media, the re-imagination of the nation-state, and the use of the urban as an essential locus for hegemonic strategies. Just as the success of global terrorism (as an example of the most extreme form of radical religious expression) cannot be understood outside of the global networks of communication and the most modern elements of technology, so too is the pressing need both to recognize and reframe the uncanny intersection of religion and the urban product of this moment of late modernity.

Notes

1 *Selections from the Travels of Ibn Batuta* (edited and translated by H.A.R. Gibb). London: George Routledge and Sons, 1929, p. 194. Also cited in Spear (2002, p. 14).

2 For a full transcript of the speech, see 'We Are Not One Because We Are Citizens Of The Same State, But It Is Because We Are One That India Is One State', *Organiser*, 9 October 1961, p. 5.
3 See note 12 for an explanation of the VHP.
4 This was one of the tenets of Hindutva as preached by Savarkar, but there are also ideas which run contrary to it, which group all Hindus living everywhere in the world as part of one nation or family. In fact, the image often evoked is that of a Hindu tree with thirty different branches that symbolize various diasporas in different parts of the world.
5 *Organiser*, 19 August 1948.
6 In a compelling argument on the rise of Hindutva hegemony, Hans Bakker pointed out that there has been a reluctance to apply the term fundamentalism to polytheistic religions such as Hinduism. Over many centuries of Islamic domination and later Western colonialism, however, Hinduism has borrowed such facets of monotheism as organized prayer, militant advocacy, and most recently, the concept of 'holy war' against other religions. Indeed, one of the fundamental projects of Hindutva has been the canonization of Rãma as the central deity of modern Hinduism, in a manner that mimics other monotheistic religions (Bakker, 1991).
7 While it is imperative to recognize the historical context of the origin of fundamentalism, it should also be pointed out that several other discursive frameworks such as nationalism, modernity, democracy, and secularism have deeply Euro-Americ-centric roots. In recent years scholars of 'non-Western' contexts have recognized, wrestled with, and successfully expanded these formerly Euro-centric definitions by illuminating previously marginalized histories, thereby exposing the narrow formulations of the aforementioned concepts. It is perhaps time to subject fundamentalism to the same discursive rigour by recognizing its various articulations in diverse contexts.
8 Regarding the constitutional articulation of secularism in India, Tejani (2007, p. 47) rightly remarked, 'Rather than being distinct from the categories of community and caste, nationalism and communalism, liberalism and democracy, Indian secularism emerged at the nexus of all of these. It was therefore a relational category arising out of a series of specific historical negotiations. As such, it had meanings that were modern, in the sense that they were contemporaneous with such negotiations elsewhere, but they were not universal in that they were closely tied to the historical context out of which they emerged and did not replicate the narratives of history as staged in the West'.
9 It should be pointed out here that the categories of Hindu architecture and Islamic architecture are deeply contested and flawed. They owe their origins to colonial framing of Indian history as divided into three distinct phases: the ancient period of Hindu and Buddhist architecture; the medieval period of Islamic building; and the modern period marked by colonial or European architecture and urbanism. Such facile distinctions have perpetuated myths of Hinduism and Islam as culturally distinct processes with little overlap and wholly outside of modernity, which is seen as a colonial import to India.
10 For more information on the excavations, see the reports of the Archaeological Survey of India edited by Ghosh (1955), Lal (1973), and Deshpande (1974).
11 For example, a map in the INTACH (Indian National Trust for Art and Cultural Heritage) register of buildings identifies a very specific part of Delhi as Indraprastha. See Nanda *et al.* (1999). Also see affirmations such as the following in Mani (1997): 'History has witnessed Delhi as the capital-city of many kingdoms

and empires. The foundation of the city of Indraprastha during the period of Mahabharata war is well known'.

12 The Hindutva political brigade is represented by the Sangh Parivar – an umbrella organization for three separate but politically similar organizations: the Vishwa Hindu Parishad (VHP), the Bharatiya Janata Party (BJP), and the Rashtriya Swayamsevak Sangh (RSS). The Sangh Parivar has often described the VHP as its socio-cultural, the BJP as its political and the RSS as its organizational wings – all working in tandem towards the establishment of India as the land of 'Hindutva'. For more information on the historical rise of the Hindu right in India, see Basu *et al.* (1993).

13 Some of the Hindu religious shrines that appeared in the *Hindu Organiser* from the year 1955 include: Parsuram Temple (25 April 1955); Sankaracharya Temple in Srinagar (1955); the Sun Temple at Martand (13 June 1955). Other articles feature Chittor as a historic site (18 April 1955); the cities of Vijayangara (5 April 1954) and Takshashila (29 February 1960) as examples of urban grandeur; the state of Himachal Pradesh for its scenic beauty and sacred landscape (18 July 1955); and even the battlefield of Kurukshetra, where the mythical battle of the *Mahabharatha* was supposedly fought (20 June 1955).

14 Refer to the articles cited in the previous note.

15 There is considerable archaeological evidence that while the base structure of the smaller Lal Kot and the larger Quila Rai Pithora was commissioned during the time of the Tomara kings and Prithviraj Chauhan, much of what is visible today was constructed during the time of Alauddin Khilji in the early fourteenth century. For example, Maulvi Zafar Hasan's report (1990, p. 41) on excavations in the area claims, 'Muslim chronicles of India aver that Alauddin Khilji repaired and fortified the walls of Raipithura's city after the invasion of the Mongols in the year 1303 … and this statement receives support from the structural remains discovered'. Also see the later article in the ASI Reports where the same view is expressed (Ghosh, 1958, p. 25). There, in speaking of the excavations at Lal Kot and Quila Rai Pithora, the archaeologist writes, 'Possibly the stonework was erected by the Rajputs and the brickwork was added by the Muslims after Delhi had been wrested by them from the former'.

16 Several newspapers reported the antagonism between the ASI and the DDA. See 'High Court Issues Notices to ASI, DDA, Delhi Police', *Statesman*, 15 June 2001, p. 4; and 'ASI Opposes DDA Center at Saket', *Times of India* (New Delhi edition), 21 June 2001, p. 2.

17 See Lok Sabha, Synopsis of Debates, 22 December 2000 (Proceedings other than Questions and Answers).

18 Several historians have analyzed this particular historiographic stance as a narrative of loss, premised on the dubious assumption that all pre-Islamic groups shared a common 'Hindu' identity, and therefore represented a united community that was wiped out and fragmented by the introduction of Islam to the subcontinent (Thapar, 2004). For example, one historian (Vaidya, 1926, Vol. III: *Downfall of Hindu India*, p. 362) writes, 'Internecine warfare has always been the bane of the Rajputs … [they] should have stayed their quarrels and combined. They did not stop their fights even against the common impending danger and they consequently were all destroyed'.

19 For a critique of how the Hindu right used images of the Rama temple at Ayodhya and conflated it with middle-class aspirations of Indians, see Raychaudhuri (2000). The author cleverly outlines how images of the new temple at Ayodhya were often conflated with commodity images such as the Indian-made Maruti car

(a classic symbol of middle-class Indian life) in an effort to make their message of communal violence seem more palatable to slightly less right-wing Hindus.

20 In this analysis, Nussbaum references James Laine's book, *Shivaji: Hindu King in Islamic India* (2003) and the controversy that erupted surrounding its publication.

References

AlSayyad, N. and Roy, A. (2005) Medieval modernity: citizenship in contemporary urbanism. *Applied Anthropologist*, **25**(2).

Appadurai, A. (2002) Spectral housing and urban cleansing: notes on millennial Mumbai, in Breckenridge, C.A. (ed.) *Cosmopolitanism*. Durham, NC: Duke University Press.

Appadurai, A. and Holston, J. (1999) *Cities and Citizenship*. Durham, NC: Duke University Press.

Bakker, H. (1991) Ayodhya, a Hindu Jerusalem: an investigation of 'holy war' as a religious idea in the light of communal unrest in India. *Numen*, **38**(1), pp. 80–109.

Banerjee, S. (2005) *Make Me a Man! Masculinity, Hinduism, and Nationalism in India*. Albany, NY: State University of New York Press.

Basu, T. *et al.* (1993) *Khaki Shorts and Saffron Flags*. Delhi: Orient Longman.

Brosius, C. (2005) *Empowering Visions: The Politics of Representation in Hindu Nationalism*. New Delhi: Anthem Press.

Deshpande, M.N. (ed.) (1974) *Indian Archaeology: A Review (1970–1971)*. New Delhi: Archaeological Survey of India.

Ganguly, S. (2003) The crisis of Indian secularism. *Journal of Democracy*, **14**(4), pp. 11–25.

Ghosh, A. (ed.) (1955) *Indian Archaeology: A Review (1954–1955)*. New Delhi: Department of Archaeology.

Ghosh, A., (ed.) (1958) *Indian Archaeology 1957–58: A Review*. New Delhi: Department of Archaeology.

Gould, W. (2004) *Hindu Nationalism and the Language of Politics in Late Colonial India*. Cambridge: Cambridge University Press.

Hansen, T. (1999) *The Saffron Wave: Democracy and Hindu Nationalism in Modern India*, Princeton, NJ: Princeton University Press.

Hasan, B.M.Z. (1990) Excavation at Delhi, in Dikshit, R.B. (ed.) *The Annual Report of the Archaeological Survey of India 1936–37*. Delhi: Swati Publications, reprint.

Jaffrelot, C. (1996) *The Hindu Nationalist Movement in India*. New York: Columbia University Press.

Kakar, S. (1996) *The Colors of Violence: Cultural Identities, Religion, and Conflict*. Chicago, IL: University of Chicago Press.

Laine, J. (2003) *Shivaji: Hindu King in Islamic India*. New York: Oxford University Press.

Lal, B.B. (ed.) (1973) *Indian Archaeology: A Review (1969–1970)*. New Delhi: Archaeological Survey of India.

Mani, B.R. (1997) *Delhi: Threshold of the Orient, Studies in Archaeological Investigations*. New Delhi: Aryan Books International.

Marty, M.E and Appleby, R.S. (eds.) (1993) *Fundamentalisms and Society: Reclaiming the Sciences, the Family, and Education*. Chicago, IL: University of Chicago Press.

Nanda, R. *et al.* (1999) *Delhi: A Built Heritage*. New Delhi: INTACH.

Nandy, A. (2007) Closing the debate on secularism: a personal statement, in Rajan, N. and Rajan S. (eds.) *The Crisis of Secularism in India*. Durham, NC: Duke University Press.

Needham, A.D. and Rajan, R.S. (eds.) (2007) *The Crisis of Secularism in India.* Durham, NC: Duke University Press.

Nussbaum, M. (2007) *The Clash Within: Democracy, Religious Violence and India's Future.* Cambridge, MA: Harvard University Press.

Pandey, G. (1990) *The Construction of Communalism in Colonial North India.* Delhi: Oxford University Press.

Pannikar, K.N. (1993) Religious symbols and political mobilization: the agitation for a Mandir at Ayodhya. *Social Scientist*, **21**(7/8), pp. 63–78.

Pellegrini, A. and Jakobsen, J.R. (2000) World secularisms at the millennium. *Social Text – 64*, **18**(3), pp. 1–27.

Rajagopal, A. (1994) Ramjanmabhumi, consumer identity and image-based politics. *Economic and Political Weekly*, **29**(27), pp. 1659–1668.

Rajagopal, A. (1996) Communalism and the consuming subject. *Economic and Political Weekly*, **31**(6), 341–348.

Rajagopal, A. (2009) Violence, publicity, and sovereignty: lawlessness in Mumbai. *Social Identities*, **15**(3), pp. 411–416

Raychaudhuri, T. (2000) Shadows of the swastika: historical perspectives on the politics of Hindu communalism. *Modern Asian Studies*, **34**(2), pp. 259–279.

Sarkar, S. (2002) *Beyond Nationalist Frames: Postmodernism, Hindu Fundamentalism, History.* Bloomington, IN: Indiana University Press.

Sassen, S. (2003) The repositioning of citizenship: emergent subjects and spaces for politics. *The New Centennial Review*, **3**(2), pp. 41–66.

Savarkar, V.D. (1969) *Hindutva.* Bombay: Veer Savarkar Prakashan.

Sinha, M. (1995) *Colonial Masculinity: The 'Manly Englishman' and the 'Effeminate Bengali' in the Late Nineteenth Century.* Manchester: Manchester University Press.

Spear, P. (2002) Delhi: a historical sketch, in *The Delhi Omnibus.* New Delhi: Oxford University Press.

Taylor, C. (2006) Religious mobilizations. *Public Culture*, **18**(2), pp. 281–300.

Tejani, S. (2007) Reflections on the category of secularism in India: Gandhi, Ambedkar, and the ethics of communal representation, c. 1931, in Rajan, N. and Rajan, S. (eds.) *The Crisis of Secularism in India.* Durham, NC: Duke University Press.

Thapar, R. *et al.* (1993) *Communalism and the Writing of Indian History.* New Delhi: People's Publishing House.

Thapar, R. (2004) *Somanatha: The Many Voices of a History.* New York: Penguin.

Vaidya, C.V. (1926) *History of Mediaeval Hindu India.* Poona: Aryabhushan Press.

Van der Veer, P. (1994) *Religious Nationalism: Hindus and Muslims in India.* Berkeley, CA: University of California Press.

Varma, R. (2004) Provincializing the global city: from Bombay to Mumbai. *Social Text – 81*, **22**(4), pp. 65–89.

Chapter 12

Excluding and Including the 'Other' in the Global City: Religious Mission among Muslim and Catholic Migrants in London

John Eade

Britain and its capital city, London, appear to be quintessentially secular spaces and, therefore, would seem to provide a sharp contrast to the other locations discussed in this volume. Britain, like other European countries, has experienced a rapid decline in religious observance, and Christian leaders there often seem to be lonely figures in their critique of secular materialism and defence of their particular moral teachings and religious practices. Yet, the continuing debate about secularization shows that appearances may be deceptive (see, for example, Wilson, 1966; Bruce, 1992; Casanova, 1994; Berger, 1999; Bruce, 2002; Davie, 2002; Davie *et al*., 2003). Although religious attendance in mainstream Christian churches has rapidly declined, belief in key Christian concepts is still widespread, and Christian traditions are deeply embedded as national rituals and imaginings (Davie, 2000; Hervieu-Leger, 2000). Global migration has also encouraged the growth of new Christian sects – for example, Pentecostalist congregations – as well as Hindu, Sikh, Buddhist, and Muslim communities. All this, combined with 'new-age' cults and a widespread embrace of 'spiritual' values, has led commentators to talk about the emergence of a 'postsecular' society (Hammond, 1985; Kyrlezhev, 2008; Morozov, 2008; Davey, 2002).

What the secularization debate has largely ignored, however, are the ways in which people shape and are shaped by these secularizing and sacralizing processes in urban contexts (although see, for example, Ahern and Davie, 1987; Davey, 2002). This neglect is surprising, since Western cities dramatically illustrate the stubborn refusal of religion to disappear from the public stage. Here religious identities are intimately bound up with the process of migration, cultural diversity, and the relationship between geopolitical events at the global and local levels. Conflicts in Iraq, Chechnya, Afghanistan, Sudan, Somalia, and the Balkans, for example, affect Western European cities and are reinforced by such iconic assaults on Western, multicultural urban centres as the 9/11 attack on New York's World Trade Center, the bombings in Madrid, and the 7/7 explosions in London.

Central to the public reaction to these events and to critiques of multicultural policies has been the trope of fundamentalism, popularly interpreted as a direct threat to 'Western' political and cultural traditions. In the ensuing 'war against terror', groups or individuals, who can be defined as fundamentalist, have attracted keen public attention, and in some cases the state has intervened using the widening range of its judicial powers. At the same time, the trope of fundamentalism has been interpreted by religious leaders to justify strategies which are quite separate from any 'war against terror'.

Again, what has attracted scant attention is the degree to which these public debates concerning fundamentalism are reflected in spatial practices that seek to maintain exclusivist communities in the face of secular materialism. London provides a fertile ground for these exclusivist attempts, since its success – and more recent weakness – as a global hub for business and financial services has been accompanied by the rapid growth of a 'super-diverse', multicultural society through the influx of 'black and Asian' migrants, refugees, asylum seekers, and, more recently, migrant workers from east and central Europe.[1] These newcomers have been central figures in the debates concerning fundamentalism, and religious missions have sought to protect them from the wider world – a process which involves both exclusionary discourses and spatial practices. In this chapter I will contextualize this process by focusing on two minority groups – British Muslims and Polish Catholics.

British Muslims and London's East End

Although the Muslim presence in Britain has a long history, the formation of substantial Muslim communities did not begin until after World War II. Large-scale migration from the 'New Commonwealth' developed during the 1950s and 1960s, mainly from South Asia, West Africa, and the Caribbean.

But during the last forty years the predominance of migrants with South Asian links has been modified to some extent by newer arrivals from the Middle East, North and East Africa, Malaysia, and the Balkans. London has attracted a high proportion of these settlers. According to the 2001 national census, almost 40 per cent of British Muslims lived in the metropolis, comprising 607,073 residents, or 8.5 per cent of London's total population (Greater London Authority, 2006).

British Muslims are highly diverse in their religious organization and practices. Yet, despite this diversity, a key question has engaged those who want to be actively involved as Muslims in Britain: how to live as authentic Muslims in a non-Muslim, secular, and highly urbanized Western country. The response invokes two further questions. The first is whether to purify Islam from its historic syncretic accretions through a return to the foundations of faith, as embodied in texts (the Quran and the Hadith), or to preserve the traditions inherited from the countries of origin. If the answer to this question is a return to the foundations of faith, then the second, contingent, question is whether to do this solely within a Muslim context, or to use the resources available within the local and national political arena.

A prominent global organization – the Tabligh Jamaat (TJ) – has sought to answer these issues through its mission to reclaim migrants to what it sees as the pure, authentic foundations of faith, and to avoid being drawn into community politics or engagement with local or central government agencies. In London's East End it has pursued these aims from a well-established base in Tower Hamlets (in a former synagogue on Christian Street!!). It has become even more influential in Birmingham, through a network of mosques and madrasas, which are mainly supported by Pakistani and Gujarati residents; and in Oldham, where it operates from several mosques run by Pakistani and/or Bangladeshi management committees (Eade and Garbin, 2002, 2006).

Despite the insistence by the Tabligh on an apolitical strategy, it has been impossible to avoid the political arena entirely. The vast majority of Muslims are concentrated in Britain's towns and cities, and their public meetings require fixed territorial bases. The establishment of these bases and any change in external use have forced the Tabligh's leaders to negotiate with local governments. That has meant becoming embroiled in a process of public meetings and negotiations with officials, where the Tabligh is subject to the glare of political, media, and neighbourhood opinion (Metcalf, 1996; Gale, 2005; Smith and Eade, 2008). A grandiose plan, for example, to build a large mosque in the neighbouring borough of Newham not only revealed the weakness of TJ's apolitical claims but also the ways in which it was a prime target for public debates about fundamentalism and the war against terror, not only in Britain but in other Western countries.

The Abbey Mills Mosque Dispute

The site on which the Tabligh want to build the mosque lies near the West Ham underground station and was once occupied by a chemical factory. The Tabligh bought it in 1990, and at first only erected several small huts for worship there. However, with the success of London's bid to host the Olympic Games in 2012 the site became a prime asset, since it adjoined the area where the main stadium and other key buildings were to be erected. Outsiders' concerns about the Muslim group's proximity to a site of such high global significance were exacerbated by the TJ's desire to make an iconic statement by employing a firm of young architects whose outline plan in 2006 included a futuristic, eco-friendly, and very large mosque.

During the ensuing, protracted debate the issue of fundamentalism was repeatedly mentioned, as different interest groups rallied either in opposition to or support of the proposal at local and more global levels. According to a local Conservative councillor, the proposed 'mega-mosque' would be a serious security threat. *The Daily Express*, a national newspaper, took up the story on 15 July 2007:

> Local Tory councillor Alan Craig said: 'It will be a horrendous security nightmare if they are allowed to build this large mosque so close to the Olympics... They have a growing and ominous track record as further young men follow Tablighi teaching about Islam and then go on to plan horrendous atrocities... The Newham councillor accused the sect of radicalising and dehumanising young Muslims and said: 'The dangerous truth about Tablighi Jamaat is coming out'.[2]

This local development was interpreted here as part of a wider plan by the 'fundamentalist Tablighi Jamaat sect' to encourage terrorism. The newspaper noted that '[t]he 7/7 suicide bombers Mohammed Siddique Khan and Shehzad Tanweer attended the European headquarters of Tablighi Jamaat at Dewsbury, West Yorkshire', and another TJ member had been involved in a recent attack on Glasgow airport. It then proceeded to comment on the size of this 'mega-mosque': '[It] will hold 12,000 people – four times as many as Britain's largest Christian building, Liverpool's Anglican cathedral'. It asked rhetorically:

> Are we sure, as a nation, that we want by far the largest place of worship in our land to be sponsored by an organisation which holds views directly opposed to our

democracy and a religion which, in many parts of the world, denies essential freedoms?[3]

A key feature of the debate was the involvement of Christian pressure groups. The Newham councillor, Alan Craig, was the leader of the Christian Peoples Alliance (CPA), which describes itself as a 'Christian political party whose members come from all backgrounds and church traditions'.[4] He was supported in Newham by two other CPA councillors, and had campaigned for a Christian Europe in June 2009 during the run-up to the European Union election:

> Our campaign is to expose attempts by the political and intellectual elites of Europe to impose the false idea that human rights are secular, not Christian. They falsely believe this dry, secular philosophy can form the basis of a universal culture to which non-Christian nations can subscribe.[5]

Not surprisingly, the CPA was supported by other Christian groups, such as Christian Voice, whose website speaks of its

> ... prophetic ministry in the 'forthtelling sense'. We attempt, with God's grace, to analyse current events in the light of scripture, proclaim God's word to those in public life and provide the information Christians need in order to pray with the mind of God and witness in these dark days.[6]

Christian Voice took up the Abbey Mills plan with relish. It contrasted the powerful local development agency's benign vision of multicultural diversity with the 'reality' of 'Islamicization':

> The London Thames Gateway Development Corporation have published a brochure setting out their vision for the Lower Lea Valley. In it they write that 'a large mosque ... *will, if realised, significantly increase the importance of West Ham as a cultural and religious destination*'. This is part of their vision to '*transform West Ham into a vibrant, mixed use centre...*'. But what will really happen if the mosque is allowed to go up in West Ham? Tablighi Jamaat, who plan to use the site as their new European headquarters, have a history of taking over towns. If the mosque is built, there is little question that West Ham will be Islamicised.[7]

Clearly, the Christian Peoples Alliance hoped to gain electoral advantage from the dispute. And their strategy was put to the test by the European elections of May 2009. In London, the Alliance competed for the first time on the same platform as the Christian Party (CP) and collected 2.9 per cent of the total vote.[8] This was not enough to gain a seat, but their exploitation of the emotive issue of Muslim fundamentalism and the Islamicization of local urban space may have played well in Newham and other East End boroughs. The eight London candidates entered by the CPA/CP were part of a total of sixty-nine fielded across England, Scotland, and Wales. And, as its website proudly announced, 'This is the first time that the electorate have been offered a Christian choice on the ballot paper across the country'.[9] Even more significantly, the CPA/CP saw itself as part of a wider European movement, and looked 'to our friends across Europe, who better established, have succeeding in returning Christian MEPs [Members of the European Parliament] to the Parliament'.[10]

These opponents of the mosque development raised the spectre of a dangerous, fundamentalist, secretive, and exclusionary group, which was seeking to change the local urban landscape through a process of 'Islamicization'. The TJ responded by recruiting a public-affairs company to both lobby central and local governments and get its message across through a website and YouTube videos. To the charge that the mosque would be exclusive, a spokesman replied that 'the whole thing behind the Tablighi Jamaat is to welcome all schools of thought, which encompasses Islam'; and as for non-Muslims, 'every faith is welcome'. Indeed, the mosque 'would be a bridge between all communities'. As for the claim that the mosque was being subsidized by the Saudi royal family, another representative declared that 'the money comes from the members of the local Muslim community'.[11] According to the *Times Online*, the Tablighi Jamaat stands 'for democracy and freedom' and acting as 'a role model to promote social and religious integration'.[12]

The contrast between the TJ's self-presentation and the reaction by outsiders to their foray into the public arena is striking. Clearly, the attempts by the TJ's representatives to reassure outsiders about its peaceful, inclusive mission met with deep scepticism, if not outright disbelief. The Newham mosque proposal, which was subsequently withdrawn, not only aroused general fears about the organization but also concerned outsiders because of its position at such a symbolically important site within the global city. If the 'mega-mosque' was built by 2012, its presence could be beamed around the world, allowing the 'Islamic village' to become for some an all-too-visible sign of fundamentalism within the secular metropolis.

The discourse concerning fundamentalism and exclusion, in other words, has had a direct impact in this case on spatial practice. However, behind the public furore there exists another discourse and set of practices bound up with the planning process. Although political opposition to the mosque clearly affected the planning application, the TJ could still have achieved its aims if it had learned to play the game. In this regard, insight into the key considerations from a planning perspective was provided by a report produced for the local council by its Environmental Management Services department (made public, significantly, by a TJ website). The report outlined the history of the site and the changes which had taken place there since it had been bought by the Tabligh. The focus was, not surprisingly, on the ways in which the Tabligh had complied not only with planning procedures, building regulations, and local development plans, but could also satisfy such environmental issues as highway and pedestrian access, transport, parking, and site contamination.

The Tabligh's leaders were slow to learn the rules of the planning game. For example, when they submitted a planning application in 2001, it received the following evaluation:

> Despite the earlier informative, a Transport Assessment and a Contamination Report did not accompany the application. The agent was continually reminded of need to submit this additional information by both letters, emails and at the scheduled monthly meetings that occurred as part of the memorandum of Agreement. The Council's Transport Consultants offered their assistance by providing a scoping report of exactly what was required. Eventually, a traffic count was submitted but this failed to provide any analytical work, or provide information on existing traffic flows, junctions, parking levels, and also implications for the wider public transport network. As such, there was insufficient information for officers to fully consider the application.[13]

Despite the failure of the submission, Tabligh members went ahead with the building works. When council officers met the TJ representatives at the site, it was explained that 'certain younger members of the Trust [had become] impatient with the delays in dealing with the planning application, and had taken it upon themselves to proceed with the development'. Although assurances were made that no more work would proceed, the officers returned a week later to find that 'further works had commenced and the extension had now almost been completed'.[14]

Clearly, failure to obey the rules of the planning game played into the hands of those who accused the TJ of not being open and trustworthy. However, council officers and political leaders continued to support the redevelopment as long as it complied with the formal procedures. As the planning report noted, if the site's owners 'make an application, do the opening work, and corrects [*sic*] any contraventions found there is nothing more that can be done under the Building Regulations. The Regulations are not designed to stop people building'.[15] Furthermore, when in November 2008 Alan Craig asked the council to reclaim the site though a Compulsory Purchase Order, the leader of the council replied:

> The Trust have advised that they are currently preparing another application so that they can continue using the site for a further temporary period... Development of the area will have to be resident-friendly and the facilities must serve everyone in the local community. There is currently no evidence that the Trust will not do [this] and therefore we are unable to CPO the site until they submit their masterplan for the site.[16]

Roman Catholics in Britain: The Decline of a 'Fortress Church'

The Christian Alliance Party, the Christian Party, and Christian Voice could all be dismissed as on the fringe of mainstream Christianity in Britain, exclusivist in their opposition to the 'Islamicization' of urban space. However, while established church leaders have adopted a more moderate, inclusive rhetoric and sought to influence the mainstream parties more discreetly, the related issues of fundamentalism and exclusivism have likewise emerged in anxious debates among Roman Catholic leaders about the implications of global migration for the integration of newcomers within the British Catholic structure, and British society more broadly.

Catholicism has a much longer history in Britain than Islam, of course. For centuries the Catholic Church acted as a crucial counter to monarchical power in England – the dominant economic and military territorial force then as now. However, the Reformation led to the reduction of Roman Catholics to an insecure minority with limited rights, and the Catholic religious hierarchy was not restored until the mid-nineteenth century. At that time immigration by Irish workers resulted in the emergence of substantial new Catholic communities in Britain's rapidly expanding industrial cities and towns. However, until the mid-twentieth century, Roman Catholicism in England took the form of a 'Fortress Church' where a religious elite sought to maintain a wide range of educational, social, and religious institutions for those descended from Irish and other immigrants (Italian, French and Polish, for example) – institutions which were mainly located in urban centres.

During the second half of the twentieth century the walls of the Fortress Church crumbled, however. Socioeconomic changes played a large part in this process: rapid upward mobility weakened urban ethnic enclaves and encouraged Catholics moving into the leafy suburbs to mix with their middle-class non-Catholic neighbours. There, they become increasingly secular or selective in their beliefs and practices and more critical of religious leadership. Wider changes within the Church also played a part, such as the Second Vatican Council reforms, the drastic fall in religious vocations generally, and the decline of educational and social resources.

But since the late 1970s a reaction to these developments has gathered pace. The appointment of John Paul II signalled a concerted effort to defend traditional Church teaching against corrosive innovation. And Cardinal Ratzinger, who, as Benedict XVI, succeeded John Paul as pope, played a key role in this reaction. Indeed, Julie Byrne has even suggested that John Paul II could be regarded as 'the first Catholic fundamentalist pope'.

> Looking at the years since he appointed Ratzinger head of the CDF in 1981, it is hard to tell where the words of Pope John Paul II ended and the new Benedict XVI's words began. But with his embrace and celebration of modernity – all that traveling, all those languages, all that interfaith dialogue – no one thought to characterize John Paul II as a fundamentalist. [However] ... [f]undamentalism is not anti-modern. It is the modern questioning itself, asking if relativism has gone too far, if God has abandoned the modern world.[17]

Before he was appointed pope, Ratzinger suggested that there was no conflict between the term 'fundamentalist' and traditional Church teaching:

> What often gets 'labeled as fundamentalism', he said during a Mass he celebrated on Monday, is merely having 'a clear faith, based on the creed of the church'.[18]

This entailed a return to Catholic essentials. As one of his cardinals recently explained: 'I think that Benedict XVI wanted to return to the fundamentals, showing the essence of Christian life – friendship with Jesus'.[19] This essentialist reading of the Church's mission presents relativism as a false religion – an expression of postmodern culture. Return to the fundamentals of faith would enable the Church to resist the advance of postmodernity, as well as the threat posed by religious sects. As another leading cardinal explained:

> The present challenges the Church faces today are connected with the spreading of the new post-modernistic culture, its features being secularism, laicism, agnosticism of the intelligentsia, relativism and religious pluralism.[20]

Although he did not mention the urban context, the features he listed are most strikingly evident in Western urban centres, especially cities such as London.

Polish Catholics in London: Old and New Poles

In May 2004 Poland and other central and east European countries joined the European Union. However, only Britain, Ireland, and Sweden subsequently opened their doors fully to these 'A8' countries, and as a result, Britain attracted a level of migration that was unprecedented in its rapidity and volume. Estimates of how many migrants came from Poland vary wildly, but around 600,000 were probably in the country by 2008. And many more had come and gone, since a striking feature of this influx, compared with the 'black and Asian' immigrations after World War II, was its highly circular nature, as A8 migrants flowed back and forth between their countries of origin and destination. Another distinctive feature of A8 migration was its geographical spread. Many did not stay in the major urban centres, but found seasonal work in the agricultural sector of both England and Scotland.

Approximately 18 per cent of the Polish migrants worked in London, primarily in the service sector (e.g., in cafes, restaurants, and pubs and in welfare and family support) and construction. They were predominantly young – between the ages of eighteen and thirty – and relatively well-educated compared with the jobs they took. They found cheap accommodation across London, and about half kept their options open as they checked for better opportunities in Britain, 'back home', or further afield.[21]

These new Poles were very different from the members of the Polish armed forces, who had stayed in Britain at the end of World War II, and who formed strong bases in London and other cities. This earlier British-Polish community became highly assimilated, and its second generation enjoyed rapid upward mobility, helped by the Catholic Church's system of schools and colleges. But this London community also eventually dispersed across the suburbs, although a substantial concentration formed in the western suburb of Ealing. This established community was replenished to some extent by partners recruited during the 1950s and 1960s, and then by the Solidarnosc activists. However, these migrations were dwarfed by the sudden influx of post-Accession workers, who posed a major challenge to the assimilative strategy of British Polish institutions, such as the Church of the 1980s. This challenge only increased when even larger numbers began arriving on student and tourist visas during the 1990s.

Sustaining the Foundations of Faith: Polish Migrants in Britain

In the eyes of Church leaders Polish migration to Britain was both an opportunity and a threat. The older settlers and their offspring represented the traditional virtues of Polish Catholicism – resistance to atheism, materialism, and relativism through conformity to the fundamental truths embodied within the universal Church. They could serve the fundamentalist mission, which both John Paul II (the former Archbishop of Krakow) and his successor hoped would redeem secular, highly urbanized Western Europe with its rapidly declining indigenous Catholic congregations.

The danger was, of course, that both the assimilated British Poles and the newcomers would be influenced by the secular values of mainstream British society, and come to question Church teachings. Parish congregations in Britain's cities and towns were revived by the sudden influx, and priests arrived from Poland to meet the new demand. However, such migration could also cause disruption to family life back in Poland, and migrants could lose touch with the Church by being drawn into the secular life of the multicultural city.

Whereas Muslims across London – and, indeed, throughout Britain – lacked a strong hierarchical structure despite the state-sponsored creation of the Muslim Council of Britain, the Catholic Church relied on an interlocking system of territorially based units dominated by a celibate, clerical elite. For centuries this system had withstood 'turf wars' of varying intensity and range, and the mass influx of Polish workers after May 2004 soon led to public debate in Britain about how to 'integrate' the newcomers within the national hierarchical structure. The debate highlighted the tension between a national structure, which encouraged the assimilation of newcomers into English-language rituals and congregations, and the Polish Catholic Mission, which supported Polish-language churches and priests across England and Wales.[22]

During December 2007, Cardinal Murphy-O'Connor, the Archbishop of London's Westminster diocese and head of the Catholic Church in England and Wales, expressed concern that 'the Poles would create some kind of a separate Church; so there isn't a one Church for the Poles and a second one for the rest of Catholics'.[23] He was quoted as expressing the hope that the newcomers 'would integrate with British parishes "as soon as possible when they learn enough of the language"'.[24]

It did not take long for the cardinal's comments to be firmly contradicted by the head of the Catholic Church in Poland, Cardinal Glemp, who reportedly 'insisted that Polish migrants should "seek out Polish pastors" and "find Polish Church centres"'.[25] The Polish Catholic Mission was seen as protecting the newcomers in a hostile environment – because the newcomers could easily lose their bearings if they had access only to English-language services. Even more dangerous was the prospect of being corrupted by the powerful forces of secular multiculturalism – a concern which was clearly articulated in a January 2008 pastoral letter from the Polish bishops to the diaspora:

> There is a great danger that the loss of inherited cultural values can lead to a loss of faith, particularly when the cultural values accepted in the new environment lack a Christian character… Only by maintaining our Polish links can we overcome our isolation and avoid being lost in a varied, multicultural, God-seeking contemporary world.[26]

The dispute was eventually resolved – formally, at least – when Cardinal O'Connor visited his counterpart in Warsaw, where he declared that 'the pastoral care of migrants must include the celebration of the sacraments in Polish "as people must have the opportunity to celebrate their faith in ways they are used to"'.[27] However, a working party was set up to consider whether the Polish Catholic Mission's separate structure in England and Wales should be revised. And it was clear that the English cardinal's focus was on those who were settling in Britain. He was reported as saying,

> I am conscious that many Polish people working in Britain would like to return to Poland. That's why they want to pray in their own language… But many of them will stay and establish families, and those people could contribute a lot to Catholic life in Britain.[28]

Behind this dispute about providing for the religious needs of Polish migrants through the predominantly urban spaces of churches and other buildings lay important differences between Catholic responses in England and Poland towards those outside the Catholic fold. The Polish hierarchy were much more exclusivist in their approach towards outsiders, while English Catholic leaders reflected the movement from a 'Fortress Church' to a more open, democratic community which was largely at ease with other Christian traditions and with multicultural diversity more generally. As one commentator in the Anglican Church's weekly, the *Church Times*, perceptively noted,

> Roman Catholic traditions in Poland, forged in dramatic historical circumstances, are unlikely to transfer easily to the pluralistic atmosphere of British Christianity. The

> Polish Church's ecumenical engagement is negligible, and it has shown little interest in global poverty and injustice, which are a vital part of Western church discourse.
>
> In social and cultural terms, meanwhile, English Roman Catholics have much more in common with Anglicans than with their fellow-Catholics from Poland. Although Poland is well supplied with renewal movements, lay people play no part in running the Church. Practices taken for granted at British churches, such as Sunday schools and post-worship socialising, are virtually unknown.[29]

Very little is known, however, about the religious beliefs and practices of the migrant workers. Many who were brought up in Communist Poland had little exposure to Church teaching, and their offspring were unreliable missionaries to Western Europe. As a Polish priest noted on his return from London:

> These young people often come from families that have no religious background and their knowledge in this field is perhaps on the level of their first communion to which they were prepared in some way but afterwards they did not keep any contacts with their parishes. In those times many people did not attend religious instruction classes, it is estimated that 80% of the pupils of vocational schools during the period of the Polish People's Republic did not attend catechesis in their parishes. Later they got married, have children and today these children go to the West. Therefore, we have got a generation of poorly religious people, who have often no catechetical foundations.[30]

Some insight can be gleaned from a survey of fifty young men and women undertaken in London during 2006. Many chose to say little or nothing about religion, and enjoyed the freedoms of a global city where they could pursue a number of different strategies, including keeping their options open (Drinkwater *et al.*, 2009). The few who did comment ranged from those who were worried about losing touch with the Catholic calendar to those who were highly critical of what they saw as a deeply conservative and out-of-touch clergy. Lucija, for example, was religious but noted:

> In Poland I feel more there is a festive day… Here you don't feel that… If you don't go to a church you won't notice that anything is special… With time you lose track of the festive days… Later you forget what is going on in the calendar because there are different Churches, different faiths… In Poland it's only Catholics and the Orthodox Church that is dominant … and here you see this Catholicism much less.[31]

London's highly multicultural, secular society made it difficult for Lucija to remember to observe the regular Catholic festivals. Even in the West London suburb of Ealing, where the Polish church held five masses a day to cater for the increased demand, Catholicism could not provide an exclusive haven from the outside world of cultural diversity. Nor was there, according to Katija, a straightforward division between a pure, devout Poland and the impure world of the global city, since Polish churchgoers in Ealing and 'back home' were not free from hypocrisy.

> I go the church in Ealing but, you know, many people go there, pray and then afterwards do that same stuff – cheat… I don't like it… Besides, this is the same thing in Poland.

These highly independent-minded newcomers tried to behave as sincerely as they would back in Poland. Marek, for example, chose to be religious and did not want just to impress other Poles.

> I am fond of Polish culture – I sustain it. I am religious and go to church, not because to show off but because I want… And there are Sundays when I don't want to … but I try to go. I think it is needed. I celebrate all important days… I don't know, I think I am the same Polish [person] that I was in Poland.

Others had begun the familiar drift away from the Church.

> I don't go to church that often now … and to confession also… I don't celebrate fast any more … or other feasts like Advent… So anything like Christmas or Easter I do not celebrate that much … so I drift away a bit from that.

Thus, recent Polish migration to Western Europe and its highly secular cities, such as London, did not neatly fit within the mission envisaged by papal understandings of 'fundamentalism' or the Polish church leaders' defence of an exclusive diasporic community. Not only was post-Communist Poland much more secular than the hierarchy desired, but its migrants found themselves in a city where Catholics were widely dispersed; and even in an area like Ealing the Church was not very visible, nor were its members necessarily 'good people'. London provided a wide variety of secular and religious options, as Sonia explained. She enjoyed travelling around England, going to concerts, and visiting different churches:

> I go to the Polish one, but there are so many others, and I really find it fascinating to go to other churches and see how they do it over there … Anglican, Baptists …

> other nationalities … but it doesn't matter really… Sometimes I got invited to St Mary's in Marble Arch … great mass – like a concert, plenty of young people … loads of positive energy, something completely different from these grannies from Radio Maryja [Polish ultra-Catholic radio station]… As I rebelled in Poland against lot of things I saw, here I see that people can accommodate that rebellion, that they can pursue their things without a feeling that he betrays something where he comes from.

The discourse about religious belief and practice once again led to considerations about the urban spaces where religion should be expressed. The Polish hierarchy resented any interference with the role played by the Polish Catholic Mission in catering for the diaspora. Its members were suspicious of the English hierarchy's desire to 'integrate' Polish migrants within the churches and other facilities provided by the English-speaking parochial and diocesan structure. Although the dispute explicitly revolved around the issue of language and national identity, there lurked broader religious concerns about the global direction of the Church. Religious mission was seen as involving a return to the foundations of faith, where the term 'fundamentalism' could be happily embraced – a move which excluded many Catholics, let alone those outside the Roman Catholic Church.

Concluding Remarks

In this chapter I have focused on migrants to Britain, who are the subject of widespread scrutiny from the media, politicians, and religious leaders. There are, of course, very important differences between British-Muslim and Polish-Catholic migrants, but by bringing them together here, a number of significant issues are raised for this volume on cities and fundamentalism.

Both Tablighi and the Polish religious leaders spoke in very broad terms about how Muslim and Catholic migrants respectively could lead a life of moral virtue by following the foundations of faith. The Western, secular, and multicultural city was seen as a challenge to this common mission, which sought to provide safe, exclusive places where the faithful could correctly perform religious traditions.

Migration was, therefore, seen as an opportunity and a threat. The opportunity lay in freeing migrants from the inadequacies of their religious upbringing in the countries from which they had come. A high proportion of British Muslims have left rural areas where there are high rates of illiteracy, low levels of religious instruction, and a long tradition of syncretic 'folk' beliefs and practices. Many Polish migrants in Britain have also come with inadequate knowledge about their faith, due to the heritage of the Communist regime. Both the Tabligh Jamaat and the Polish Catholic hierarchy saw the opportunity of returning migrants to a pure observance of religious fundamentals and, thereby, resisting the threat posed by the enticements of secular materialism and the breakdown of social and cultural ties with families and communities 'back home'.

In physical terms, this parallel religious mission would be achieved through the mosques and churches located in areas of high migrant concentration. Religious mission inevitably led, therefore, to issues of urban spatial practice. The Abbey Mills mosque provided the Tabligh with the opportunity to develop a highly strategic site within the global city, while the Polish Catholic Mission had the chance to expand its existing array of Polish-language churches across London and other urban centres.

Insiders and outsiders interpreted these religious missions by drawing on the trope of fundamentalism. In the case of the Abbey Mills site, its development was interpreted by critics as part of a 'fundamentalist' strategy, which encouraged rigid exclusivism and attracted followers who were violently hostile to Western society. The Tabligh leaders responded by vigorously rejecting the fundamentalist label and claiming to be open to mainstream society.

Pope John Paul II and Benedict, on the other hand, provocatively embraced the trope of fundamentalism, but changed its popular meaning to refer to the traditional foundations of faith. Although the Polish hierarchy did not refer to fundamentalism during the dispute over the religious needs of Polish migrants in London and elsewhere, they shared the two popes' hostility towards religious pluralism and the blandishments of secular Western cities. In order to protect Poles abroad, they sought to maintain exclusive physical and spiritual resources for the diaspora through the Polish-language facilities provided by the Polish Catholic Mission. However, interviews with some Polish migrants revealed a more inclusive process. They were already looking beyond the high walls of a Fortress Church and enjoying, with their English-speaking co-religionists, the much wider range of resources provided by the multicultural global city.

The impulse towards a purificatory exclusivism was countered, therefore, by the everyday forces of cultural diversity and engagement with mainstream society. In the case of the Abbey Mills mosque development, inclusivism was encouraged by the Tabligh's need to negotiate with the local councillors and officials and to reassure outsiders through the use of the media and public-relations advisers. Polish religious leaders, in turn, were drawn into negotiation with the English hierarchy, which was far less fearful of mainstream British society and more open to lay opinion. Furthermore, the English hierarchy also provided a much wider range of resources through its own structure of parishes, schools, religious congregations, and voluntary organizations than the Polish Catholic Mission could hope to provide – something which probably added to the Polish hierarchy's anxiety.

Religious mission, in other words, required a physical base from which to operate. Muslim and Catholic activists needed to have territorial bases from which they could seek to convert insiders and outsiders to their particular version of faith. British Muslims lacked the coordinated structure of parishes and dioceses, as well as a priestly hierarchy like that of the Catholic Church. However, mosques and madrasas provided territorial bases for congregational prayer and religious instruction. One of the advantages of developing the Abbey Mills mosque for the Tabligh was the iconic statement it would make both to other Muslims and to outsiders at a globally important site – hence, the repeated attempts by its opponents (Muslims as well as outsiders) to stop the development by mobilizing support at local and more global levels.

Thus, although exclusivist religious missions needed territorial bases from which to operate and, therefore, had to deal with London as a multicultural global city, they looked beyond the metropolis and the nation to a world which was to be redeemed by a return to the foundations of faith. Mosques and churches were the sites of a redemptive, purificatory mission where the global and local joined in a struggle between exclusivist and inclusivist forces. Despite the furore surrounding migrants and the exclusivist strategies pursued by certain religious leaders, the pressure towards inclusion generated by everyday negotiation with others, including the state, promises to overcome attempts to keep migrants within the walls (physical and mental) of embattled religious communities.

Notes

1 Although Sassen's *The Global City* (1991) presented a pioneering and highly influential comparison between New York, London, and Tokyo during the late 1980s and early 1990s, various commentators have since pointed to its limitations. In particular, its use of the term 'global city' suggests that certain cities have reached some end-state; as an alternate, 'globalizing cities' indicates an ongoing process open to cities around the world. Despite this critique, for the sake of convenience, I will continue to refer to London as a global city and leave open the similarities between it and other globalizing cities.

2 See http://www.express.co.uk/posts/view/13553/. Accessed 2 June 2009.

3 *Ibid.*

4 See http://www.cpaparty.org.uk/. Accessed 2 June 2009.
5 See http://www.cpaparty.org.uk/resources/CPA_Manifestofin3.pdf. Accessed 2 June 2009.
6 See http://www.christianvoice.org.uk/about.html. Accessed 2 June 2009.
7 See http://www.christianvoice.org.uk/mosque1.html. Accessed 2 June 2009.
8 See http://news,bbc.co.uk/shared/bsp/hi/elections/euro/09/html/ukregion_39.stm. Accessed 24 June 2009.
9 See http://www.cpaparty.org.uk/?page=news&id=311. Accessed 24 June 2009.
10 http://www.cpaparty.org.uk/?page=news&id=311. Accessed 24 June 2009.
11 See http://www.youtube.com/watch?v=_1wo-A0u9LU&feature=channel. Accessed 2 June 2009.
12 See http://www.timesonline.co.uk/tol/. Accessed 2 June 2009.
13 Newham Borough Council, 5 July 2002, 'Retention and Expansion of Extensions to Temporary Mosque', application No. P/01/1375. Available at http://www.magamosquenothanks. com/pdf/RTZSite.pdf. Accessed 27 June 2009.
14 *Ibid.*
15 *Ibid.*
16 Cohen, T. (2008) Application for London mega-mosque can go ahead. *Religious Intelligence News*, 2 November 2008. Available at http://www.religiousintelligence.co.uk/news/?NewsID=3204. Accessed 27 June 2009.
17 Byrne, J. (2005) Pope Benedict XVI and the not-so-new Catholic fundamentalism. Office of News and Communications, Duke University, 22 April. Available at http://www.dukenews.duke.edu/2005/04/fundamental_oped.html. Accessed 9 August 2008.
18 *Ibid.*
19 Two years of Benedict XV1's Pontificate: Wlodzimierz Redzioch talks to Cardinal Paul Poupard. *Sunday* (*Niedziela*), 4 August 2008. Available at http://sunday.niedziela.pl.artykul.php. Accessed 7 August 2008.
20 Wlodzimierz Redzioch talks to Cardinal Tarcisio Bertone. *Sunday (Niedziela)*, 4 August 2008. Available at http://sunday.niedziela.pl.artykul.php. Accessed 7 August 2008.
21 See Eade, J., Drinkwater, S. and Garapich, M. 'Class and Ethnicity – Polish Migrants in London'. Research Report for the RES-000-22-1294 ESRC project. Available at www.surrey.ac.uk/Arts/Cronem/polish/Polish _Final_Report_Web.PDF.
22 A separate agreement had been established with the head of the Roman Catholic Church in Scotland.
23 Polish migrants were 'creating a separate church in Britain'. Available at http://www.eni.ch/featured/article.php?id=1572. Accessed 3 June 2009.
24 *Ibid.*
25 http://www.catholicherald.co.uk/articles/a0000231.shtml. Accessed 3 June 2009.
26 http://www.catholic.org/international/international_story.php?id=23485. Accessed 25 June 2009.
27 http:/www.catholicherald.co.uk/articles/a0000231.shtml. Accessed 25 June 2009.
28 *Ibid.*
29 Luxmore, J. (2008) Can Poles integrate in the UK? *Church Times*, No. 7563, 29 February. Available at http://www.churchtimes.co.uk/content.asp?id=52489. Accessed 25 June 2009.
30 Thinking about Polish Londoners. *Sunday* (*Niedziela*), 4 August 2008. http://sunday.niedziela.pl.artykul.php. Accessed 7 August 2008.
31 This and subsequent interview quotations come from Drinkwater, Eade, and Garapich (2008).

References

Ahern, G. and Davie, G. (1987) *Inner City God: The Nature of Belief in the Inner City*. London: Hodder and Stoughton.

Berger, P. (ed.) (1999) *The Desecularization of the World: The Resurgence of Religion in World Politics*. Washington DC: Ethics and Public Policy Centre and Grand Rapids, MI: W. Eerdmans Publishing.

Bruce, S. (ed.) (1992) *Religion and Modernization: Sociologists and Historians Debate the Secularization Thesis*. Oxford: Clarendon Press.

Bruce, S. (2002) *God is Dead: Secularization in the West (Religion and Spirituality in the Modern World)*. Oxford: WileyBlackwell.

Casanova, J. (1994) *Public Religions in the Modern World*. Chicago, IL: University of Chicago Press.

Davey, A. (2002) *Urban Christianity and Global Order: Theological Resources for an Urban Future*. Peabody, MA: Hendrickson Publishers.

Davie, G. (2000) *Religion in Modern Europe: A Memory Mutates*. Oxford: Oxford University Press.

Davie, G. (2002) *Europe: the Exceptional Case. Parameters of Faith in the Modern World*. London: Darton, Longman & Todd.

Davie, G., Heelas, P. and Woodhead, L. (eds.) (2003) *Predicting Religion: Christian, Secular and Alternative Futures*. Aldershot: Ashgate.

Drinkwater, S., Eade, J. and Garapich, M. (2008) Class and Ethnicity: Polish Migrant Workers in London. ESRC study RES-000-22-1294. Available at http://www.surrey.ac.uk/Arts/CRONEM/polish/reports.htm. Accessed 30 March 2010.

Drinkwater, S., Eade, J. and Garapich, M. (2009) Poles apart? EU enlargement and the labour market outcomes of immigrants in the UK. *International Migration*, **47**(1), pp. 161–190.

Eade, J. and Garbin, D. (2002) Changing narratives of violence, struggle and resistance: Bangladeshis and the competition for resources in the global city. *Oxford Development Studies*, **30**(2), pp. 137–149.

Eade, J. and Garbin, D. (2006) Competing visions of identity and space: Bangladeshi Muslims in Britain. *Contemporary South Asia*, **15**(2), pp. 181–193.

Gale, R. (2005) Representing the city: mosques and the planning process in Birmingham. *Journal of Ethnic and Migration Studies*, **31**(6), pp. 1161–1179.

Greater London Authority (2006) *Muslims in London*, available at http://www.london.gov.uk/gla/publications/equalities/muslims-in-london.pdf. Accessed 25 June 2009.

Hammond, P. (ed.) (1985) *The Sacred in a Post-Secular Age*. Berkeley, CA: University of California Press.

Hervieu-Leger, D. (2000) *Religion as a Chain of Memory* (translated by S. Lee). Cambridge: Polity Press.

Kyrlezhev, A. (2008) The postsecular age: religion and culture today. *Religion, State and Society*, **36**(1), pp. 21–31.

Metcalf, B. (ed.) (1996) *Making Muslim Space in North America and Europe*. Berkeley, CA: University of California Press.

Morozov, A. (2008) Has the postsecular age begun? *Religion, State and Society*, **36**(1), pp. 39–44.

Sassen, S. (1991) *The Global City*. Princeton, NJ: Princeton University Press.

Smith, M.P. and Eade, J. (eds.) (2008) *Transnational Ties*. Brunswick, NJ: Transaction Books.

Wilson, B. (1966) *Religion in a Secular Society*. Harmondsworth: Penguin.

Index